Introduction to Telecommunications

Introduction to Telecommunications

Converging Technologies

Kimberly Massey

Stanley J. Baran

San Jose State University

Mayfield Publishing Company

Mountain View, California

London • Toronto

Library of Congress Cataloging-in-Publication Data

Massey, Kimberly B.

Introduction to telecommunications: converging technologies / Kimberly B. Massey, Stanley J. Baran.

p. cm.

Includes bibliographical references and index.

ISBN 0-7674-1119-6

1. Telecommunication. I. Baran, Stanley J. II. Title.

TK5101.M34 2000

384—dc21

00-055034

Manufactured in the United States of America

10 9 8 7 6 5 4 3 2 1

Mayfield Publishing Company
1280 Villa Street
Mountain View, California 94041

Sponsoring editor, Holly Allen; *production editor,* Carla White Kirschenbaum; *manuscript editor,* Kay Mikel; *text and cover designer,* Ellen Pettengell; *design manager,* Jean Mailander; *illustrators,* Judith Ogus, Lineworks, and Alice Thiede; *art editor,* Robin Mouat; *manufacturing manager,* Randy Hurst; *photo researcher,* Brian Pecko. The text was set in 9.75/12 Meridian Medium by Thompson Type and printed on 45# Somerset Matte by Phoenix Color.

Cover image: Benjamin Shearn/FPG International

For Madelyn Baran and Bruce Friedberg,
great family, great friends

Contents

Preface

Ours is the Information Age. We live and work in an information society. Every day we read about the newest "killer ap" or the most recent "largest merger in telecom history." The television networks are in decline. Overall television viewing is starting to slip. But the amount of time people spend with media is on the rise. Record companies fret over online piracy and local radio stations can be heard around the globe on the Web. An alphabet soup of new and emerging technologies constantly enters the language. AM, FM, VHF, and UHF have been joined by CD-ROM, DARS, DAT, DBS, DSL, DVD, DVR, DMX, FTP, GEO, LEO, and MEO, HDTV, DTV, IBOC, LAN and WAN, LMA, MATV, MUD, MMDS, PLC, ISP, SMATV, SCA, T1 and MP3, TELNET, TNC, URL, and e-everything. Mergers turn phone companies into cable companies, Internet Service Providers into broadcasters, and radio stations into profit centers for liquor companies and book publishers. Real world events like tragic schoolyard killings and elections with dwindling numbers of voters thrust the media and their effects into public consciousness and discourse. For students, especially those wanting careers in the telecommunications industries, there is always something new to know and understand. Likewise, there is always something new to know and understand for everyone living in our modern telecommunications world.

This remarkable explosion of telecommunications activity is fueled by convergence, the erasure of the distinctions among media. Is Web radio the Web or radio? Just what are family members doing when they sit in their living room, in front of the television set, watching a satellite-delivered game show originally produced and distributed by a cable network, while playing along on an interactive Internet link displayed in a picture-in-picture screen embedded in the lower right hand corner of the larger, HDTV screen? Are they watching television, cable television, or satellite television? Are they television viewers or Internet users? When the phone company delivers to people's home television, video-on-demand, radio and audio, high speed Internet access, and fax, along with multiple telephone lines and long distance and local phone service, is it really the *phone* company?

But convergence is not the only modern force shaping and reshaping the telecommunications environment. The audience for the various media is increasingly fragmented, divided into ever smaller, more discrete, and more homogeneous bits that can be gathered and sold to advertisers, producing media that are hypercommercialized. Ownership of telecommunications companies is becoming concentrated into fewer and fewer hands, and that ownership is quite likely global. Some observers see these as natural forces; others see them as damaging to media and audiences alike. Either way, they are here.

Perspective

It is the reality of these on-going changes and their effects in all aspects of the telecommunications environment that provides the foundation for this book. These forces may be inevitable, but the outcome of their operation is not. What we do—as telecommunications professionals and as citizens in general—with the ever-changing communication technologies will determine not only their future, but ours. The pages that follow focus on these forces of convergence, audience fragmentation, hypercommercialization, concentration, and globalization. How these forces will impact—both positively and negatively—the telecom industries and those who work in them as well as the people who use and rely on those industries is addressed in every chapter.

Thus, you will find throughout this text hints, sometimes more than hints, of a number of contemporary conceptions of telecommunications. Among them are our adherence to the ritual perspective of mass media, our reliance on the cultural view of mass communication, and our belief in the need for critical thinking.

The ritual perspective of mass media sees media as much more than mere senders of information. The media are central to how we know ourselves and our world. They sit at the center of our politics, our economy, and the organization of most of our experiences. For this reason, people working in telecommunications have important responsibilities, often forgotten in the reality of doing a job and making a living. Audiences and users must understand this centrality, and aspiring media professionals need to understand this responsibility. You will see these beliefs reflected in the nature of the boxes and chapter-ending exercises, which ask students to imagine themselves in situations not uncommon in the everyday life of working media professionals. They also push students to consider responses and solutions beyond those that first come to mind.

The cultural view of mass communication demands that we examine the media, not as isolated industries and technologies, but as integral participants in the construction and maintenance of the culture that binds us. We must understand not only how telecommunications technologies work and how the industries on which they are based operate, but also the nature of their relationship with their audiences and users. You will see this reflected in the attention given to regulation, self-regulation, ethics, and effects, not only in the chapters dedicated to these issues, but throughout our discussion of the different technologies and industries.

Critical thinking, understanding the operation of the specific media and the telecommunications industries in general, not as natural developments, but as the product of the interplay of science, economics, and power, is reflected in the organization of the individual chapters. The chapters that focus on specific media and industries progress through their origins (to better understand how and why we have arrived where we are today) to a thorough examination of those media and industries as they currently operate to a detailed study of not only how convergence, audience fragmentation, hypercommercialization, concentration, and globalization promise to change those media and industries, but how they might alter the relationship between them and the people who use them. Another application of critical thinking is the subject of Chapter 8, Conceptualizing and Counting the Audience. We believe that teaching students how to count audiences—ratings, shares, AQHs, focus groups—is important. But we believe just as strongly that to do so without asking them to consider just who and what the audience is and its role in the overall functioning of telecommunications ultimately cheats students. In addition to learning the tools of audience measurement and how to use them, students should also know why they use those tools and what that use means, not only to themselves as professionals, but to those people who are counted.

Many, if not most, of the readers of this book will be students seeking careers in the telecommunications industry. We applaud their desire, and have tried to honor it with what we believe is a thoughtful, complete, and challenging approach that emphasizes the vast difference between the business they want to enter and all others, a difference in challenges and opportunities, yes, but also a difference in responsibility.

Features

The pedagogical features of this book are designed to support this perspective. Convergence (largely aided by digitalization and the Internet) will define all aspects of telecommunications for many years to come. Therefore, discussions of the Internet, World Wide Web, and digital technologies occur throughout the text, not only in the chapter dedicated to these technologies.

Recognizing that students come to their introductory telecommunications classes with a variety of experiences and aspirations, we have constructed four different sets of boxes designed to address those varying backgrounds and goals. Although we believe that all students can benefit from all the boxes, *Careers* and *Tech Talk* are aimed primarily at those students bent on industry careers. *Careers* appears in the industry-oriented chapters, offering insight into the kinds of jobs available, the demands of these jobs, and the personal characteristics necessary for success in them. *Tech Talk* offers detailed, technical explanations and graphic representations of the technical aspects of the technologies under discussion.

Two other boxes, *Critical Issues* and *The Real World,* are present in every chapter. They should be of interest to all students, but are geared primarily to those who will live in a telecommunications-rich environment, rather than those who will work in a telecommunications industry. Nonetheless, they raise issues of race, gender, class, power, ethics, and obligations that are designed to make students think—about themselves, about the people around them, and about the media they consume and enjoy. *Critical Issues* focuses on telecommunications and power—who has it and how they use it. Chapter 3, for example, offers "Women in Network Television" and Chapter 8 presents the controversy surrounding television's supposed abandonment of older viewers in "One Dollar, One Vote." *The Real World* puts students in the shoes of either industry pros or affected consumers and asks them to determine how they would act if they were to find themselves in certain situations. Students are then asked to consider how they might have responded differently if they had occupied a different position in the given situation. For example, the box in Chapter 6, "Paper Tiger Television," describes this provider of controversial cable access programming and asks students, first in their role as general manager of an access channel, then as member of a local franchising authority, and then as owner of a large MSO, to decide whether or not they would carry Paper Tiger's shows. In many cases, boxes in later chapters also ask students to go back to earlier boxes and reconsider their responses now that they have more information.

The different chapter-ending exercises offer to and demand of students different things. Chapter Review gives a quick, to-the-point rearticulation of what students have just read, and Questions for Review reinforces those ideas, facts, and concepts that are most crucial. Questions for Critical Thinking and Discussion poses questions—sometimes about controversies, sometimes about personal media use—that flow from the chapter. Most questions are designed to be answered individually, but others ask students to work cooperatively to construct their responses. Finally, Engaging Telecommunications directs students to the online presence of the major players discussed throughout the chapters. Students are asked to determine how these entities—media companies, regulators, critics, professional associations—represent themselves and their activities. In many instances, the identified Web site is interactive, in which case students are encouraged to enter into dialogue with site authors.

Organization

Introduction to Telecommunications: Converging Technologies is divided into three parts. Part I, Introduction to Telecommunications, consists of an introductory chapter, Introduction to Telecommunications, and five industry-specific chapters—Radio and Audio (Chapter 2), Television (Chapter 3), Noncommercial Broadcasting (Chapter 4), Corporate and Industrial Telecommunications (Chapter 5), Cable (Chapter 6), and The Internet (Chapter 7). Each follows the format described on pages xxi–xxii, and each details the medium designated by its title.

Part II, The Practice of Telecommunications, contains four chapters specific to the business of telecommunications. Chapter 8, Conceptualizing and Counting the Audience, begins with a discussion of the audience's role in the telecommunications process and then proceeds to a lengthy overview of the different audience evaluation techniques, focusing on radio, television, and the Internet. Chapter 9, Advertising, looks at the role of commercials and the sale of commercial time on radio, television, cable, and online. Its final section highlights the contemporary advertising environment—its regulation and effectiveness and criticisms of it. Chapter 10, Programming, focuses on programming and programming practices in radio, television, and cable, at both the local and national levels. Finally, Chapter 11, Global Telecommunications, studies international (country to country) and global (worldwide) media. Different media systems and the philosophies that support them are detailed, as is the ongoing controversy over cultural imperialism.

Part III, Telecommunications Rights and Responsibilities, presents three chapters. Rights and Regulations (Chapter 12), offers a discussion of the First Amendment—the freedom it provides and the responsibility it carries—and a discussion of how regulatory agencies operate in a free telecom environment. Then, a number of contemporary regulatory issues are presented, specifically deregulation, copyright, the regulation of cable and digital television, and control of the Internet. Chapter 13, Ethics and Self-Regulation, makes the argument that the freedom granted to telecommunications professionals makes industry self-regulation and the ethical conduct of individual careers all the more important. Not only are several ethical dilemmas posed, but three distinct, systematic ethical decision-making strategies are detailed. Chapter 14, The Effects of Telecommunications, presents the effects debate in general—the arguments for and against the presence of media influence—and it examines several effects issues that vex many in and outside the telecommunications industries, including violence, stereotyping, media's influence on politics, prosocial media effects, and the impact of the Internet on democracy and self-governance.

Instructor's Support

The Instructor's Resource Manual to accompany this text was written by Bob Mendenhall of Southwestern Adventist University. For each chapter there are learning objectives, detailed chapter outlines, key terms and concepts, lecture suggestions and resources. Additionally, there are approximately 60 test items (multiple choice, true/false, short answer and essay) for each chapter and a computerized test bank available for both PC and Macintosh. A Web site with monthly updated material provided for lecture materials can be found at www.mayfield.com/massey.

Acknowledgments

Naturally, an undertaking of the magnitude of this book required the work and dedication of more than its two authors. The Mayfield staff, with whom we both have a long and warm relationship, operated at its usual high level of skill and efficiency. Our editor, Holly Allen, is as intelligent, committed, and forward-thinking a media educator as any we've met in the academy. Our production editor, Carla Kirschenbaum, brought an infectious grin, a sweet laugh, and cold competence to our interactions. Robin Mouat's amazing creativity and vision transformed our facts and figures into stunning, imaginative graphics. Brian Pecko, our tireless photo researcher, searched long and hard for the unique images in the text.

The nature of this text required us to count on the experience and resources of many telecommunications professionals. But because we have both been in the university for some time, we were able to call on (and renew friendships with) several former students now making good livings "in the biz." Most helpful were Kevin Knudson and Leonard Ybarra. Bob Mendenhall also provided a sharp eye and great interpersonal skill in pointing out places where we could have been more precise in our writing.

Numerous reviewers, too, lent us their expertise. Among those who vetted our work (to its great advantage) were Henry Aldridge, Eastern Michigan University; Joseph Chuk, Kutztown University; Bill Davie, University of Southwestern Louisiana; Michael Gerhard, Ball State University; Joe Johnson, San Diego State University; Barbara Moore, University of Tennessee, Knoxville; Michael Murray, Western Illinois University; Paul Prince, Kansas State University; Karla Robinson, University of Kentucky; Joe Tenerelli, Indiana State University; and George Whitehouse, University of South Dakota.

Finally, there is our daughter Simmony Margaret Baran. Nowhere in the following pages is her spirit absent, because we wrote this book in the belief that she will live in the world we and our students leave for her. If we can help students—as practitioners and as audience members and users—produce a more humane, more democratic, more just telecommunications industry, the world she and her children will inhabit will be that much better.

—KM & SJB

Chapter One

Introduction to Telecommunications

When television exploded onto the scene in the 1950s, many observers predicted the demise of the book as a mass medium. "Why should people read stories when they could sit at home and watch them?" they asked. But books survived the video onslaught—as did newspapers, magazines, radio, and movies—by finding and fulfilling new functions. Books made serious fare, recipes, and self-help their domain. Newspapers provided depth, commentary, and more local coverage than did other media. Radio provided mobile and portable music and talk. Movies prospered through a combination of overwhelming production values and provocative themes.

But media survive and prosper today not so much by finding and fulfilling new functions—a process known as **functional displacement**—but by adopting and sharing one another's functions. We are all familiar with this **convergence,** the erasure or blurring of distinctions among different forms of communication media. Music videos, for example, draw from radio *and* television. VCR and DVD players blur the distinction between movies and television. The **Internet,** a global network of interconnected computers, combines the best and worst of virtually all our media, print and broadcast.

Convergence is reshaping the audience too. "Convergence technologies are different," writes *Wired* magazine's New York editor, Randall Rothenberg (1998). "Unlimited, ubiquitous, real-time, full-motion, personalized, integrated media fragment the monolithic audience into myriad pieces, subverting the rules by which conventional communication companies operate" (p. 207). Convergence is even changing how the federal government interacts with the telecommunications industry. In its document, "A New Federal Communications Commission (FCC) for the 21st Century" (<http://www.fcc.gov>), the FCC admitted that, "As the traditional lines dividing communications industries blur and eventually erode, the traditional ways of regulating or monitoring these industries will also have to change. The FCC must think about the complex issues resulting from converging communications markets."

There is a rush toward convergence on a scale unimagined even by the best science fiction writers of just a few decades ago. The fruits of this technological melding, merging, and marrying will change our future, whether we plan careers in the media or lives as consumers of their content. All of us—media professionals and everyday citizens—will live in the world that results from these technological advances.

What that world will look like, as well as our ability to know and find our place in it, will depend on how we choose to use these newly converged technologies and the **mediated communication**—communication conducted through the use of electronic and digital technology—they provide and permit. And how we use the existing and emerging telecommunications technologies will be determined by our knowledge and understanding of them.

Traditionally, **telecommunications**—the technologies and industries that facilitate mediated communication—has included television, radio, audio and video technology, and computer technologies such as the Internet and the **World Wide Web** (a tool that uses sophisticated transfer protocols as a means of accessing audio, video, graphic, and text files on computers connected to the Internet). But telecommunications now also encompasses the technologies that frequently deliver these familiar media: telephone, satellite, and cable. It is no longer appropriate, however, to exclude newspapers, magazines, or even books from membership in the world of telecommunications. The reason is convergence.

Convergence: Two Examples

Readers can now create digital books, either by accessing their Internet-connected computers or by dialing up their satellite provider and having entire books downloaded into files that exist in the spine of what appears to be a normal book. Silicon Valley's NuvoMedia began selling its Rocket **e-book** in late 1998, and other manufacturers (SoftBook Press, for example) introduced their own versions at the start of 1999. The Rocket e-book weighs less than a pound and a half and is roughly the size of a standard paperback. It holds more than 4,000 pages of text—enough for 12 books—and 1,100 titles

The Rocket e-book offers a good example of convergence. It marries computer, Internet, satellite, and video technology to produce a new way of accessing written material.

were available at the start of 2000. In the case of the Rocket e-book, readers download their selections from bookseller Barnes & Noble's Web site.

The pages of these e-books are actually flat computer screens that enable a single page to have one collection of content one moment and different content in the next. Charles Mann (1998), contributing editor of *Atlantic Monthly,* describes the operation of the typical e-book. Readers can open one, he writes,

> and see a list of works on the title page—the collected works of William Faulkner, say. To read *The Sound and the Fury,* one would tap the title, and the text, stored in the circuitry of the spine, would flow noiselessly onto the pages. If the tale of the Compsons seemed too involuted, one could tap *Light in August* and actuate it instead, or *Absalom, Absalom!* Given the advances in computer memory, it should be feasible to store entire libraries in every electronic book. Readers could use a stylus to underline favorite passages and write in the margins; the next time the text was generated, the marginalia would appear as well. (p. 61)

Is an e-book a book, a computer, a collection of video screens that display satellite material, or the Internet? Regardless of your answer, people who want to read Faulkner or any other author in an e-book will have to be familiar with the use of a number of technologies beyond the traditional bound paper book. And those who will want to provide that content, that is, those who wish to meet this audience demand, will work in fields much different from publishing as it has been understood for five centuries.

Technology also exists, and in fact is already in use, that permits publication of up-to-the-moment, individually personalized newspapers for readers while they are traveling. IBM Corporation of the United States and Spiegel Verlag of Germany, publishers of *Der Spiegel* magazine, have developed a system that uses digital satellite signals to deliver full-size, full-color newspapers to computer-based desktop printers onboard trains in transit. In addition, IBM and Spiegel offer "intelligent templates that will permit front pages to be customized to the interests of individual readers. Information about such interests . . . could be collected at the time of ticket purchases. With appropriate advertising or editorial content available, customization can be accomplished" for each and every individual passenger on a train (Rosenberg, 1998, p. 34). Are readers flipping through one of these publications looking at a newspaper, accessing satellite-delivered content, or reading a computer printout?

This second example raises issues beyond convergence. The IBM/Spiegel effort demonstrates the operation of four other forces that are fundamentally reshaping telecommunications: fragmentation of the audience, concentration of ownership and conglomeration, globalization of media, and hypercommercialism. It's important to note that these forces do not operate independently. For example, the same advances in technology that encourage convergence also make it possible to fragment the audience and enable large conglomerates to form and communicate internationally. In addition, reaching the individual segments of the fragmenting audience is one explanation often given for the rush toward concentration, and corporations' need to pay for their newly acquired media outlets explains hypercommercialism.

Fragmentation of the Audience

The telecommunications audience is becoming more fragmented, and its segments more narrowly defined. IBM and Spiegel's ability to customize their in-transit "newspaper" to each individual reader may be an extreme example, but it demonstrates the point.

The introduction of television began this trend, but audience fragmentation has reached its highest level with the advent of digital technologies. Before the coming of television, radio and magazines were national media serving national audiences. Television displaced both of them, forcing these media to find new functions. Unable to meet the needs of a mass audience as effectively as did television, radio and magazines adopted similar strategies, first for survival and then for prosperity. They began targeting smaller audiences whose needs they were able to identify and then meet with specific content. Because the different parts of this fragmented audience tended to be homogeneous, or alike in some important characteristic, they became more attractive to specific advertisers, ensuring the financial well-being of radio and magazines. Where once we had national radio networks and mass circulation magazines such as *The Saturday Evening Post,* we now have format-based local radio and narrow-interest consumer magazines such as *Surfing* and *Model Builder.*

Cable television, satellite, and home video technology such as VCR and DVD have wrought the same effect on television. Before cable television, viewers had a choice of three commercial networks—ABC, CBS, NBC—possibly one noncommercial television station, and in some larger markets maybe an independent station or two. Viewers now have literally thousands of viewing options (remember all those videotapes and DVDs). Therefore, television's audience has been fragmented. Incapable of attracting a mass audience, as NBC, CBS, and ABC had for decades, each channel now serves a more specific group of viewers. Fox Family targets kids. The History Channel attracts older, more affluent viewers. Even the large, national commercial broadcast networks aim their content at ever-narrower parts of the audience. Fox appeals to younger viewers, CBS to older viewers, and ABC and NBC tend to fight over 24- to 48-year-old professionals.

As implied by the IBM/Spiegel onboard newspaper, digitalization allows even more dramatic fragmentation of the audience. The Internet, by its mere existence, further fragments the audience—those who spend time online spend less time with other media. Throughout this book you'll see example after example of digitalization that permits ever-finer divisions of the audience. Digital sound recording, for example, makes it possible for anyone to become a "recording company," serving even the tiniest group of listeners. Digital television permits broadcasters to send many different signals on one channel, and because of the interactivity it allows, digital television can create the ultimate fragmentation of the audience—individual audience members designing their own content, in essence becoming an audience of one for that given piece of content.

The telecommunications audience has traditionally been viewed as a large, varied group about which the different media industries knew only the most basic demographic information. The shift toward smaller, narrower, more specific, and better known audiences, therefore, raises the question of whether this change is for better or worse. The pessimistic view is that the national culture that binds us will fragment even more as people are increasingly divided into narrowly targeted **taste publics**—groups of people or audiences bound by little more than their interest in a given form of media content. Many observers already identify the loss of shared experience as one reason for what seems to be a growing absence of civility and respect in people's interactions with one another. There is also the worry that some fragments of the audience—the poor, for example—will go unserved because they are of little interest to media executives and advertisers.

The optimistic view is that a fragmented, better served audience means that all groups and all interests will be considered. The national, uniform culture of the traditional mass media not only ignored many people and groups but served up a common experience that may not have been particularly

Critics of increased audience fragmentation lament the loss of a common, shared national media experience. Others argue that the big television networks of years past may have contributed to a shared national experience, but not necessarily a good one. For example, the portrayal of women in the hit 1960s NBC series *I Dream of Jeannie* was anything but enlightened.

enriching. Television of the 1940s, '50s, and early '60s makes the point. It may have provided the United States with a common, unifying experience, but it was one that at best ignored minority groups and women and at worst denigrated them. The racist *Amos 'n Andy* and the sexist *I Dream of Jeannie* were national network television hits. The availability of a fragmented audience will encourage someone to find a way to serve its various parts (and the advertisers who want to reach them). That means more variety and, quite possibly, a greater likelihood that people will encounter a more varied array of content.

Steven Brill, creator of cable channel Court TV and the media watchdog magazine *Brill's Content,* suggests another beneficial outcome of the growing fragmentation of the audience. As greater numbers of outlets develop to serve (and be supported by) a greater number of audience segments, "the barriers to entry for alternative media aren't as bad as they used to be. I think the ultimate salvation of media is that good media . . . can drive out bad media" (as quoted in Ledbetter, 1998, p. 46). Writing in the *Columbia Journalism Review,* Tom Goldstein (1998) presented a variety of optimistic comments on the beneficial influence of the fragmenting audience. He quotes *Newsweek* columnist Robert Samuelson, "As media multiply, the (media) elite becomes less exclusive. Smaller audiences give them less prominence and market power," distributing influence more evenly among different media voices (p. 53). Goldstein also offers the comments of respected television critic Michael Arlen, who marveled at the power of the newly emerging "cacophony of market-crazed citizens all over the world, with their insistence on two-way communication and their appetite for fragmentation of broadcast authority" (p. 53).

Concentration of Ownership and Conglomeration

In addition to convergence and audience fragmentation, concentration is also reshaping telecommunications as ownership of media companies is increasingly being concentrated in fewer and fewer hands. In fact, it is convergence that is, in part, fueling this concentration. As media consultants Michael Wolf and Geoffrey Sands (1999) wrote of the dwindling number of independently owned media outlets:

> The survivors will not just be broadcast networks, or cable news networks, or newspapers, or Websites. They will be entities that encompass all of them, and there will be only a few of them. They will deliver news anywhere, at any time you want it, through your TV, your laptop, even your Palm Pilot, as portable devices allowing online access revolutionize the way people get their news and entertainment. When you watch your favorite cable channel, the announcers will urge you not just to keep watching but to pick up its magazine, tune to its radio station, and log on to its Website. (p. 110)

Author Stanley Bing (2000) wrote of the logical outcome, 50 years in the future, of convergence-driven concentration:

> Our programming mixtures reach us through a variety of pipelines all owned by one of four Great Big Media Companies. These are all exactly alike in their collection of assets, each of them owning broadcast, narrowcast, die-cast, retro-cast and cybercast, broadband, narrowband, audio, video, satellite and an upload-and-download phalanx of option-driven interfaces. Each of our Great Big Media Companies has thousands of brands that make us feel all warm and toasty and provide an emotional connection to a past that nobody can actually remember. (p. 58)

The result of this intermarriage of telecommunications companies is **oligopoly**—the concentration of control of these industries into an ever-smaller number of companies. As a result, a shrinking number of large companies is coming to own not only a greater proportion of the world's media outlets but also a bigger chunk of the communication infrastructure companies that service them, that is, the telephone, satellite, and cable companies. For example, in 1999 Viacom bought CBS for $37 billion, creating a company owning 34 television stations, 173 radio stations, a national television network, cable television networks such as MTV, USA, Nickelodeon, Showtime, Comedy Central, and the Sundance Channel, publishing houses such as Simon & Schuster, the world's largest outdoor advertising company, and Paramount Pictures, one of the world's leading movie studios. Three years earlier Disney bought Capital Cities/ABC, and Time Warner bought Turner Broadcasting. News Corporation, owner of the Fox Television Network, reaches 35% of all U.S. homes with its 22 owned and operated television stations and virtually the entire country with its broadcast and cable networks.

Concentration is a reality in nonbroadcast telecommunications as well. In June 1998 Disney bought Internet search engine Infoseek, and in 1999 NBC bought a share of search engine Snap! As 1998 closed, America Online (AOL) paid $4.2 billion for the Internet's premier browser, Netscape. And just after 1999 opened, @Home, the high-speed Internet access company that uses cable television lines to serve its subscribers, bought Web search engine Excite for $6.7 billion, both now controlled by major stockholder AT&T. In July of 1998 that same AT&T, the largest U.S. communications company and owner of cable television giant TCI, and British Telecom, Britain's biggest telecommunications provider, merged their international operations into a single $10 billion unit. One day after the AT&T and British Telecom deal was made public, Bell Atlantic announced its intentions to purchase GTE Corporation, the largest independent U.S. local and long-distance telephone company, for $52.8 billion. These, however, are small change compared to the third-largest corporate merger in world history, MCI's $115 billion buyout of rival telecom Sprint in 1999, and history's biggest merger, the $190 billion union in 2000 of German telecom giant Mannesmann and British wireless telephone provider Vodafone.

Media observer Ben Bagdikian reported that in 1997 the "number of media corporations with dominant power in society" was 10 (in March 2000 *Harper's* set the number at just 6), and this "new communications cartel" has the power to "surround almost every man, woman, and child in the country with controlled images and words." This places in their hands, Bagdikian argues, the "ability to exert influence that in many ways is greater than that of schools, religion, parents, and even government itself" (as quoted in Goldstein, 1998, p. 52). To critic Eric Effron (1997a), the "serious consequence" of concentration is "the story not told, the questions not asked, the power not challenged" (p. 47). Journalist and social critic Bill Moyers calls this concentration of media ownership "the central issue that faces us as a democratic society" (as quoted in Moore, 1999, p. A11).

There is no more obvious example of this convergence-fueled unification than what was at the time the largest corporate merger in world history, the $184 billion deal through which the world's largest producer of "traditional" media content, Time Warner, and the world's largest-by-far Internet provider, America Online, were united in January 2000. Speaking specifically of the merger's impact on the Internet, *Washington Post* media writer Howard Kurtz

(2000) wrote, "The creation of the world's largest media and cyberspace company seems to muddy the original dream of the Internet as a democratizing force that would enable thousands of individual voices to compete with major media organs. But the rapid expansion of newspaper, magazine, and network (television) sites on the Web has given it a decidedly corporate cast" (p. 7). Kurtz quoted Mark Crispin Miller, director of New York University's Project on Media Ownership, "Is there something good about one very large entity being able to reach everybody?" and James Ledbetter, New York bureau chief of *Industry Standard* magazine, "The wonders of the information age are in the hands of very few companies" (p. 7). *Post* reporters Joel Garreau and Linton Weeks (2000) were equally pessimistic. They wrote, "If AOL becomes just another traditional, arbiter-of-taste, top-down media empire like Time Warner, what we're facing is a stunning number of ways to be inundated with Martha Stewart—on your computer, your TV, your refrigerator screen" (p. 8). Figure 1.1 provides an overview of the last few years of concentration of telecommunications ownership.

Closely related to concentration is **conglomeration,** the increase in the ownership of media outlets by larger, nonmedia companies. "The threat is clear," wrote media critic Steven Brill (2000):

> The bigger these conglomerates get, the less important their journalism gets and the more vulnerable that journalism becomes to the conglomerate's other interests. . . . These mega-companies, therefore, present a new, sweeping, and unprecedented threat to free expression, independent journalism, and a vibrant, free marketplace of ideas. Their sheer enormity makes it almost routine that they are covering a subject involving one of their own divisions or some competitor to one of their enterprises. And their involvement in so much other than journalism threatens to water down the values that would assure that they deal with those conflicts honorably. (pp. 26–27)

According to many observers both inside and outside the media industries, the true impact of conglomeration was on display in the media's behavior during President Bill Clinton's impeachment crisis, a performance widely assailed as questionable at best. For example, at a 1999 *Columbia Journalism Review* forum titled "How the Press Is Shaping (or Misshaping) Politics," historian Arthur Schlesinger argued that the media's failures were "associated with a change in the structure, a competitive structure, of the press." New York University law professor Burt Neuborne concurred. "The press," he said, "has been subsumed into a market psychology, because they are now owned by large conglomerates, of which they are simply a piece. And they [news organizations] are expected to contribute their piece of the profit to the larger pie. You don't have people controlling the press anymore with a fervent sense of responsibility to the First Amendment. Concentrating on who's sleeping with whom, on sensationalism, is concentrating on essentially irrelevant issues" (as quoted in Konner, 1999, p. 6).

There are, however, less dire observations on this growing oligopoly and conglomeration. Many telecommunications professionals argue that

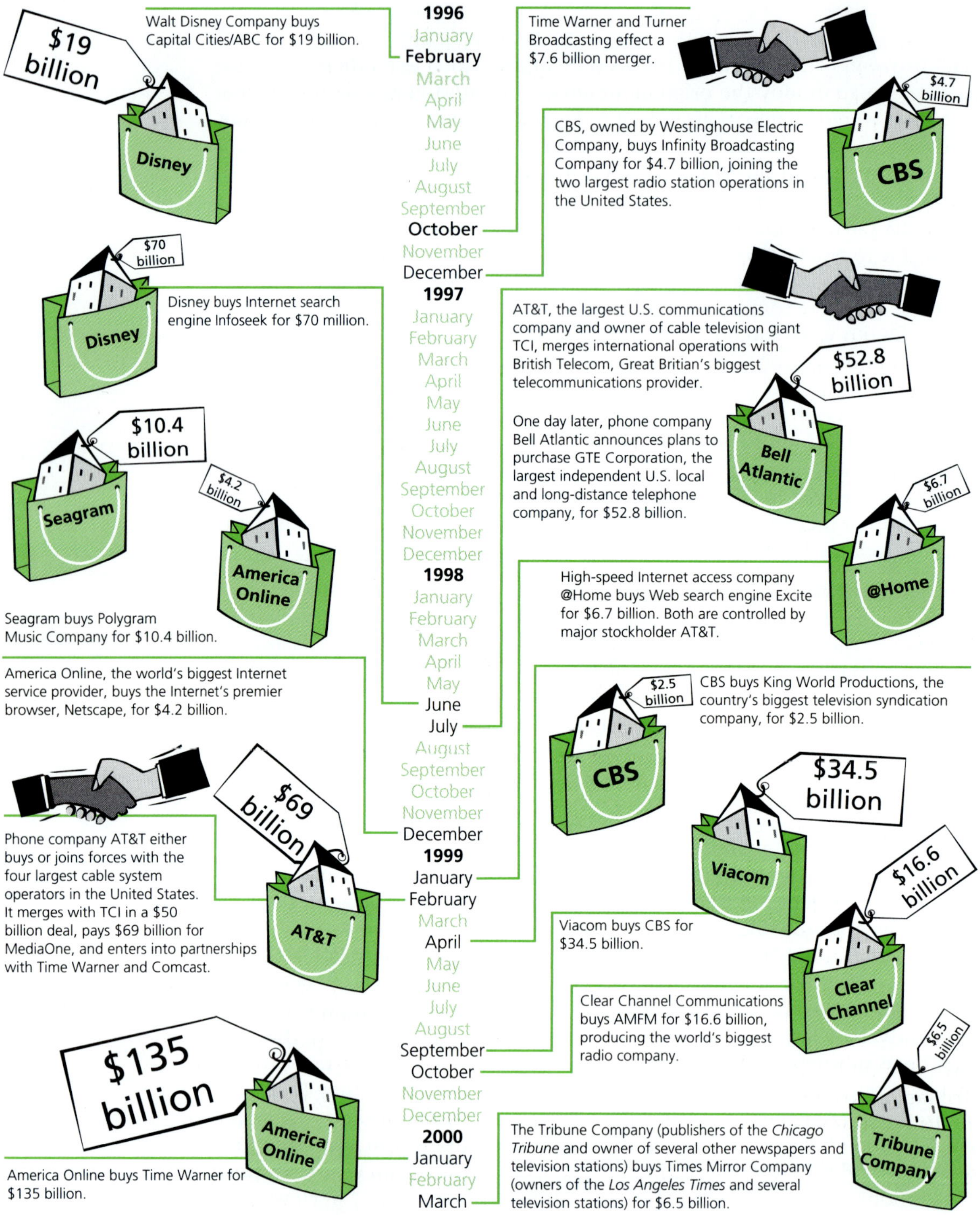

Figure 1.1 Merger (and Acquisition) Mania
Concentration of ownership of telecommunications companies reached epidemic levels in the last few years of the 1990s and the start of the 2000s. Here is a list of some of the notable events in this process.

concentration and conglomeration are not only inevitable but necessary in a telecommunications environment that is increasingly fragmented and internationalized. Companies must maximize their number of outlets to reach as much of the divided and far-flung audience as possible. If they do not, they will become financially insecure, and that is an even greater threat to free and effective mediated communication because advertisers and other well-monied forces will have increased influence over them.

Another defense of concentration and conglomeration has to do with **economies of scale;** that is, bigger can in fact sometimes be better because the relative cost of an operation's output declines as the size of that endeavor grows. For example, the cost of collecting the news or producing a television program does not increase significantly when that news report or television program is distributed over 2 outlets, 20 outlets, or 100 outlets. The additional revenues from these other points of distribution can then be plowed back into even better news and programming. In the case of conglomeration, the parallel argument is that revenues from a conglomerate's nonmedia enterprises can be used to support quality work by its media companies.

Globalization

Globalization is closely related to concentration and conglomeration because it is primarily large international conglomerates, or **transnational corporations (TNC)**, that are making most of the media acquisitions. We've already mentioned the AT&T and British Telecom merger. News Corporation, also just mentioned, is an Australian TNC owning 789 different businesses in 52 countries, and it had $1.32 billion in income in 1997 ("World-Class," 1998, p. 3).

Globalization manifests itself in two ways. First, ownership of telecommunications companies is increasingly international; and second, media content is increasingly borderless—that is, it reaches a worldwide audience. When you go online via America Online or buy discs from RCA Records and Arista, you're doing business with German TNC Bertelsmann. When you went to the theater to see *Titanic,* you watched a movie from two big studios, Paramount, owned by U.S. company Viacom, and 20th Century Fox, owned by Australia's News Corporation. When you rented *Jurassic Park,* you watched a video from Universal Pictures, owned by Canadian TNC Seagram, also owners of the USA cable television network and Putnam books. Japan's Sony owns Columbia Pictures, TriStar Pictures, Sony Pictures Classics, Columbia TriStar Television, Columbia TriStar Home Video, and Loews Theaters. It also owns recording labels Columbia Records, Sony, Chaos, Epic, Relativity, and TriStar Music. Even watching *Sesame Street* puts you in touch with a big TNC, Germany's EM.TV & Merchandising, owners of the Jim Henson Company.

Critics of globalization fear that as telecommunications companies become not only bigger but more distant from the audiences they serve diversity of expression will suffer. The question they ask is, "Will distant, anonymous corporations, many owning nonmedia holdings as well as media businesses, use their power to shape news and entertainment content to suit their own ends, reducing the diversity of opinion and expression available to audiences?" There have been examples of just that. For instance, HarperCollins, a subsidiary of News Corporation, dumped a book from the last Governor of Hong Kong, Chris Patten, even though the publisher had given him a hefty advance and early reviews of the manuscript were glowing. Internal memos were later discovered that made it clear that once News Corporation executives, including Chairman Rupert Murdoch, saw the text of Patten's *East and West* they thought it too critical of the Chinese government. Patten faulted Beijing for its lack of commitment to democracy. But News Corporation had significant business dealings with the Chinese government and had ambitions of even more. News Corporation had already pulled BBC World Television from its Asian television service because that respected news source had aired a speech critical of China's human rights record.

The world's largest untapped audience again figured into a scenario that frightened critics of globalization. Just weeks after his company bought CBS, Viacom CEO Sumner Redstone attended the celebration of the 50th anniversary of communism in China where he announced, "Journalistic integrity must prevail in the final analysis. But that doesn't mean that journalistic integrity should be exercised in a way that is unnecessarily offensive to the countries in which you operate" (as quoted in Baker, 1999, p. 6B). Globalization's opponents fairly asked the question, "What message has the boss just sent his new employees at historically respected CBS News? What stories will go untold, what questions unasked, what power unchallenged?"

Yet defenders of increased globalization point to the same factor fueling concentration—the need to reach a fragmented and widespread audience—as encouraging this trend. They also add the growing economic clout of emerging democracies (and the need to reach the people who live in them) and the increasing intertwining of the world's economies as additional reasons globalization is necessary for the economic survival of telecommunications businesses.

Hypercommercialism

The costs involved in acquiring numerous or large media outlets, domestic and international, must be recouped somehow. Selling more advertising on existing and new media and identifying additional ways to combine content and commercials are the two most common strategies. This leads to what media critic Robert McChesney (1999b) calls **hypercommercialism.** McChesney explained, "Concentrated media control permits the largest media firms to increasingly commercialize their output with less and less fear of consumer reprisal" (pp. 34–35). The rise in the number of commercial minutes in a typical broadcast or cable show is evident to most viewers. The American Association of Advertising Agencies reported that in 1999 there were 16 minutes and 43 seconds of advertising in an average network television prime time hour, a 21.8% increase from 1991. ABC's *Norm,* for example, contained 10 minutes and 19 seconds of commercials during its 30-minute broadcast (Jessell, 2000).

The sheer growth in the amount of advertising is one troublesome aspect of hypercommercialism for many observers, but even more so is the increased mixing of commercial and noncommercial media content. For example, television station WDSI, a Fox News television station in Chattanooga, Tennessee, informed advertisers in June 1999 that it would produce a series of three "positive news segments" about their companies, airing them on its morning, midday, and 10 o'clock news programs. In addition, the station would throw in several promotional spots for advertisers to ensure that viewers tuned in to the "news" featuring the "positive news segments," all for only $15,000 (Cooper, 1999, p. 17). MTV explicitly informs advertisers that it provides editorial coverage and promotional tie-ins only to those movie studios that purchase large amounts of advertising on MTV (Shapiro, 1998). NBC invited its own evening news anchor Tom Brokaw to its *Today Show* four times in 6 weeks to talk about his book, *The Greatest Generation.* NBC's prime time news magazine, *Dateline NBC,* devoted a full hour to the best-seller; later the *Wall Street Journal* revealed that the network owned a 25% interest in the book. Many radio stations now accept payment from record companies to play their songs, an activity once illegal and called **payola.** It is now quite acceptable as long as the "sponsorship" is acknowledged on the air.

Again, as with globalization and concentration, where critics see damage to the integrity of the media themselves and disservice to their audiences, defenders of hypercommercialism argue that it is simply the economic reality of today's telecommunications world.

The Inevitability of Convergence

Whether we deem audience fragmentation, concentration, globalization, and hypercommercialism beneficial or problematic, they are realities. They are not only reshaping the world of telecommunications with which we—as practitioners or as audiences—interact, but they also combine to hasten convergence, further altering telecommunications. Take globalization and concentration as examples. Imagine that one company owns newspapers in the United States; a worldwide online service; U.S., Canadian, Australian, and French television stations; book publishers in the United States and in Great Britain; magazines in Asia, North America, Europe, and Africa; and a film company with international distribution. This creates very strong organizational and economic incentives to maximize the use (and therefore the value) of its content, whether news, education, or entertainment, by using a single piece of content across as many channels of delivery (called **platforms**) as possible and by reducing the format distinctions among those channels. This process is called **synergy,** and it is the impetus behind several recent mergers and acquisitions in the telecommunications industry. In 1997, for example, computer software titan Microsoft paid $1 billion for a 6% interest in cable

Critical Issues *People Left Behind in the Wake of Convergence*

The telecommunications revolution is an established part of our culture. Its achievements fill the front pages of our newspapers. Television news magazines chronicle its marvels. Movies like *The Net, The Matrix,* and *You've Got Mail* cash in on its popular appeal. But not everybody travels the information superhighway, and among those who do, not everyone travels with the same degree of efficiency and luxury. For all its promise and glamour, the newly converged telecommunications technologies are leaving many people behind. Those folks are at the wrong end of the technology gap.

The **technology gap** is the widening disparity between the communication technology haves and have-nots. For example, the typical Internet user is better educated and financially better off than the nonuser. She or he has the money to buy the hardware and software needed to access the Internet and World Wide Web as well as to pay for that connection. These users have the kinds of jobs that grant them the opportunity and the time to spend learning and using the Internet. These are people with the necessary sophistication to do both. These are not all, in fact these are not most, U.S. citizens.

For example, in 1998 the National Telecommunications and Information Administration (NTIA), a division of the Department of Commerce, issued a report titled "Falling Through the Net II: New Data on the Digital Divide." The NTIA reported that,

> despite (a) significant growth in computer ownership and usage overall, the growth has occurred to a greater extent within some income levels, demographic groups, and geographic areas, than in others. In fact, the "digital divide" between certain groups of Americans has *increased* between 1994 and 1997 so that there is now an even greater disparity in penetration levels among some groups. There is a widening gap, for example, between those at the upper and lower income levels. Additionally, even though all racial groups now own more computers than they did in 1994, Blacks and Hispanics now lag *even further behind* Whites in their level of PC-ownership and online access. (McConnaughey, 1998, p. 2)

The Commerce Department revisited the digital divide a year after the NTIA study was released. It showed little improvement, indicating that only one in four African American or Latino households owned a computer with Internet access, compared with half or more of Asian American and White households (Glass, 1999).

Those without access to the new telecommunications technologies are doubly denied. First, they do not have access to what it offers and, as such, their opportunities for information and entertainment are diminished by that much. Second, because

---

television operation US West. Microsoft's goal in this and other similar purchases (it already owned part of cable giant Comsat Corporation and, at the time, was negotiating for a one-third stake in TCI Cable) was to make the delivery and content of cable and the Internet indistinguishable.

Audience fragmentation has a similar impact on convergence. A telecommunications company that has difficulty reaching a mass audience can reach its component parts through a combination of various media. Another factor also supports the trend toward convergence—the audience itself. People are increasingly comfortable receiving information and entertainment from a variety of sources, and nontraditional technologies such as the World Wide Web and e-books frighten fewer and fewer consumers. Not all people, however, are equally comfortable with or equally served by the modern converging

ours is a democratic political and social system that holds sacred the idea of "one person, one vote," those people lacking the necessary technology and skill to participate in the political and social dialogue will be denied their vote. Another important belief in our social and political system is that people live their lives and govern themselves best with full access to information. But people without the requisite technology obviously have diminished access to the information it makes available, disadvantaging them and the system as a whole.

Then there is the problem of what people find when they do enter the information superhighway. According to a study conducted in early 2000 by the Children's Partnership, a nonprofit advocacy group, millions of Americans who are on the Net—those with low income, low education levels, and limited literacy and English language skills—are poorly served by the medium. The organization's Wendy Lazarus explained, "The good news is that as we start to see more under-served people get connected, we see the next set of questions being raised. [The biggest] question is whether they are finding the applications and services they want and need" (as quoted in O'Brien, 2000, p. 1C). Lazarus argued that the answer to that question is "No." She explained, "We found a strong desire among people [at or below the federal poverty line] for practical, local information about their neighborhoods that seems to fly in the face of the way the Internet is moving in terms of national portals" and the Net's increasing commercialism (as quoted in Shiver, 2000a, p. A12).

With telecommunications technologies increasingly converging into digitally produced, programmed, and distributed synergies, this problem becomes even more troublesome. Many people today are not even being served by "traditional" media. Access to cable television is lowest among the urban working class and poor people. Many newspapers, uninterested in these same people because they don't have the demographic profile sought by advertisers, do not promote their papers in poor neighborhoods and, in some large cities, do not even deliver there. For this same reason, few consumer magazines are aimed at less well-off people. If the computer technology gap creates an even wider **information gap** between these and other citizens, they and democracy will surely suffer.

What will become of these people left in technology's wake? What kind of society might result if the technology and information gaps are not closed? Why should you, as an individual, care about these gaps? What can *you* do to close them?

technologies, as can be seen in the box titled People Left Behind in the Wake of Convergence.

Studying Telecommunications

One of media theorist Marshall McLuhan's famous expressions is, "First we shape our tools, thereafter they shape us" ("The Wired Diaries," 1999, p. 97). Bill Davidow, a Ph.D. electrical engineer, Silicon Valley venture capitalist, and former head of Intel's microcomputer division, makes a similar argument. Stating that *tools make the rules,* Davidow (1998) wrote,

Marshall McLuhan

> What will the society of the future, designed by computers, look like? Hundreds of millions of computers are working on the answer today. Millions are redesigning government, business, and commerce, and some are probably engaged in redesigning man. Man has, of course, empowered computers to carry out these processes, but once they are in place, increasingly they proceed on their own, carrying out their mission with little or no human interaction.
>
> We tend to view tools as passive and to believe that since man created them, he controls them. Nothing could be further from the truth.

But Davidow is no **technological determinist;** that is, he does not believe machines *necessarily* make the rules. Accepting **Kranzberg's law**—technology is neither good, nor bad, nor neutral (Ogilvy, 1998, p. 188)—he argues that *the way people use technology* permits people to be the rule-makers. Davidow continued,

> In the Renaissance, the scientific revolution gave man the ability to dominate nature. In the centuries that followed, it gave him the knowledge required to create inanimate sources of power and to raise crops in the face of adversity. It freed him from the constraints imposed by geography by providing him with rapid forms of transit, and permitted him to create new materials to meet his ever changing needs. Man ruled nature and only man ruled man. It would indeed be

Gutenberg and his printing press.

> ironic if one of the final contributions of the "great information evolution" would be that the tools wrenched much of the control from man. . . . This is precisely what will happen if we view tools as passive and assume that they will be used to empower existing institutions and make them better. It is certain that man's efforts will be devoted to making what exists better, but the tools will have a different "plan." (1998, p. 4P)

Before we can effectively use telecommunications technologies, that is, before *we* can rule *them,* we must

- Be familiar with them,
- Know their various uses—their limitations *and* potential, and
- Know how to use them to overcome those limitations and realize that potential.

In fact, a better question than "How can we use telecommunications technologies?" is "How will we know *how to use* telecommunications technologies?" We study telecommunications to answer this question. Let's look at the printing press to examine why.

When Johannes Gutenberg invented the mechanical printing press in the mid-1400s, it was his intention to use his device in service of the Catholic Church. Until this time, Bibles had to be hand copied in a long and laborious process. Gutenberg decided to print Bibles. Once people had ready access to his machine-printed Bibles, he reasoned, the Word of God could spread and more people would enter the Church. But his machine had just the opposite effect. Once the Bible was placed in the hands of more and more people, and once these readers could interpret its content without the Church's supervision and direction, numerous interpretations became possible. Many of those interpretations were not consistent with Church doctrine. Therefore, rather than helping the Church expand, the printing press led in part to its fracture—it contributed to the Protestant Reformation. To use Davidow's argument, Gutenberg saw the printing press as passive and assumed it would be used to empower an existing institution (the Church) and make it better, but others with access to Gutenberg's invention had a different plan.

It's not enough to be familiar with a technology. Who could have been more familiar with the printing press than its inventor? We are all familiar with radio, television, and virtually all the telecommunications technologies mentioned so far in this chapter. But *how well* do we know them, and how well do we understand their limitations and potential? How well can we—either as media professionals or as media content consumers—use these technologies to meet personal, economic, societal, and cultural goals?

After Gutenberg's "failure," others implemented *their* plans for the printing press. Publishers and their readers used print to expand empires, power the rise of democracy, entrench capitalism, and help build a middle class (Eisenstein, 1979; Inglis, 1990; Innis, 1972).

How will we use telecommunications technologies? Will we let them make the rules, or will we rule them? Will the Internet become just more television, offering users entertainment, weather, sports, and commercials? Or will it become a soapbox from which all voices can be heard, all shades of expression aired? What will we do with digital television? Will we use it simply as high-definition television with clearer, sharper images of the same old sitcoms and action adventure shows? Or will viewers demand additional channels and greater interactivity? Will radio become the haven of last resort for foul-mouthed shock jocks, or will it be an effective local voice in a world of ever-expanding international media?

Getting Started

Before we begin our investigation of the different telecommunications technologies and the issues surrounding them, we need to familiarize ourselves with a few central terms and concepts, some narrow in focus, others quite broad.

We've already seen that mediated communication is communication that is facilitated or otherwise conducted by electronic and digital technology. But we should specify exactly what it is we mean by *communication* itself. **Communication** is the process of creating shared meaning. We can talk to each other all we want, but if we do not understand one another (share meaning), there is no communication. CNN can produce the most remarkable documentary on the health crisis in Honduras after Hurricane Mitch, but communication does not occur if we have no basis for interpreting the story.

Mass communication is the process of creating shared meaning between those producing mass media messages and their audiences, and **mass media** are those technologies (and the industries that utilize and support them) that carry messages to a large number of people.

Converging telecommunications technologies increasingly permit communication between media outlets and very small numbers of people (remember audience fragmentation), or, in the case of the Internet and the telephone, even allow single individuals to communicate with solitary others. Because of this, many people prefer the expression mediated communication to mass communication. For this same reason, telecommunications, rather than mass media, is increasingly used to refer to the various communications technologies and the industries that employ and sustain them.

To fully understand telecommunications itself, we need to be familiar with a number of other terms and concepts. The **information superhighway** is an illustrative, collective expression used to describe worldwide digital data networks. Where it may once have been used specifically to describe the Internet and the World Wide Web, because of convergence it now encompasses not only these technologies but the many media that can and do travel it. Cable television and satellite-delivered cellular phones are examples. **Digital** refers to the conversion of signals (audio and video) into 1s and 0s (digits) logged in millisecond intervals in a computerized translation process that permits near-instantaneous and noise-free replication, transmission, and reception.

Included in the telecommunications companies on the information superhighway are the **broadcasters,** those radio and television stations that transmit their signals (digital or otherwise) over the airwaves. These broadcasters frequently link themselves together into **broadcast networks** for the purpose of reaching large national or regional audiences. Those stations that belong to a network are called **affiliates;** those that remain apart from a network are called **independents.** But many people receive radio and television programming in their homes via **cable,** the delivery of content to consumers by wire. When the local NBC affiliate comes to our home screens through the cable, it is still a television station, but it has been **retransmitted** by the cable operator. When we watch content produced specifically for delivery by cable, for example MTV, HBO, and the Comedy Channel, we are watching **cable channels.** Regardless of how content reaches us, those who create it are referred to as **producers;** those who schedule and organize it are called **programmers;** and those who deliver it to the various outlets used by audiences are **distributors.**

Among the most powerful travelers of the information superhighway are the **telcos,** or telephone companies. The phone companies were long content to be **common carriers;** that is, they simply carried the messages of others with no power to shape or restrict them. Today, however, telephone companies are increasingly combining with cable and Internet technologies (both also originally conceived as common carriers) to create new roles and responsibilities for themselves. In other words, they are experiencing convergence.

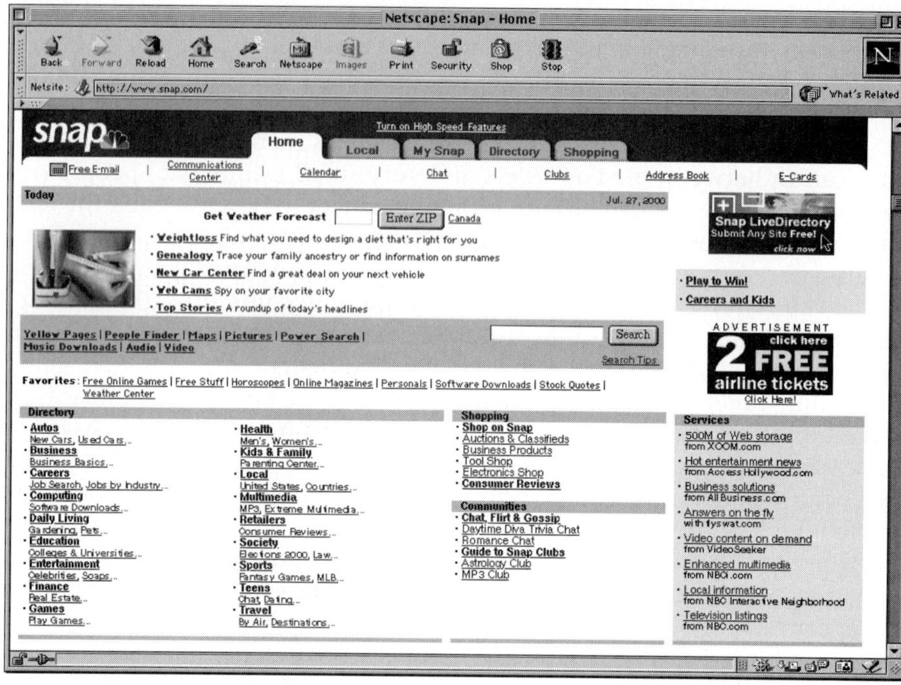

Internet traditionalists dislike portals. NBC-owned Snap! embodies two problems for these critics. First, Snap!'s corporate/big-media ownership is contrary to the Internet's free-wheeling freedom. Second, its many preselected links threaten to steer users away from more robust use of the Net.

Conflicting Telecommunications Cultures

This convergence of media traditionally operated as common carriers (the **telecom model of telecommunications**) with those traditionally based on the **broadcast model of telecommunications**—that is, those relying on advertising for financial support—leads to many of the conflicts and questions you will study in the following chapters. The broadcast model, for example, may not demand direct payment from audience members, but those media companies must expend more effort to reach the people of greatest interest to advertisers than they do to reach the entire body of potential audience members (Chapter 8). Common carriers prosper by delivering greater amounts and a variety of services to as many people as possible who are willing to pay for those services. Just think of the array of services your telephone and cable companies (Chapter 6) and your Internet provider (Chapter 7) offer. But what happens when these technologies converge, forcing a convergence of operating models?

Internet traditionalists, and many everyday users, cringe at the advertising that clutters their screens and slows their download time (Chapter 9).

They also fear the limiting tendencies of **portals,** one-stop brand-name entry points to the Web that offer users news, chat, search functions, original content, links to specific sites, advertising, and e-commerce. Many people believe these portals send users to an ever-narrower, sponsor-paid-for set of Internet sites. The Internet, in this view, ceases to be a common carrier and becomes, in effect, the equivalent of a television network, directing users/viewers to a few preselected pieces of content.

We revisit this and similar controversies and conflicts throughout this text. But it is important to remember that the technologies and industries we are learning about are in a period of remarkable transformation, fueled primarily by convergence, and that convergence affects virtually all aspects of the telecommunications media and industries with which we interact, either as practitioners or audience members, or both.

Chapter Review

Traditionally, when a new medium arrives, older media must find new functions or face being displaced. This is functional displacement. But the more modern form of media change today is convergence, the erasure or blurring of distinctions among different forms of media. Convergence is but one force reshaping telecommunications. Among the others are fragmentation of the audience, the increasingly narrow segmentation and classification of media consumers; concentration of media ownership into an ever-smaller number of companies; globalization, the internationalization of both media ownership and reach; and hypercommercialism.

Current arguments either defend these trends for bringing even better media and media content to audiences or lament them for eroding the quality of media and their content. Individuals in the telecommunications industries and those who make up their audiences can best judge for themselves the correctness of these differing views. But to do so requires a knowledge of each individual medium, the telecommunications industries, and the communication they encourage and shape. To gain this knowledge, we must familiarize ourselves with telecommunications technologies, know their various uses, and understand how to use them to realize their potential and overcome any limitations they may have.

This is not an argument for technological determinism, the belief that machines and their development drive political, economic, and cultural change. Rather, it is an acknowledgment of Kranzberg's law. That is, because our lives are increasingly involved with these technologies and their content, it is *we* who will determine if they are used for good or bad.

Increasingly, telecommunications technologies are converging on the information highway. Broadcasters, broadcast networks, affiliate and independent stations, cable television, satellite content providers, and the telcos are joining together as producers, programmers, and distributors. But convergence of technologies brings with it a parallel convergence of models of financial support. As media become increasingly indistinguishable, the broadcast and common carrier models of financial support inevitably conflict.

Questions for Review

1. What is the distinction between functional displacement and convergence?
2. What are the arguments in support of and against the idea that increased audience fragmentation will produce better media content?
3. What are the arguments in support of and against the idea that concentration of ownership will produce better media content?
4. What factors encourage increased globalization of telecommunications? In what two ways does that globalization occur?
5. What is hypercommercialism? Why is it occurring?
6. What are oligopoly, synergy, and economies of scale? How are they related?
7. What factors render convergence inevitable?
8. What is technological determinism?
9. What is Kranzberg's law?
10. What is mass communication? What is mass media?
11. What is the broadcast model of telecommunications' financial support system?
12. What is a common carrier?

Questions for Critical Thinking

1. Can you give an example of convergence from your own experience that might have surprised you? If so, how was your use of that medium or content affected by its newly converged form?

2. Are critics of concentration of ownership making too much of the fact that some corporations own ever-larger numbers of media and nonmedia companies? Does it matter if Disney or News Corporation own all those different media and nonmedia companies? Won't the desire to earn profits keep them from engaging in action that would appear to be unethical to potential audience members?

3. If you were to ride the train and the conductor brought you a newspaper whose front page was specifically detailed to you and your interests, would you welcome this as a fine service or see it as an invasion of your privacy? Defend your answer.

4. Do you welcome or fear globalization? Will it usher in a true global village, uniting the world and its people, or will it reduce the world's diversity of expression? If you see globalization as homogenizing and limiting different expression, can you argue that this is not a bad thing? Would you argue that we need to limit those things that separate us?

5. Do the technology and information gaps worry you? Why or why not? On which side of the digital divide do you reside?

Engaging Technology

1. Are you a technological determinist? For an interesting, media-based discussion of the interplay of technology and the societies that shape and are shaped by it, visit the Association for Media, Communication and Cultural Studies' site at http://www.aber.ac.uk/~dgc/tecdet.html> and read Daniel

Chandler's discussion of technological determinism.

2. Regularly updated listings of who owns what in telecommunications reside at <http://www.cjr.org/owners/>. From this site you can link to articles on media ownership from *Columbia Journalism Review* as well as to other sites devoted to the issue.

3. The U.S. National Telecommunications and Information Administration regularly tracks technology use and gaps in that use between different groups of people. Access its Web site at <http://www.ntia.doc.gov/ntiahome/>. From there you can link to a variety of sites, including the NTIA reports on the digital divide.

4. One group that believes conglomerate ownership distorts the news is Project Censored. Visit its site at <http://www.projectcensored.org>, and read through the list of stories Project Censored says are underreported due to corporate self-interest. How many of their "most underreported" stories had you heard about? Do you find Project Censored's arguments persuasive as to the value of these stories and the reasons they were ignored? Why or why not?

Chapter Two

Radio and Audio

Gerald Nachman has been a journalist for 40 years, covering entertainment and broadcasting for some of the country's biggest and best newspapers. He is also the author of *Raised on Radio* (1998), a loving, personal memoir of radio's glory days. So it was no surprise that he was stunned when *Time* magazine reviewed the most significant events and people of the 20th century in its November 1998 magazine and ignored radio in its section on entertainment. "There's a huge legacy of radio in contemporary television," he said (as quoted in Miller, 1998, p. 5C). After all, the video formats with which we are familiar—situation comedies, talk shows, game shows, soap operas, newscasts, action adventure shows, and sportscasts—were all begun on radio. The reason for *Time*'s oversight, Nachman believes, is that television "just swallowed radio whole" (p. 5C). But, Nachman reminds us that radio gave the world *CBS News,* soap operas, *Gunsmoke, The Adventures of Ozzie and Harriet, Father Knows Best, Dragnet, Face the Nation, The Lone Ranger, Perry Mason, Truth or Consequences, Twenty Questions,* Groucho Marx's *You Bet Your Life, What's My Line, The Guiding Light,* and a show about "a frantic, endearingly spacy wife trying to compete with her husband" called *My Favorite Husband,* which became *I Love Lucy* on television (p. 5C).

Another person who loves radio is Quincy McCoy, senior editor of the magazine *Gavin*, an important radio business publication. Unlike Nachman, McCoy's concern for radio is not that its past will be forgotten but that its future will render it incidental and unimportant. Listeners, he said "are getting less music, less choice, and more research. Radio was safe enough as it was, so imagine it more homogenized" (Zoll, 1998, p. 16). The threat to radio's well-being, says McCoy, resides in concentration and convergence. These two forces are "really hurting radio's creativity, spontaneity—everything that makes it exciting" (p. 16).

Despite the fact that radio's Golden Age has passed and its future is uncertain, it remains a popular medium. According to the National Association of Broadcasters (2000), there are two radios for every person in the United States, 77% of all U.S. citizens 12 years old and over tune in daily to the radio, and the average listener spends 3 hours and 30 minutes each workday with the medium. This listenership is attractive enough to advertisers that they buy more than $15.2 billion a year of radio airtime (Ditingo, 1999).

Preview

In this chapter we follow radio's rise from wireless telephony through its Golden Age to the medium we know today. We detail the work of scientists interested in the physics of waves and sound, and we relate that work to the advances of Marconi and DeForest. We see how radio's network, economic, and regulatory structures came to be, and how they produced the medium's Golden Age. Television's impact on radio is discussed, as is how radio changed not only to survive but to prosper.

We discuss the operation of contemporary radio, especially its value as an advertising medium and its fragmented, specialized, personal, and mobile nature. These characteristics of radio may be altered by deregulation (and the convergence and audience fragmentation it permits) and new technology. Therefore, we examine these factors in depth, paying particular attention to satellite, cable television, digital recording, and Web radio.

Technological and Economic Roots

American inventor Samuel Morse made long-distance communication a reality with his 1844 introduction of the telegraph. Morse's telegraph messages were composed of electrical impulses—dots and dashes—carried on wires strung from poles. The next challenge was to produce the same long-distance communication but *without* wires, called **wireless telegraphy.** Not long after Morse's advance, Alexander Graham Bell exhibited his new invention, the telephone, at the 1876 Philadelphia Centennial Exhibition. Bell, a teacher of the deaf, developed the telephone to provide an effective hearing aid and in that process demonstrated that voices could be transmitted. This produced a

Guglielmo Marconi

second challenge for radio's early innovators, **wireless telephony,** that is, producing long-distance communication not only without wires but with voices instead of impulses.

Several different wireless systems were tried. Mahlon Loomis received the first U.S. patent for the wireless in 1872. Like many now-forgotten inventors of his time, Loomis thought the secret to wireless communication was in sending signals through the water and the ground, but this **conduction system** proved too limiting. What was needed was a system that would allow signals to be reliably sent and received through the air, a **radiation system.** The term "radio" comes from this practice of "radiating" signals through the air.

Scottish physicist and mathematician James Clerk Maxwell predicted that radio waves existed, mathematically demonstrating in 1864 that a radiation system was possible using electromagnetically sent signals. Twenty-three years later German physicist Heinrich Hertz experimentally demonstrated that Maxwell's equations were correct, sending and receiving wireless signals over short distances in his laboratory. Today, the standard measurement of audio (and other) wave frequency is the **Hertz,** honoring this scientist. In the early 1890s, soon after Hertz's advance, French physicist Édouard Branly developed a much improved receiver, and before that decade ended English physicist Sir Oliver Lodge perfected the principle of tuning, allowing the transmitter and receiver to operate on the same **wavelength** (the distance between successive positive or negative peaks of a wave).

Guglielmo Marconi, son of a wealthy Italian diplomat and his Irish wife, avidly followed these advances in the scientific literature. By 1896 his improvements on the work of Maxwell and Hertz allowed him to send and re-

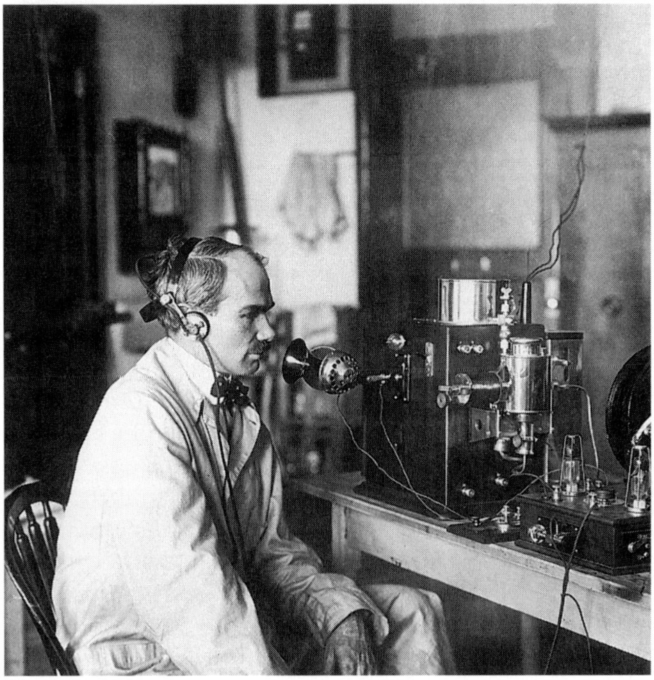

Lee DeForest

ceive telegraph code over distances as great as 2 miles. When his native Italy expressed disinterest in his invention, Marconi used his mother's contacts in Great Britain to find support and financing for it there. England, with a vast global empire and the world's largest navy and merchant fleets, was naturally interested in long-distance wireless communication. With the financial and technical help of the British, Marconi successfully transmitted a radio signal across the English Channel in 1899 and across the Atlantic in 1901. The wireless transmission of signals had become a reality, and, as a result, Marconi is remembered as the "Father of Radio." Marconi was satisfied with wireless telegraphy, but other scientists still sought the secret to wireless telephony. Several important developments would soon make this transmission of voices by wireless possible.

In 1903 Reginald Fessenden, a Canadian, invented the **liquid barretter,** the first audio device permitting reception of wireless voices. On Christmas Eve in 1906 Fessenden made the world's first public broadcast of voices and music from Brant Rock, a small Massachusetts coastal village. His listeners were a few specially equipped ocean-going ships and newspaper offices.

Also in 1906, American Lee DeForest invented the **audion tube,** a vacuum tube that improved and amplified wireless signals. Now the reliable transmission of clear voices and music was a reality. But DeForest's second important contribution was that he saw radio as a means of *broadcasting.* DeForest's technical advances combined with his vision of radio as a broadcast medium led radio historian Michael Adams (1996) to call him "the single most important individual in the twenty year evolution from two-way wireless telegraphy . . . to the technical perfection and use of the radiotelephone for entertainment broadcasting" (pp. 99–100).

The early pioneers, including Marconi, saw radio as a means for conducting point-to-point business and commercial communication, for example, from ship to ship or ship to shore. But in the 1907 prospectus for his radio company, DeForest wrote, "It will soon be possible to distribute grand opera music from transmitters placed on the stage of the Metropolitan Opera House by a Radio Telephone station on the roof to almost any dwelling in Greater New York and vicinity. . . . The same applies to large cities. Church music, lectures, etc., can be spread abroad by the Radio Telephone" (as quoted in Adams, 1996, pp. 104–106). In fact, from 1907 to 1912 DeForest did conduct a series of highly publicized broadcasts of opera in New York. Soon, countless "broadcasters" went on the air. There were so many "stations" and so little order that havoc reigned. Some broadcasters were large corporations looking to dominate the medium for profit, and some were hobbyists, playing with the medium for the sheer joy of it. Yet the promise of radio was such that the medium continued to mature until World War I, when the government ordered the immediate cessation of all radio communications.

Early Advances in Audio Recording

The decades just before the turn of the 20th century saw the beginning of sound recording. In 1877 prolific inventor Thomas Edison patented his "talking machine," a device for duplicating sound that employed a needle and a hand-turned, grooved cylinder. As the cylinder was cranked, it would rotate, causing the needle to pass along its groove and hit bumps that were made there by a similar needle. This mechanical movement of the needle was converted into electrical energy that activated a diaphragm in a loudspeaker, producing sound. The drawback of this system was that only one "recording" could be made of any given sound; that is, the cylinder could not be duplicated. That problem was solved in 1887 by German immigrant Emile Berliner. He developed the gramophone, which used a flat, rotating, wax-coated disc that could easily be copied or pressed from a metal master. Berliner's equally important contributions to audio recording were the development of a sophisticated microphone and, through his company Victor Records, the importation from Europe of recordings by famous opera stars. Now people not only had a reasonably priced record player but had records to play on it.

Columbia Phonograph Company improved on Berliner's innovation with its introduction of the two-sided disc in 1905, and by 1920 the record player—either as the gramophone or the phonograph—was a standard feature in U.S. homes. There were more than 2 million machines and 107 million recordings sold in 1919 alone. Public acceptance of the new medium was enhanced even more by the development of higher fidelity electromagnetic recording in 1924 by Bell Laboratory's Joseph P. Maxwell.

The Coming of Broadcasting

The idea of broadcasting—transmitting voices and music at great distances to a large number of people—was around well before the development of radio. Bell's telephone company had a subscription music service in major cities in the late 1800s, delivering music to homes and businesses by phone wires. The public anticipated and, after DeForest's successful opera broadcasts, was eager for music and voices at home. According to a legend now questioned by many historians, Russian immigrant David Sarnoff, then an employee of American Marconi, recognized this desire and in 1916 sent his superiors what has become famous as the "Radio Music Box Memo." In this memo Sarnoff wrote of

> a plan of development which would make radio a "household utility" in the same sense as the piano or phonograph. The idea is to bring music into the house by wireless. . . . The receiver can be designed in the form of a simple "Radio Music Box" and arranged for several different wavelengths, which should be changeable with the throwing of a single switch or pressing of a single button. (as quoted in Sterling & Kitross, 1990, p. 43)

The introduction of broadcasting to a mass audience was delayed in the first two decades of the 20th century by patent fights, lawsuits, and war. DeForest and Fessenden were financially destroyed by the legal conflict, and use of the airwaves was forbidden to nonmilitary broadcasters during the United States' involvement in the armed conflict in Europe. When World War I ended, an expectant audience awaited what had become a much improved medium. In a series of developments that would be duplicated for television at the time of World War II (see Chapter 3), radio was transformed from an interesting technological idea into an entertainment and commercial giant. First, to aid the war effort, the government took over the patents relating to radio, continuing to improve it for military use. Thus, the refinement and development of radio's technical aspects continued throughout the war. Then, when the war ended, the patents were returned to their owners, but the bickering was renewed.

Some were concerned that the medium would wither as patent holders refused to cooperate with one another. Others were fearful that a foreign company (British Marconi) would control this vital resource. With the government's encouragement, in 1921 the competing companies merged. American Marconi, General Electric, American Telephone & Telegraph, and Westinghouse each controlled a vital radio patent. They joined to create the Radio Corporation of America (RCA). The creation of RCA helped stave off direct government control of the new medium. Twenty-eight-year-old David Sarnoff, author of the Radio Music Box Memo, was RCA's commercial manager. The way for the medium's popular growth was paved. Its success was guaranteed by a public that, because of the phonograph, was already attuned to music in the home and, thanks to the just concluded war, was awakening to the need for instant, wide-ranging news and information.

On September 30, 1920, a Westinghouse executive, impressed with press accounts of the number of listeners who were picking up broadcasts from the garage radio station of company engineer Frank Conrad, asked him to move his operation to the Westinghouse factory and expand its power. Conrad did so, and on October 27, 1920, experimental station 8XK in Pittsburgh, Pennsylvania received a radio license from the Department of Commerce to broadcast as KDKA. On November 2 KDKA made the first commercial radio broadcast, announcing the results of the presidential election that sent Warren G. Harding to the White House.

The Coming of Regulation

Not only did radio give television many of its most recognizable formats and popular programs, as we saw at the beginning of this chapter, it also transferred to the newer medium its regulatory philosophy and structure.

As its encouragement of the creation of RCA demonstrated, the government had a keen interest in the development, operation, and diffusion of radio. At first its interest was in the use of point-to-point communication. In 1909 almost all the passengers on the wireless-equipped *Republic* were saved as the ship sunk off the east coast of the United States. The vessel had used its radio to call for help from nearby ships. Soon after, Congress passed the Wireless Ship Act of 1910, requiring that all ships using U.S. ports and carrying more than 50 passengers have a working wireless and operator. The wireless industry did not object, as the legislation boosted sales. Then the *Titanic* struck an iceberg in the North Atlantic in 1912. When it was later learned that hundreds of lives were needlessly lost because some ships in the area had left their radios unattended, Congress passed the Radio Act of 1912, which not only strengthened the rules regarding onboard wireless but also required that wireless operators be licensed by the Secretary of Commerce and Labor.

The Radio Act of 1912 established definitions of authority between federal and state governments, provided for allocating licenses and fining violators, and assigned frequencies for station operation. The government was in the business of regulating what was to become broadcasting.

Where the 1910 act had boosted sales, the 1912 act imposed control. Therefore, many in the wireless industry objected. Two legal challenges to the government's regulatory authority eventually followed, *Hoover v. Intercity Radio Co., Inc.* in 1923 and *U.S. v. Zenith Radio Corp. et al.* in 1926. The broadcasters won both, effectively negating the federal government's power to license. President Calvin Coolidge ordered an end to government regulation of radio despite his publicly stated belief that chaos would descend on the medium.

Coolidge proved prophetic. The industry's years of flouting the 1912 act had led it to the brink of disaster. Radio sales and profits dropped dramatically as listeners tired of the chaos. Stations arbitrarily changed frequencies, power, and hours of operation, and there was constant interference between stations, often intentional, making it difficult or even impossible to clearly hear signals. Leaders of the commercial radio industry petitioned Commerce Commissioner Herbert Hoover who, according to historian Erik Barnouw (who ti-

The wireless-telegraphy room of the *Titanic.* Despite the heroic efforts of wireless operator Jack Philips, scores of people died needlessly in the sinking of that great ocean liner because the ships in its vicinity did not man their receivers.

tled his book on radio's early days *A Tower in Babel,* 1968), "encouraged firmness" in his efforts to regulate and control the competitors. The government's response was a series of four National Radio Conferences involving industry experts, public officials, and government regulators. These conferences led to the Radio Act of 1927. Order was restored and the industry prospered, but the broadcasters had made an important concession to secure the saving intervention. The 1927 act authorized broadcasters to *use* the channels but not to own them. The airwaves belonged to the public, and broadcasters were simply the caretakers of this national resource.

The act further stated that when a license was awarded the standard of evaluation would be the "public interest, convenience, or necessity"; that is, broadcasters would be required to serve the listeners in the local areas for which they were licensed. The Federal Radio Commission (FRC) was established to administer the provisions of the act and based this **trustee model** of regulation on two premises (Bittner, 1994). The first was the philosophy of **spectrum scarcity.** This is the idea that because broadcast spectrum space is limited, and therefore not everyone who wants to broadcast can, those who are granted licenses to serve a local area must accept regulation. The second was the issue of influence—because broadcasting reaches virtually everyone in the society, it is a power so great that it requires official oversight.

The Communications Act of 1934 replaced the 1927 legislation, substituted the Federal Communications Commission (FCC) for the FRC, and cemented its regulatory authority, which continues today. Chapter 12 expands this discussion of telecommunications regulation.

Advertising and the Networks

While the medium's regulatory structure was evolving, so were its financial bases, and these, too, would carry over to television when it eventually appeared. The formation of RCA had ensured that radio would be a commercial, profit-based system. The industry supported itself through the sale of receivers; that is, it operated radio stations in order to sell radios. Once everybody had a radio, however, radio purchases would slow down. The solution was advertising. On August 22, 1922, AT&T's New York station WEAF accepted the medium's first paid commercial, a 10-minute spot for Long Island brownstone apartments. The cost was $50.

The sale of advertising helped speed establishment of the national radio networks. Before commercial sponsorship of programming, AT&T had linked stations into **chains,** or networks, that offered **toll broadcasting.** This system made it possible for anyone to buy airtime to deliver a message over a wide geographic area. AT&T was using radio as a kind of high-powered long-distance telephone service. Westinghouse, RCA, and General Electric quickly realized that linking groups of stations, or **affiliates,** was economically wise (the same piece of content could be "stretched" over many cities). They could deliver larger audiences, realizing greater advertising revenues, which would allow them to hire bigger stars and produce better programming, which would attract larger audiences, which could be sold for even greater fees to advertisers. RCA set up its 24-station network, the National Broadcasting Company (NBC), in 1926. A year later it bought AT&T's stations and launched a second network, NBC Blue (the original NBC was renamed NBC Red). The Columbia Broadcasting System (CBS) was also founded in 1927, but it struggled until 26-year-old millionaire cigar maker William S. Paley bought it in 1928, making it a worthy competitor to NBC. The fourth network, Mutual, was established in 1934 largely on the strength of its hit western, *The Lone Ranger.* Four midwestern and eastern stations came together to sell advertising on it and other shows. Soon Mutual had 60 affiliates. Mutual differed from the other major national networks in that it did not own and operate its own flagship stations (called **O&Os**, for owned and operated). By 1938 the four national networks had affiliated virtually all of the country's large stations and the majority of its smaller operations. They grew so powerful that the government forced NBC to divest itself of one of its networks in 1943. It sold NBC Blue to Life Saver candy maker Edward Noble, who renamed it the American Broadcasting Company (ABC).

The fundamental basis of broadcasting in the United States was set:

- Radio stations were private, commercially owned enterprises rather than government operations.
- Governmental regulation was based on the public interest.
- Stations were licensed to serve specific localities, but the most lucrative and heavily listened to hours were programmed by national networks.
- Entertainment and information were the system's basic content.
- Advertising formed the basis of broadcasting's financial support.

The Golden Age

The networks ushered in radio's Golden Age. Although the Great Depression (1929–1939) damaged the phonograph industry, with sales dipping to as few as 6 million records in 1932 (from 107 million in 1919), it helped boost radio. Phonographs and records cost money, but once a family bought a radio, a whole world of entertainment and information was at its disposal—free of charge. The number of homes with radios grew from 12 million in 1930 to 30 million in 1940, and half of them had not one but two receivers. Despite the hard economic times, ad revenues rose from $40 million to $155 million over the same period. Among them, the four national networks broadcast 156 hours of network-originated programming each week. New genres became fixtures during this period. Among these rising stars were comedies (*The Jack Benny Show, Fibber McGee and Molly*), audience participation shows (*Truth or Consequences, Kay Kyser's Kollege of Musical Knowledge*), children's shows (*Little Orphan Annie, The Lone Ranger*), soap operas (*Oxydol's Ma Perkins, The Guiding Light*), and dramas (Orson Welles' *Mercury Theater of the Air*).

News, too, became a radio staple. Prior to 1933, many newspapers refused to carry radio station schedules except as paid advertising, and wire services like the Associated Press and United Press declined to make their stories available to radio stations. Both the wire services and the newspapers feared competition from the powerful new medium. The **Biltmore Agreement,** named after the New York hotel in which the press–radio talks took place, settled the dispute, restricting radio to brief news reports at specific, limited times of the day. This ensured newspapers' dominance as a journalistic medium. Still, news had come to the airwaves and would prove to be one of the medium's most valued services during the subsequent years of global conflict, quickly rendering the limits of the agreement obsolete.

Radio and Audio Recording in World War II Radio's Golden Age shone even more brightly after Pearl Harbor was bombed by the Japanese in 1941, propelling the United States into World War II. Radio was used to sell War Bonds, and much of radio's content was aimed at boosting the nation's morale. The war increased people's desire for news, especially from abroad. It also caused a paper shortage, reducing advertising space in newspapers. Because the FCC licensed no new stations during the war years, the 950 existing broadcasters

Kay Kyser's Kollege of Musical Knowledge was not only a very popular national audience participation radio show, it was also a popular early television program as well, moving to the NBC television network in 1950.

reaped all the broadcast advertising revenues as well as additional ad revenues that would have otherwise gone to newspapers. Yearly ad revenues doubled by war's end in 1945.

The technical development of sound recording benefited from the war as well. Until WWII, recording in the United States was done either directly to master metal disc or on wire recorders, literally magnetic recording on metal wire. But soldiers brought a new technology back from occupied Germany, a tape recorder that used an easily handled plastic tape on a reel. Other advances in sound recording soon followed. In 1947 Columbia Records introduced a new $33\frac{1}{3}$ rpm (rotations-per-minute) long-playing plastic record perfected by Peter Goldmark. A big advance over the previous standard of 78 rpm, it was more durable than the older shellac discs and played for 23 rather than $3\frac{1}{2}$ minutes. Columbia offered the technology free to all other record companies. RCA refused the offer, introducing its own 45 rpm disc in 1948. It played for only $3\frac{1}{3}$ minutes and had a huge center hole requiring a special adapter. Still, RCA persisted in its marketing, causing a speed dispute that was settled in 1950 when the two giants compromised on $33\frac{1}{3}$ rpm as the standard for classical music and 45 rpm as the standard for pop.

Television Arrives When the war ended and radio licenses were granted again, the number of stations grew to 2,000. Annual ad revenues reached $454 million in 1950. Then came television. From a high of 97% in 1945,

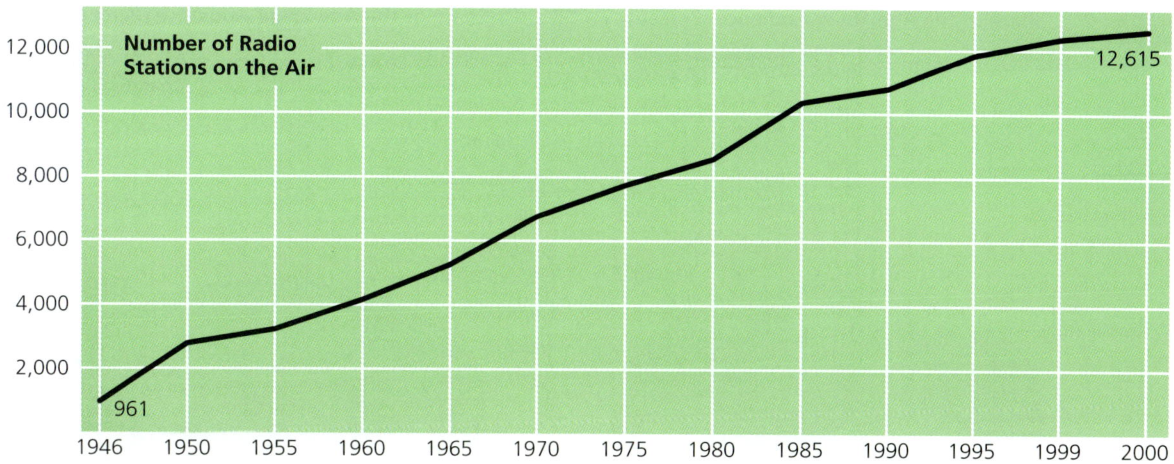

Figure 2.1 Growth of Radio After Television's Arrival
Not only did radio not die after the coming of television, the number of stations has shown constant and steady growth. *Sources: Broadcasting & Cable,* various issues; M Street Corporation.

radio network affiliation dropped to 50% by the mid-50s as stations "went local" in the face of television's national dominance (Figure 2.1). National radio advertising income dipped to $35 million in 1960, the year television found its way into 90% of U.S. homes. If radio was to survive, it would have to find new functions.

Contemporary Radio

Radio has more than survived; it has prospered by changing the nature of its relationship with its audience. The easiest way to understand this is to see pretelevision radio as television is today—nationally oriented, broadcasting an array of recognizable entertainment program formats, populated by well-known stars and personalities, and consumed primarily in the home, typically with people sitting around the set. Posttelevision radio is local, fragmented, specialized, personal, and mobile. Where pretelevision radio was characterized by the big national networks, today's radio is dominated by **formats,** that is, a station's particular sound (Figure 2.2).

There are 12,615 radio stations operating in the United States today: 4,783 commercial AM stations, 5,766 commercial FM stations, and 2,066 noncommercial FM stations.

Who is radio's audience? Teens 17 years old and younger average 2 hours, 19 minutes of radio listening a day; men ages 18 and over average 3 hours, 29 minutes; and women ages 18 and over, 3 hours, 17 minutes. Teenagers and adults listen an average of 22 hours a week. The majority of us (more than 60%) get our first news of the day from radio. Most listening is done

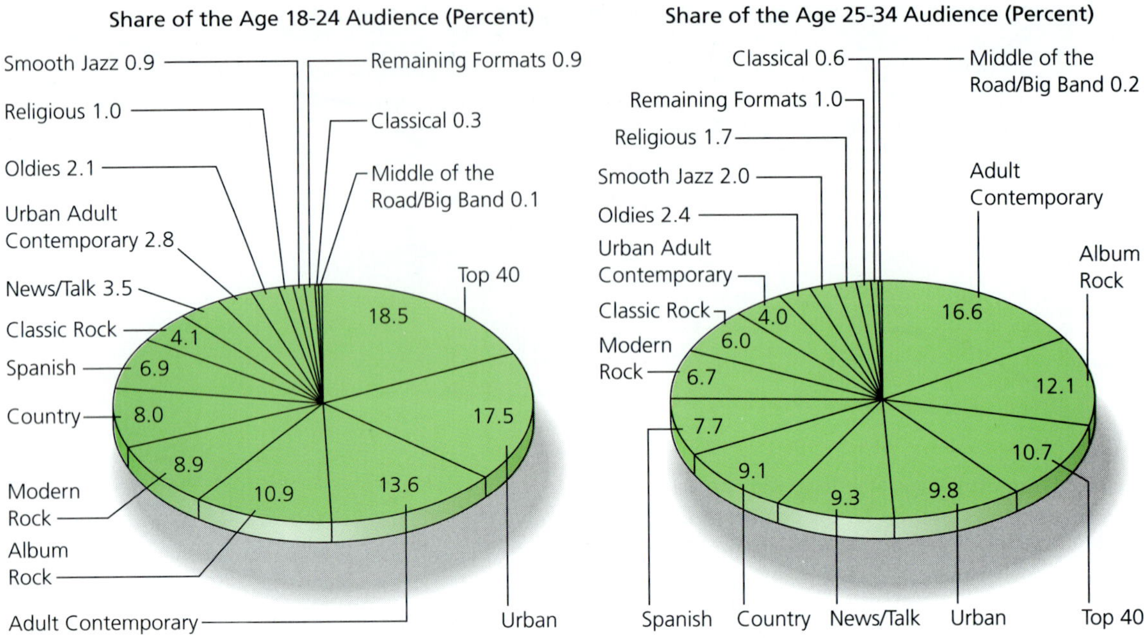

Figure 2.2 Most Listened To Formats, Ages 18–24 and 25–34
Different formats are designed to appeal to different listener demographics. Here are the most popular formats for two desirable age groups. Can you identify differences in listenership? Can you explain those differences? *Source:* Radio Advertising Bureau, <http://www.rab.com> (June 1, 2000).

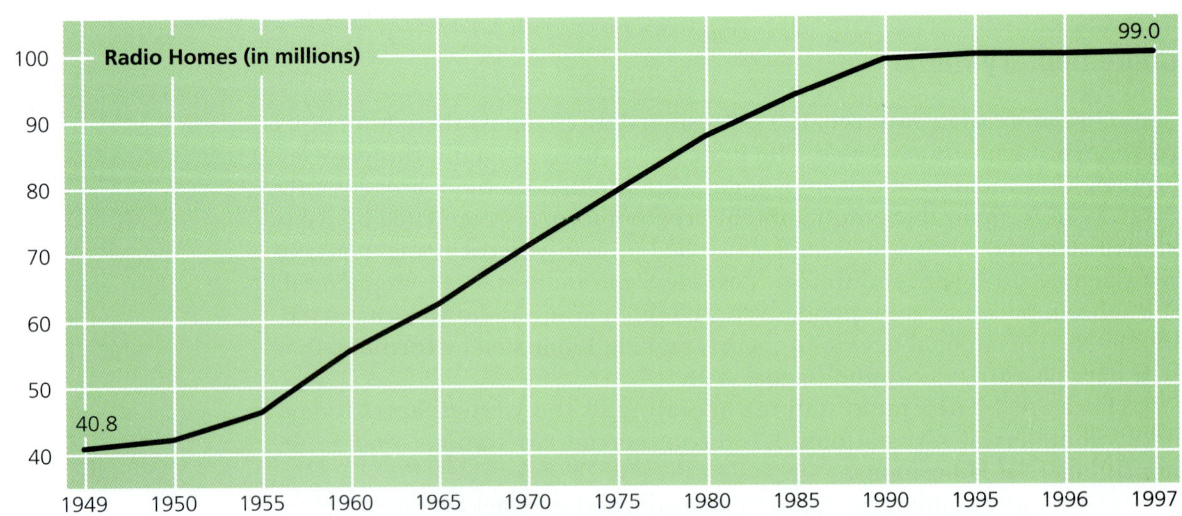

Figure 2.3 U.S. Radio Audiences in the Television Era
Not only did radio's audience not diminish in the television era, it, like the number of stations, has shown constant and steady growth. *Sources: Broadcasting & Cable Yearbook, 1997, 1998, 1999, 2000; Broadcasting & Cable*, various issues.

Tech Talk *Cycle, Hertz, and Frequency*

We've seen that radio comes to us as the modulation of either a carrier wave's amplitude (AM) or frequency (FM). The wave and its manipulation, then, are of great importance in understanding radio.

A **cycle** is the movement of an individual wave past a specific point. One undulation of a wave from a high point to a low point back to a high point (or from a low point to a high point to a low point) is one cycle.

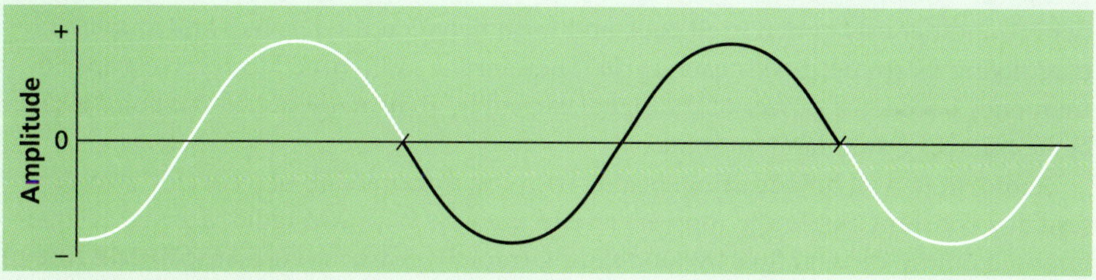

The number of cycles that occur in one second is a transmission's **frequency,** and frequency is measured in Hertz. One Hertz is one cycle per second. A kilohertz is 1,000 cycles per second. A megahertz is 1 million cycles per second.

away from home (Figure 2.3). There are 576.5 million radios in the United States, a 26% increase over the 1980 total. Although the large majority of radios, 367.4 million, are in people's homes, most listening happens outside the home on the country's 186.5 million car, van, truck, and RV radios and its 22.6 million workplace sets (*New York Times,* 1998, p. 398).

FM, AM, and Noncommercial Radio

Although FMs constitute only 55% of all commercial stations (to AM's 45%), the FMs attract fully twice as many listeners. This has to do with the technology behind each. The FM signal can be broadcast in stereo, and it also has better fidelity than the signal provided by AM stations. As a result, people attracted to music, the medium's staple, gravitate toward FM (of course, only after radio programmers, recognizing its better fidelity, moved their music to FM). People favoring news, sports, and information tend to find themselves listening to the AM dial.

AM signals travel farther than FM signals, which makes them better suited for rural parts of the country. But rural areas tend to be less heavily populated, and most AM stations serve fewer listeners. The FCC approved stereo AM in 1985 but refused to set technical standards, counting on the marketplace to do so. Broadcasters and listeners alike were unwilling to gamble on what might be a losing standard, which effectively doomed this ad-

Tech Talk *AM and FM Waves*

Whenever electrons are set in motion, radio waves occur. When you turn on a light or even a toaster, for example, electrons are activated and radio waves are generated. But what differentiates these random waves from those that give us radio is that radio waves are modulated, that is, shaped. The problem faced by radio's technical pioneers was to develop radios that could receive only the wanted waves and screen out the unwanted signals.

Edwin Armstrong discovered that a received radio signal fed back into an amplifier instantaneously cycled itself back and forth between the receiver and amplifier, magnifying its strength with each cycle. Therefore, a radio tuned to receive a specific frequency boosts the power of the signal carried by that frequency to the point it can be heard over other, random signals.

Radio, in fact all broadcasting, begins with the transmission of a **carrier wave,** sent at a constant amplitude (power) and frequency. With AM radio, a signal is superimposed on the carrier wave by modulating (rapidly changing) the **amplitude** (the height and depth) of the wave. This **amplitude modulation** gives AM its name.

When there is no sound being transmitted, the carrier wave assumes a uniform shape:

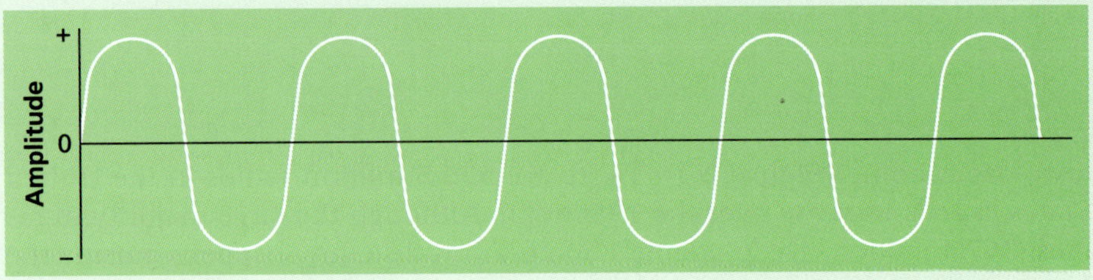

But when voices or music are added, the height and depth (amplitude) of the wave is shaped, or modulated:

vance. Even today, with a technical standard in place, relatively few people have AM stereo receivers, and there seems to be little demand for news, sports, and information in stereo. The boxes Cycles, Hertz, and Frequency and AM and FM Waves offer more technical explanations of these phenomena.

FM came about as a result of work begun in 1923 by inventor-innovator Edwin Armstrong. Already instrumental in both the early development of radio and in RCA's economic and technological success, by 1935 he was demonstrating FM, as well as stereo radio, to RCA's David Sarnoff. But RCA rejected this potential competitor to its AM domain to focus instead on television. So Armstrong turned to GE, and together they put the first FM station, W2XMN, on the air in 1939. The war, however, interrupted the diffusion of FM, and at

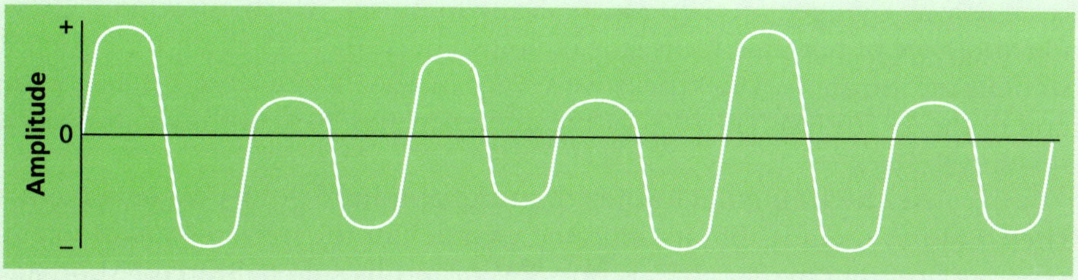

The AM receiver detects these alterations. But even though the selected signal is more powerful than the interfering ones, those other signals still exist everywhere in the air. That's why we sometimes get static when we listen to AM during an electrical storm, for example, or when we run a microwave near our receiver. It was Armstrong, again, who solved this problem. He invented FM, or **frequency modulation.** In FM it is not the height of the wave that is altered but its frequency:

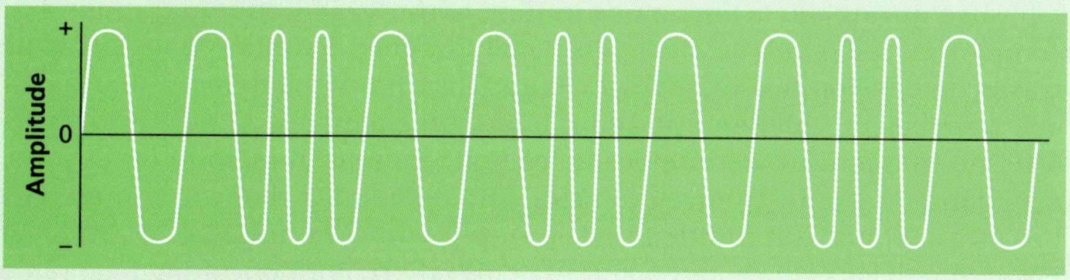

Because FM receivers detect changes in wave frequency rather than alterations in wave height, all radio waves of different amplitude (heights), naturally occurring or otherwise, are ignored.

In contemporary radio, AM operates on radio spectrum frequencies between 535 and 1705 kHz (kilohertz), and FM operates on those between 88 and 108 MHz (megahertz).

war's conclusion the FCC rededicated the part of the broadcast spectrum that had been used for FM to the hot new medium, television. This effectively rendered the half million existing FM radios useless. The commission assigned new spectrum space to FM, but it was slow to recover.

In 1961 the FCC approved FM stereo broadcasts. In 1962 it imposed a partial freeze on the allocation of new AM licenses, making it permanent in 1968. And from 1963 to 1967 the FCC put its nonduplication rule into action. Until that time, most FMs were part of **AM/FM combos,** wherein two stations had common ownership, often simultaneously broadcasting identical content. The AM licensees were content simply to keep the FM stations out of the hands of potential competitors. Under the **nonduplication rule,** however, holders of an AM and an FM license in the same market were forced to broadcast different content at least 50% of the time. The AMs were already successful, so the content was changed on the FMs, typically to music, which attracted a growing audience of portable transistor radio listeners. In 1945 there were 50 FM stations on the air; in 1960 there were 785; in 1965 there were 1,300. By the end of the 1970s FM had 70% of the audience and 70% of the medium's ad revenues.

Because the frequency rather than the amplitude of its wave is modulated, FM has room within its allocated channel for the transmission of totally separate program matter. This is how FM brings us stereo. Stereo FM broadcasters in effect send two distinct signals, and we tune to a frequency between them, giving us stereo. But this available channel space can also be used for **subsidiary communications authorizations (SCAs)** such as Muzak, financial information, and other subscription services. This practice is called **multiplexing.**

Many of the stations that audiences tune in are noncommercial—they accept no advertising. When the national frequency allocation plan was established during the deliberations leading to the 1934 Communications Act, commercial radio broadcasters persuaded Congress that they alone could be trusted to develop this valuable medium. They promised to make time available for religious, children's, and educational programming. At that time, therefore, no frequencies were set aside for noncommercial radio to fulfill these functions. At the insistence of critics who contended that the commercial broadcasters had not fulfilled their promise with AM, the FCC set aside all FM frequencies between 88.1 and 91.9 megahertz for noncommercial radio in 1945. Today, the country's 2,066 noncommercial stations not only provide local service but many offer national network quality programming through affiliation with National Public Radio (NPR), Public Radio International (PRI), or a number of smaller national networks (for example, Pacifica Radio). Chapter 4 offers a fuller discussion of noncommercial radio in its various forms.

Radio's Local Nature

No longer able to compete with television for the national audience in the 1950s, radio was forced to reinvent itself as a local medium. Because it cost much more to run a local television station than a local radio station, advertising rates on radio tend to be much lower than on television. Therefore, many local advertisers choose radio over television for the placement of commercials because they can reach local audiences for less money than they can with television. We discuss radio advertising in detail in Chapter 9.

Radio's Fragmented Nature

Radio stations are widely distributed throughout the United States, with virtually every town, even those with only a few hundred residents, having at least one station. The number of stations licensed in an area is a function of both its population and its proximity to other towns. Tiny Long Beach, Missouri, has one FM station. White Bluff, Texas, has one AM station. Chicago has 19 AM and 30 FM stations, and New York City has 17 AM and 28 FM stations. This fragmentation—many stations serving different areas—makes possible contemporary radio's most important characteristic, its ability to specialize.

Radio's Specialized Nature

With the demise of network radio, individual stations could no longer program the expensive, star-filled genres of the medium's Golden Age. The problem faced by the industry in the 1950s was how to program a station with interesting content, and do so economically. A disc jockey playing records was the best solution. Stations soon learned that a highly specialized, specific audience of particular interest to certain advertisers could be attracted with specific types of music, and format radio was born. Of course, choosing a specific format means accepting that many potential listeners will not tune in, but in format radio the size of the audience is secondary to its composition.

Broadcasting & Cable Yearbook 2000 recognizes more than 90 different formats, from the most common—Country, Top 40, Album-Oriented Rock, All Talk—to the somewhat uncommon such as Ukranian and Bluegrass. Many stations, especially those in rural areas, offer **secondary services** (formats) at particular times on certain days. For example, a Country station may broadcast a Religious format for 10 hours on Saturday and Sunday.

Format radio offers stations many advantages beyond low-cost operations and specialized audiences that appeal to demographically conscious advertisers. Faced with falling listenership or declining advertising revenues, a station can simply change disc jockeys (DJs) and discs. Neither television nor the print media have this content flexibility. When confronted with competition from a similarly formatted operation, a station can further narrow its audience by specializing its formula even more. Many midsize and large markets have Album-Oriented Rock (AOR), Hard Rock, Alternative Rock, Classic Rock, Heavy Metal, and Soft Rock stations. There are Country, Contemporary Country, Outlaw Country, Album Country, Spanish Country, and Young Country stations.

Music format radio requires a DJ. Someone has to spin the discs and provide the talk. The DJ as we know it today is often credited as the invention of Todd Storz, who bought KOHW in Omaha, Nebraska, in 1949. He turned the radio personality/music formula on its head. Before Storz, radio announcers would talk most of the time and occasionally play music to rest their voices. Storz wanted more music, less talk. He thought radio should sound like a jukebox, offering the same few songs people wanted to hear, played over and over again. His Top 40 format, demanding strict adherence to a **playlist** (a predetermined sequence of selected records) of popular music for young people, up-tempo pacing, and catchy production gimmicks, became hugely popular among posttelevision popular music stations. Gordon McLendon of Dallas's KLIF refined the Top 40 format and developed others, such as the Beautiful

In the 1950s, Gordon McLendon of Dallas's KLIF refined the Top 40 format and developed others, including the Beautiful Music format. Therefore, he is often considered one of the pioneers of format radio, along with Top 40 inventor Todd Storz.

Music format, and is therefore often considered, with Storz, one of the two pioneers of format radio.

Radio's Personal Nature

With the advent of television, radio changed its relationship with its audience. Families no longer gathered around the radio set, as they once had, listening to their favorite stars. Today, we listen to the radio alone, we select personally pleasing formats, and we listen as an adjunct to other personally important activities. Radio personalities talk to us personally as well. They play our requests, wish us happy birthday, engage in contests with us—they become our friends.

Radio's Mobile Nature

Radio's mobility accounts in large part for its personal nature. We can listen anywhere, any time. We listen at work, while exercising, while sitting in the sun. By 1947 the combined sale of car and alarm clock radios exceeded that of traditional living room receivers, and in 1951 the annual production of car radios exceeded that of home receivers for the first time. It has continued to do so every year since. Career opportunities in this profitable medium are discussed in the box titled Working in Commercial Radio.

The Business of Radio

Radio's distinctive characteristics serve its listeners, but they also make radio a thriving business. In fact, Lawrence K. Grossman (1998), former president of NBC News and PBS, reports that radio "now reigns as the most profitable of all media. Radio's recent tidal wave of corporate consolidation, its cheap production costs, and its high cash flow have transformed it into the darling of Wall Street" (p. 61).

Radio as an Advertising Medium

Advertisers appreciate radio's specialization because it gives them access to homogeneous groups of listeners to whom products can be pitched. Since the entrenchment of specialized formats, there has not been a year in which annual **billings**—dollars earned from the sale of airtime—have declined. In 1998 total radio billings were $15.2 billion (a 12% increase over 1997). Advertisers bought local time worth $11.9 billion (80% of all billings), national spots worth $2.75 billion (for example, Prestone Antifreeze buys time on several thousand stations in cold weather areas), and network time worth $557 million (Ditingo, 1999). The cost of time is based, more or less, on the ratings, an often controversial reality in radio as you will see in Chapter 8.

Radio is an attractive advertising medium for reasons other than its delivery of homogeneous audiences. Radio ads are inexpensive to produce and therefore can be changed, updated, and specialized to meet specific audience demands. They can also be specialized to different times of the day. For example, a hamburger restaurant may have one version of its commercial for the morning audience in which its breakfast menu is touted and a different version for the evening audience driving home, dreading the thought of cooking dinner. Radio time is inexpensive to buy, especially when compared to television. An audience loyal to a specifically formatted station is presumably loyal to those who advertise on it. Radio is the listeners' friend, it travels with them and talks to them personally. We will take a detailed look at these issues in Chapter 9.

Deregulation and Ownership

The business of radio is being dramatically altered by deregulation and changes in ownership rules. To ensure diversity on U.S. airwaves, the FCC had long limited the number of radio stations one person or company could own to one AM and one FM locally and seven AMs and seven FMs nationally. These

Careers *Working in Commercial Radio*

For most of us the most visible (audible?) position at a radio station is the DJ or talk show host. But there are several other career options in radio, all indispensable to the smooth functioning and financial success of a commercial radio station. Most stations, regardless of their location in large, medium, or small markets, share these job titles and descriptions. In larger markets the duties associated with a given position tend to be discrete; that is, the person with that job title does only that job. As markets and, therefore, station budgets and staffs get smaller, people with specific job titles often have to take on additional duties that might be someone else's at a large-market station. Finally, most stations divide their professional staff into management, programming, technical, and sales categories.

Management deals with running the station. People in management are concerned about ensuring that the station meets its goals, pays its people, completes its **log** (an FCC-required record of all program elements run throughout a broadcast day), meets FCC legal requirements, and recruits and retains personnel. Management, in fact, the entire station, is overseen by a general manager (GM), sometimes called a station manager. The GM's primary function is fiscal—to make sure the station makes money. Because of this focus, the large majority of GMs build their radio experience in sales.

To ensure that the station prospers financially, the GM's responsibilities include management of personnel, oversight of legal matters, supervision of sales, labor relations at the station, managing the facility, and ultimate responsibility for programming. Managers spend time in community relations. They have the last word in all disputes and disagreements at the station and are responsible for setting station policy.

The GM has help in the form of payroll and personnel staff, and the programming director (PD) and traffic manager are considered management functions at many stations. The traffic manager runs the traffic department, which prepares the log and has the important responsibility of scheduling commercials, ensuring that commercial scripts and tapes are available, and guaranteeing that the orders for spots sold by the sales staff successfully make their way from sales to production to air.

Programming deals with the sound of the station and is managed by the program director (PD, see Chapter 10). The PD's primary function is to monitor the station's sound and evaluate its competitiveness and profit potential in the market. The PD constructs and maintains the station's format, se-

numbers were revised upward in the late 1980s, and controls were almost totally eliminated by the Telecommunications Act of 1996, leading to what FCC Chair William Kennard called "the most dramatic increase in consolidation in the broadcast industry in our history" (as quoted in Markels, 1999, p. 70). This concentration, he said, threatened the commission's public interest goal of "promoting competition and promoting diversity" (p. 70).

Today, thanks to **deregulation,** which eliminated most restrictions on how many stations one person or company may own, a concentration of ownership has evolved. Depending on the size of the market, a single entity can own as many as eight stations in one market. As a result, **duopoly**—ownership of two or more stations of the same service (AM or FM) in a single market—exploded. In the 2 years following the act's passage, more than 4,000

lects music (in larger markets, the PD might be assisted by a music director), maintains the music collection, or library, schedules the music or talk and news, and interprets the research data that tell him or her how well the station's sound is doing. PDs, who almost always come up through the DJ or announcing ranks, direct the station's staff of announcers, schedule air shifts, and at some stations even have oversight of the sound of news and public affairs.

When the PD does not supervise the sound of news and public affairs, that task rests with the news director. This position entails running the station's journalistic endeavors through management of news, sports, and traffic reporters. This is accomplished primarily through the process of story assignment—placing the right person on the right story.

Another crucial element of a station's programming operation is the promotions director, the person in charge of organizing the promotion and positioning of the station. The promotions director—in fact, even the sales and news people—often have copywriters available to them to write scripts for on-air presentation.

Technical concerns itself with the operation of the station's equipment. The chief engineer runs this part of a station, and the primary job of this position is to keep the station on the air and to ensure that it meets all FCC technical requirements. In larger markets chief engineers may have a staff that could include a production director who has oversight of the station's production facilities, equipment, and scheduling. In smaller markets a station might not even have an engineer on staff, preferring to hire a consulting engineer who might serve this function for a number of other stations.

Sales sells. The executive in charge of sales is the sales manager. This position monitors local, national, and retail sales (Chapter 9), develops sales projections for the station and sales quotas for the sales staff, assigns accounts to the sales people who are most often called account executives (AEs), and trains and motivates the AEs. In larger markets sales managers are usually assisted by local and national sales managers. Sales managers tend to come up through stations' sales departments. They may experience the greatest amount of pressure of all station personnel (their jobs, not to mention everybody else's livelihoods, depend on their performance), but they also can make a lot of money, sometimes even more than the GM. In fact, as noted previously, the most common route to station management in contemporary broadcasting travels through the sales department.

commercial radio stations changed hands (Duncan, 1998). Much of the buying was done by already large media companies, permitting the growth of group owners such as Clear Channel Communications with 830 stations and Cumulus Media with 248. As a result, "in 25 of the 50 largest markets in the U.S., three companies have at least 80% of the listeners. In the other 25, they have at least 50 percent" (Boothroyd, 1999, p. 24).

This concentration is a source of concern for many radio professionals. As radio historian Mike Adams (1999) wrote, "We are left at the end of the 'broadcast century' with a corporatized version of radio broadcasting, one that serves the bottom line and the lowest common denominator, rather than enlightens and challenges" (p. 4A). Critics point to several troubling aspects of so many stations being owned by so few companies.

Critical Issues *KDIA-AM in Oakland*

KDIA-AM had served Oakland, California, for more than 50 years. Called variously a Black or Urban station, its primary audience was the African American community of that major city on the eastern shore of San Francisco Bay. KDIA, a music-oriented AM station, survived and prospered through a mix of rhythm and blues, news, community service, and local public affairs. Reporter Harrison Chastang, of San Francisco's KPOO-FM, said, "KDIA was once a place where people could get their records played and get their message out. A lot of issues and concerns of the Oakland community were aired on KDIA" (as quoted in Zoll, 1998, p. 16).

But in December 1997, the station was sold to the Disney Company. Midnight vigils and other listener protests could not overcome the $6 million that Disney paid. The FCC does not consider community input in its deliberations when approving a license transfer. The buying party must meet only well-defined economic and technical standards.

Disney replaced the station's urban sound with its automated, satellite-fed Radio Disney format, which relies on Top 40 pop music and Disney-themed features to attract children and young teens.

What did Oakland gain and lose in this change? Certainly, Oakland, with its large African American population, has other stations programming the "urban sound." And just as certainly, many children of all races enjoy Disney and its high-quality entertainment. Southern California–based Disney would not have spent the money if it did not think it could draw these young listeners.

Critics of the sale, however, point to a loss of much that made KDIA, specifically, and radio in general, important to them. KDIA was local; the new station is not. KDIA was specialized; its audience was the Oakland African American community. The new station is also specialized, but its audience is a narrow demographic attractive to a specific set of advertisers, not a definable community with a shared interest in its well-being. KDIA was personal; listeners knew the DJs and reporters—and they knew their listeners. The new station is programmed some 600 miles away by people who deliver identical content to stations across the country.

Defenders of the new Disney station argue that the market should decide the wisdom of the sale. If KDIA had been drawing good ratings, it would have been making a profit for its owners, perhaps one large enough to have made the sale to Disney unattractive. And if Disney meets its new audience's needs, it will earn a profit. In addition, in the contemporary telecommunications environment, other media can serve people's demand for local, specialized, personal content. The East Bay abounds in local and regional magazines and newspapers, cable delivers several African American–oriented channels (both audio and video), and there is always the Internet.

Since deregulation, stations have become commodities to be bought and sold. Why shouldn't KDIA's owners profit from the station they built? It is their hard work and investment of time and money that made the station worth $6 million. And why shouldn't Disney do what is not only good business but legal? The hard work of Disney's employees and the company's investment of time and money put it in the position to converge its production, satellite, and broadcast operations in the Oakland market.

What do you think? Put yourself in the position of KDIA's owners. Would you say "No" to $6 million? KDIA's listeners helped make the station worth all that money. How would you balance your obligation to your listeners with the legal and proper operation of the marketplace? Do you even have an obligation to do so? Now put yourself in the position of KDIA's listeners. How would you feel if the station with which you grew up abandoned you and your interests? Why not simply find another station that speaks to you, or use other media to replace what you lost when KDIA was sold?

1. *Loss of national diversity.* The FCC reported that in 1997 and 1998 "the number of radio station owners has dropped 11.7%, to 4,507, and the total number of radio outlets has declined 2.5%" (McConnell, 1998, p. 27). Even though there has been an increase in the number of U.S. radio stations in recent years, one out of every five owners in 1996 has now disappeared (Kiesewetter, 2000).

2. *Loss of regional diversity.* With a single owner able to claim up to eight stations in one city, many areas, despite what appears to be a large number of individual stations, actually are being served by very few *different* radio operations. In San Francisco, for example, Chancellor Media of Texas and CBS Radio of New York own 11 of that city's top 20 commercial stations. In 1988 San Francisco and Oakland's 30 commercial stations were owned by 23 different companies. In 1998 they were owned by 10 (Zoll, 1998). The box titled KDIA-AM in Oakland offers an example of this loss of regional diversity.

3. *Rising overcommercialization.* As radio stations become just one more profit center among a large company's national or international collection of businesses, there is increased pressure on individual station managers to maximize profit at the expense of service. Because deregulation has lifted all limits on the number of commercials a station can air, the amount of time devoted to advertising at an average big city operation has grown from about 12 to 14 minutes an hour to more than 20 minutes per hour (Zoll, 1998, p. 18).

4. *Loss of variety in programming.* As out-of-town conglomerates come to own larger numbers of stations, they find great efficiency in standardizing their "product" across all their holdings. As a result, stations across the country, and even in a given market, start to sound the same. *Variety*'s radio writer Christopher Stern (1999b) wrote, "Radio's eccentric ways are becoming a thing of the past as a huge wave of consolidation has turned music stations into cash cows that focus on narrow playlists aimed at squeezing the most revenue from the richest demographics" (p. 5).

In addition, the change in ownership rules and subsequent flurry of station buying have created a situation where "stations that sold for less than $200,000 two decades ago now sell for 100 times that much. And corporations, fraught with worries about paying massive debt service [payments on loans taken to buy additional stations], take fewer and fewer chances on programming" (Kava, 1999, p. 22A). This trend toward standardization is exacerbated by the profusion of **automated stations,** which are stations that are computer-programmed. These operations tend to have no local personnel other than engineers and a sales staff. They contract with a content provider who delivers the same material to hundreds of stations by satellite. The local station simply has to insert commercials where indicated.

5. *Disappearance of news.* Deregulation also erased the requirement that stations program news and public affairs. As a result, many do not. Others air only packaged, syndicator-provided headline reports. Still others contract

with news operations such as Metro Networks and Shadow Broadcast Services. These companies provide local traffic, weather, sports, and news to stations for free, selling commercial time around these reports themselves. Joe McConnell, a news bureau chief for Metro, defends the practice, arguing that "we're putting news back on the radio. Stations have been dropping radio news since deregulation, and we provide a way for stations who may not have the resources or who don't want to spend the money on news to have it" (as quoted in Zoll, 1998, p. 19). Put more starkly by Metro founder David Saperstein, "If it wasn't for Metro, stations would've dropped news altogether" (as quoted in Rosenwein, 1999b, p. 97). But Lawrence K. Grossman (1998) counters that the syndicated news companies "employ few if any reporters and do not subscribe to the AP or other wire services. Instead, they merely cannibalize local newspapers and cable news channels" (p. 61). Moreover, to service multiple stations in an individual market, a single announcer, using different names, will read the same news on reports prepared for different stations.

Recognizing that radio's local nature is jeopardized by the new ownership rules and the concentration it permits, the FCC began proceedings in late 1998 to solicit public opinion in support of its investigation into opening up spectrum space for low-power, community radio stations, eventually issuing its plan for what is called **microbroadcasting** in January 1999. FCC Chair Kennard said that the commission's proposal will "maximize the use of the spectrum for the American public" and "give voice to the voiceless" (as quoted in McConnell, 1999b, p. 100). Kennard went on to say that microbroadcasting "could create a whole new class of voices using the airwaves . . . opportunities for churches and community groups . . . so many of whom feel that they are being frozen out of opportunities to become broadcasters" (as quoted in Kava, 1999, p. 1A). Later, he told *Broadcasting & Cable* magazine that the new low-power stations "are about opening doors of opportunity for additional community uses of these public airwaves, particularly as consolidation closes those doors for so many new entrants" (as quoted in McConnell, 1999a, p. 27). And there are potentially *many* new entrants. Within a few months of the FCC's announcement, it received 13,000 inquiries about starting low-power FM service.

Resistance to the possibility of licensing hundreds of new stations came from commercial and noncommercial broadcasters alike. The commercial stations feared interference from the new stations and crowding of the spectrum, making the transition to digital radio more difficult and costly. The noncommercial stations worried that the audience for their advertiser-free fare would be hopelessly fragmented among themselves and the low-power stations. In fact, it was a coalition of these same commercial and noncommercial broadcasters that had, 20 years before, convinced the FCC to phase out licensing of low-power radio stations.

Despite this opposition, the FCC voted in January 2000 to authorize the new service, making possible 1,000 new low-power FM community stations. Licenses would be awarded based on applicants' "service to the community and proposed programming" (as quoted in Shiver, 2000b, p. 5A) and their ability to meet these criteria: "be non-profit operations that serve the educational, instructional, or cultural needs of the community [and] demonstrate a commitment to programming created locally" (as quoted in McConnell, 2000, p. 10). The first of the licensed microbroadcasters were to go on the air in late 2000.

What adds particular heat to the debate over the consolidation of ownership of radio into fewer and fewer hands, its effects on content, and microbroadcasting as the remedy is a statistical reality known quite well to both fans and critics of concentration—a steady decline in overall radio listening. "Only 15.4 percent of the population age 12 and older *regularly* listens, down from 17.5 percent in 1989," reports Thom Moon, research director of radio industry analyst Duncan's American Radio (Kiesewetter, 2000, p. E2). Moon attributes this drop in regular listenership, in part, to "much higher commercial loads and not as much programming innovations," both products of increased concentration of ownership (p. E2).

New Technology, Audience Fragmentation, and Convergence

Radio faces an upheaval today not unlike what it confronted with the coming of television. As an industry, radio is immensely profitable for its owners, listened to by almost everyone (although not as regularly as in the past), and subject to serious criticism from many quarters. This state of affairs is directly attributable to deregulation and the concentration it has allowed. The new technologies, and the audience fragmentation and convergence that they allow, will shape radio's future.

Satellite and Cable

The convergence of radio and satellite has aided the rebirth of the radio networks. Music and other forms of radio content can be distributed quite inexpensively to thousands of stations. As a result, one "network" can provide very different services to its very different affiliates. ABC, for example, maintains several different radio networks under its own name as well as those of Disney and ESPN. Together, the ABC networks have 3,050 affiliate stations reaching an audience of 50 million listeners (Rosenwein, 1998). Westwood One, which bought the NBC Radio network in 1987 and added it to its already large and varied networking and program syndication operations, counts among its affiliates 60% of all the commercial stations in the United States. The low cost of producing radio programming, however, makes establishment of other, even more specialized networks possible. Satellites, too, make access to syndicated content and formats affordable for many stations. Syndicators can deliver news, Top 10 shows, and other content to stations on a market-by-market basis. They can also provide entire formats. Local stations, if they wish, need do little more than insert commercials into what appears to listeners to be a

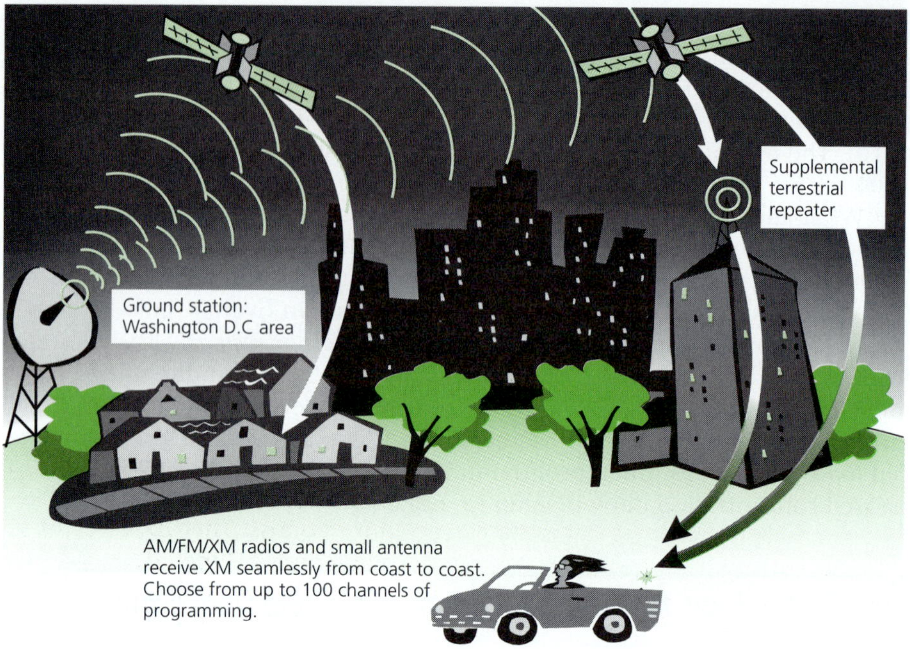

From its studios and uplink facility near Washington, D.C., XM Satellite Radio can beam up to 100 channels of digital quality music, news, and entertainment directly to two powerful satellites to people from coast to coast in their cars and at home. Car and home receivers look exactly like typical AM/FM radios, but they utilize a special satellite radio antenna.

local broadcast. Radio networks and syndicators are discussed in more detail in Chapter 10.

Satellite has another application as well. Many listeners now receive "radio" through their cable televisions. This service is called **Digital Music Express (DMX),** and the technology also exists for direct home and even automobile delivery of audio by **digital audio radio service (DARS).** The FCC has already allocated spectrum space for the service, but its diffusion is slowed by uncertainty over listener demand and disagreement over technical standards. The two licensed American DARS providers, XM Satellite Radio Inc. and Sirius, are aiming at a date in late 2000 to launch their satellites and services. Ford and General Motors offer DARS receivers in their 2000 model cars. Another DARS provider, WorldSpace Corporation of Washington, D.C. already has agreements with four receiver manufacturers—Hitachi, JVC, Panasonic, and Sanyo—who are making sets that range in price from $250 to $350. These receivers offer standard AM and FM radio as well as shortwave communications in addition to DARS. Initially transmitting via its satellite AfriStar only into Africa and the Middle East, WorldSpace launched satellites to serve Asia, Latin America, and the Caribbean in late 1999, reaching 123 different countries. WorldSpace is banking on its early entry into DARS to give its technology a head start at becoming the fledgling industry's worldwide standard.

Both DMX and DARS have the potential to fragment radio's audience even further. As DMX and DARS tend to be subscription or listener, rather than advertiser, supported, this fragmentation is the kind that the radio industry fears. Rather than dividing the audience into even more demographically identifiable, salable segments, DMX and DARS threaten to spread the listening audience across even more media, depleting the number of people tuning into commercial radio.

Digital Technology

In the 1970s the radio and audio industries began the change from analog to **digital recording.** That is, sound went from being preserved as waves to conversion into 1s and 0s logged in millisecond intervals in a computerized translation process. When replayed at the proper speed, the resulting sound was not only continuous but pristine—no hum, no hiss. The CD, or compact disc, was introduced in 1982 using digital coding on a 4.7 inch disc read by a laser beam. In 1986 Dire Straits' *Brothers in Arms* became the first million-selling CD. In 1988 the sale of CDs surpassed that of vinyl discs for the first time, and today CDs account for more than 80% of all music sales.

Digital audio tape (DAT) offers digital-quality sound purity on minicassettes to buyers of manufactured tapes and to home tapers as well. DAT was developed in the early 1970s and initially released in Japan in 1987. Although DAT tapes must be played on DAT recorders that cost about $1,000, the price of those machines is coming down. Because DAT makes high-fidelity taping of complete digitally recorded albums on digitally broadcast radio stations and taping of borrowed CDs possible, it is a threat to the recording industry's profits. At the insistence of the recording industry, the federal government kept DAT out of the United States. But in 1992 Congress passed a "taping tax" to be paid to music rights holders. This tax is currently 2% of the wholesale price of DAT recorders and 3% of the price of blank DAT tapes. The tax helps compensate the recording industry for revenues lost to home taping (see Chapter 12 for more detail). Radio stations, especially those broadcasting digitally, look favorably on DAT. Because DAT allows people to cleanly record entire albums off-air, numerous stations promote their broadcasts of full albums as a service to listeners—and as a way to attract listeners to their stations during times when people typically aren't listening, for example, late Sunday nights.

Digital technology, in the form of computers and the Internet, offers other challenges and opportunities to the radio and audio industries. The way the recording industry operates is likely to be altered by the Internet. Traditionally, a record company signs an artist, produces that artist's music, and promotes the artist and music through a variety of outlets, but primarily through the distribution of music to radio stations. Listeners learn about the artist and music through radio and go to a record store and buy the music. Today, the

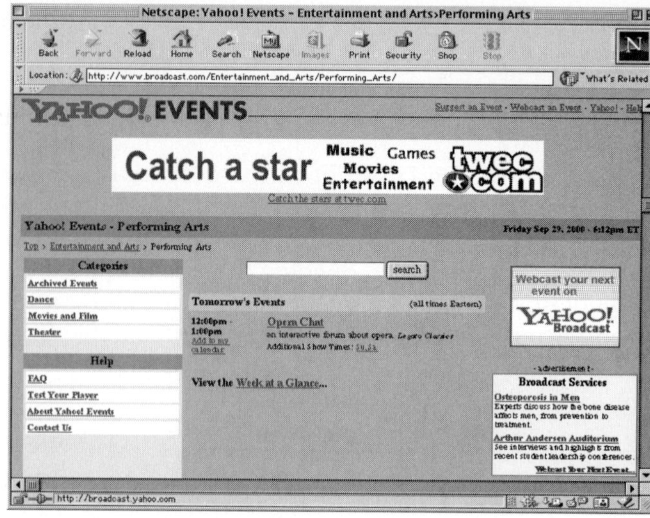

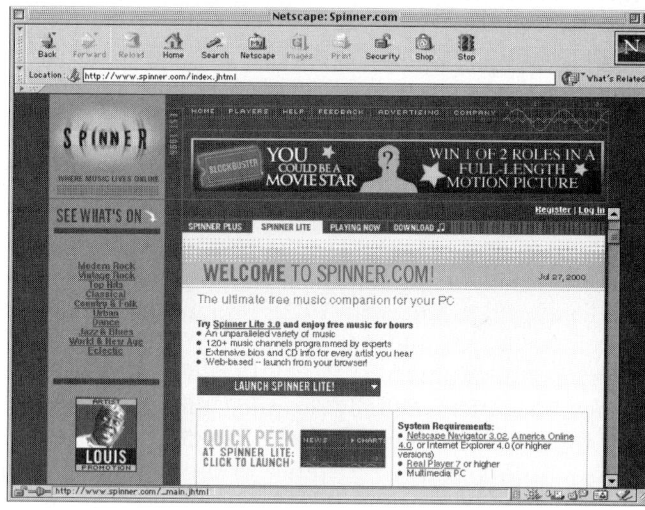

CDNow, Spinner, and Yahoo!Broadcast are only a few of the many Web sites devoted to the online sale of recorded music.

Internet and the Web are increasingly being used for music promotion (potentially disrupting the radio station–recording company relationship) and distribution. N2K Entertainment, for example, maintains a number of online music sites, among them Music Boulevard <http://www.musicboulevard.com>, Rocktropolis <http://www1.rocktropolis.com>, and Jazz Central Station <http://www.jazzcentralstation.com>. With the appropriate software, listeners can sample, download, and tape music from these and similar sites. As improvements in the Internet transmission of high-quality audio continue (see the discussion of audio compression technology in Chapter 7), these and similar sites will be used for direct online digital release of music, recorded directly onto CD or DAT, bypassing the record companies and radio stations altogether. In addition, an increasing number of performers themselves, among them Prince (<http://www.love4oneanother.com>) and Todd Rundgren, are using the Internet and the Web to communicate directly with their fans in the production, promotion, and distribution of their music. For more on this topic, see the box titled Hello, It's Todd Rundgren.

Digital Radio

Several technologies exist for the digital broadcast of traditional, terrestrial radio stations. No one technology has yet been deemed the standard by the FCC, and broadcasters are understandably hesitant to invest in the conversion from analog at this time. The favored technology is called **in-band-on-channel (IBOC),** which employs digital compression to "shrink" both digital and analog signals, allowing them to occupy a single channel. This permits stations to broadcast in both formats without acquiring additional spectrum space and without losing existing listeners. One industry plan, backed by CBS and USA Digital, would enable simultaneous broadcast of analog and digital signals for 12 years, at which time analog would disappear. Other companies that have invested time and money in competing digital delivery technologies want the FCC to take its time before selecting a standard. The commission, however, began issuing experimental IBOC station licenses in August 1999.

Web Radio

Radio's convergence with digital technologies is no more pronounced and potentially profound than in **Web radio,** the delivery of "radio" directly to individual listeners over the Internet. Nearly 10,000 "radio stations" exist on the Web today in one of three forms.

1. *Radio simulcast.* These traditional over-the-air stations are simulcast on the Web. Sometimes the Web-delivered signal is little more than the original broadcast re-created online. More often, however, the simulcast includes additional information, for example, the lyrics of the song being played, biographical information on the artist, and concert dates. According to BRS Radio Consultants (1999), in August 1999 there were more than 4,200 U.S. commercial, nearly 1,200 U.S. noncommercial, and over 3,000 non-U.S. radio stations delivering their normal broadcasts over the Web. Simply search the Web using a station's call letters to see if your favorite is online.

2. *Web-only radio.* **Bitcasters** are "radio stations" that can be accessed only over the Web. The 150 Web-only stations deliver two distinct services. The first simply delivers digital sound via the Web. Sites like <http://www.spinner.com> offer listeners a choice of more than 100 different channels of music, everything from jazz to Celtic to New Age, all free of commercials and DJ chatter. All users have to do is click on the type of "station" they want to hear, and they can hear it. There are also more narrowly targeted Web-only stations. If you want All Grateful Dead All the Time, visit <http://www.deadradio.com>.

3. *Interactive Web-only radio.* Another type of Web-only station permits users to interactively personalize the Webcasts they receive. At <http://www.

The Real World *Hello, It's Todd Rundgren*

In the 1970s and early 1980s rock 'n' roller Todd Rundgren had a string of hits. "Hello, It's Me," "We Gotta Get You a Woman," "Open My Eyes," "I Saw the Light," and "It Wouldn't Have Made Any Difference" are just a few of his million-sellers. But then he became artistically and economically invisible. That is, he ceased to be commercial. That meant no radio airplay for his new music, no record company contract, and no MTV.

Rundgren is now using the Internet and the Web to communicate directly with his fans, bypassing the record companies, MTV, and radio. He is creating his own resurrection, as he calls it.

The home page at <http://www.tr-i.com/subscribe.html> begins with a question and an explanation.

> How about a subscription to the resurrection?
>
> So if I were to go to a record label and ask for a deal, they would make a guess as to how many people would buy the record and give me an advance based on that number, in effect lending me some fraction of the money that my fans would eventually (2 or so years later) pony up.
>
> It occurred to me that with the aid of some modern advances I could go directly to my audience, ask them if they would commit to buying the music, and then deliver it to them as it is produced, thus eliminating the middlemen.
>
> Additionally, I wouldn't be constrained by the limitations of manufacturing, which would allow me to complete and deliver ideas that might not normally appear on a CD. In fact, I could consider extending the experience beyond music to include interactive presentations, live online events and other experiments. I just need to pitch the concept to my fans.
>
> Which is precisely what I'm about to do.

The site then offers listeners a number of options on what Rundgren calls his Members Only TR-i (Todd Rundgren interactive) site. The first is The Online Package. It includes "a personal key" to the site that permits members to preview and download new music as it's created, access "rarities and oddities" available to members only, send e-mail that Rundgren may answer discussing the works in progress, get information and discounts on products and events, ask for and have made custom tapes and CDs, engage in online chats with the artist, attend an online performance, and get the jacket for the resulting disc in the mail. All this for only $25.

The site continues, "Since you're unlikely to see me on MTV," Rundgren also distributes The Video Option, permitting members to view the premiers of at least three new videos, download high-qual-

ImagineRadio.com>, for example, listeners choose the type of music they wish to hear from a list of 20 categories. Listeners can specify certain artists they prefer over others and whether they want news and other informational content. Listeners can even skip tunes they don't want to hear.

To access Web radio, users must have some form of file compression software such as Real Player. It permits **streaming,** which is simultaneous downloading and playing digital audio or video data. Real Player is available for free downloading at <http://www.real.com>. But even streaming software can't give all users the full Web-radio experience. Users who have dial-up access to the Internet experience a hum when listening, a sound analogous to that heard on AM. Perfectly clear sound is possible only with very fast cable or DSL modems (see Chapter 7).

Todd Rundgren, tired of constant conflict with his recording labels, was among the first musical artist to turn to the Internet for the distribution, sale, and even creation of his music.

ity versions of these and others from his archives, view "behind-the-scenes" and "the making of . . ." videos, and later, in the mail, receive a VHS tape of these pieces.

Members can even buy The Literary Option, "since it takes so long to write a book." This selection digitally delivers half of Rundgren's autobiography before it is published on paper.

Would you consider entering such an arrangement with Rundgren or another artist of interest to you? Why or why not? How different might the work produced be, free of the influence (Rundgren would say interference) of radio station programming directors and record company executives? The radio station/record company relationship has always been a mutually beneficial one: stations get the best of a company's music and artists for free, and the record companies get free promotion and airplay of that same music and those same artists. Does Rundgren's innovation threaten that relationship? Do you think there will ever be enough activity on sites like Rundgren's to pose a real threat to the way radio stations and record companies operate? Will stations be forced to broaden their playlists to meet listener demand created by the Internet and the World Wide Web?

Chapter Review

Radio's roots are in science and technology. The search for wireless telephony and wireless telegraphy led to advances by James Maxwell, Heinrich Hertz, Édouard Branly, and Oliver Lodge, culminating in Guglielmo Marconi's success around 1896. Marconi's wireless was in turn improved by Reginald Fessenden and Lee DeForest. At about the same time sound recording was also being perfected. Thomas Edison patented a cylinder sound recorder in 1877, and Emile Berliner introduced the gramophone 10 years later. Both technologies soon found their way into U.S. homes, with broadcasting becoming firmly established at the end of World War I.

Advertising had become the base of radio's economics by the 1920s. Regulation of the airwaves came in the form of the Radio Acts of 1910, 1912, and

1927. The Communications Act of 1934 formalized the core concepts of the spectrum as a national resource and radio as a public service medium. The new industry's organizational structure was also formed early, as national networks CBS, NBC Red and Blue, and Mutual soon provided the content for most of the stations in the United States. These came together in the 1930s to create radio's Golden Age, a period of influence and prosperity that lasted until the end of World War II. The introduction of television changed almost every aspect of radio, including the transfer of the existing FM spectrum to use for that new medium.

The more than 12,000 radio stations operating in the United States today are either commercial FM, commercial AM, or noncommercial FM stations. New government rules, plus the new youth music of the 1960s, have made FM the more popular of radio's two technological forms. Whatever its form, radio survives today because it is local, fragmented, specialized (dominated by formats), personal, and mobile. These characteristics make it an attractive medium for advertisers, and recent deregulatory moves have made radio even more successful commercially. Deregulation has, however, led to concentration on such a scale that some observers worry that radio's future is in jeopardy. These critics lament radio's loss of national and regional diversity, rising overcommercialism, loss of programming variety, and disappearance of news. Radio is also being changed by technological advances. Satellites can deliver music directly to homes and cars and makes possible the proliferation of radio networks. Advances in digital recording make for better fidelity but also promise to alter the traditional relationship between record companies, radio stations, and listeners. Nowhere is this more evident than in Web radio, the delivery of "radio" directly to listeners' computers.

Questions for Review

1. Who were James Clerk Maxwell, Heinrich Hertz, Édouard Branly, Sir Oliver Lodge, Guglielmo Marconi, Reginald Fessenden, and Lee DeForest?
2. What is the difference between conduction and radiation systems of broadcasting?
3. What is meant by wireless telephony and wireless telegraphy?
4. Who is Emile Berliner, and what were his major contributions to sound recording?
5. What is the significance of KDKA and WEAF?
6. How do the Radio Acts of 1910, 1912, and 1927 relate to the Communications Act of 1934?
7. What were the five defining characteristics of the U.S. broadcasting system as it entered radio's Golden Age?
8. What is the Biltmore Agreement? What were some of its provisions?
9. How did WWII and the introduction of television change radio?
10. What are the nonduplication rule and AM/FM combos?
11. What are SCAs and multiplexing?
12. What does it mean to say that radio is local, fragmented, specialized, personal, and mobile?
13. What are formats and secondary services?
14. What is duopoly?
15. Critics of concentration in radio fear its impact on the medium's quality and ability to serve the public. What are four of the critics' specific concerns?

16. What is microbroadcasting? Why is it receiving so much recent attention?
17. How have cable and satellite affected the radio industry?
18. What are DMX, DARS, DAT, and IBOC? What technologies make each possible?
19. What is an automated radio station? What technology makes it possible?
20. What are the three different types of Web radio?
21. What are a wave's cycle and frequency?
22. What is a carrier wave? What happens to it when its amplitude is modulated? When its frequency is modulated?

Questions for Critical Thinking

1. Would you have favored a noncommercial basis for our broadcasting system? Why? If you are familiar with the noncommercial system of another country, how would you describe its content?

2. Are you primarily a commercial AM, commercial FM, or noncommercial FM listener? Which are your favorite formats? Why?

3. Do you depend on radio for news? If your answer is "No," did you ever do so? How would you characterize today's radio news?

4. How much regulation do you believe is necessary in U.S. broadcasting? If the airwaves belong to the people, how can we best ensure that license holders perform their public service functions?

5. Have you ever had a favorite radio station change formats after being sold? How did it make you feel? How free do you think new owners should be to change formats when its listeners helped make the station profitable and valuable in the first place?

Engaging Technology

1. The convergence of the Internet and radio has produced Web radio. Visit <http://www.broadcast.com>, <http://wmbr.mit.edu/stations/list.html>, or <http://www.radio-directory.com> for instant links to virtually all the simulcast and Web-only stations currently in operation. Listen for a while to a Web-only and to a traditional simulcast station. Which do you prefer? Why? Do you think Web radio will ever replace traditional radio for you? Why or why not?

2. Radio buffs and those who take a more serious interest in the medium's traditions should visit 100 Years of Radio History at <http://web.lconn.com/rtb/history.htm>. Which radio stars are you familiar with from television and film? Which programs can you identify from your own television watching that had their start on radio?

3. Access the FCC's Web page at <http://www.fcc.gov> and work your way to its pages that deal with radio. What kinds of issues does the commission seem to concern itself with? Do you think the level of government rule-making regarding radio is too much or too little?

Chapter Three

Television

David and Marshall Fisher (1996) begin their book *Tube: The Invention of Television* with this observation on the medium's significance:

> Television . . . needs no excuse, no proclamatory words establishing its importance. Not a day goes by but nearly everyone in this country turns to or is affected by it. As other nations struggle toward the status of "developed countries," the first thing their people want, before air conditioners or cars or computers, before they even emerge above the poverty line, is a television set. (p. xv)

Television has changed the way teachers teach, governments govern, religious leaders preach, and the way we organize the furniture in our homes. It has changed the nature, operation, and relationship to their audiences of books, magazines, movies, and radio. The computer, with its networking abilities, may one day exceed television's power as a medium of mass communication, but today television defines even the computer's future: Will the promise of the Web drown in a sea of commercials, like television? Can online information services deliver faster and better information than television? Even the computer screens we use look like television, and we await better Internet video, WebTV, online video conferencing, and the new and improved video computer game.

But television is at a historic crossroad in its technological and economic development. Viewing levels in the United States are at an all-time high, with at least one set on for an average of 7 hours and 40 minutes a day in a typical U.S. household. But at the same time we see frequent reports of alarming declines in the viewing audience for traditional network television. Where ABC, NBC, and CBS drew 66% of all television viewing in 1985, they attract only 43% today. In the 1998–99 season the national networks averaged only single-digit prime time ratings for the first time in their history (Carman, 1998). It's logical, then, to ask what will become of the networks. To what technologies and channels are viewers turning? How will advertisers react to these changes? How will the relationship between the medium and its audiences be altered by this upheaval?

Preview

This chapter traces the development of television from early experiments with mechanical scanning to the electronic marvel that sits in virtually all our homes. We follow its rapid transformation into a mature medium after and because of World War II, as well as several events early in its existence that changed and cemented its character: the quiz show scandal, *I Love Lucy,* McCarthyism, and reliance on the Nielsen ratings.

The remarkable reach of television accounts for its attractiveness as an advertising medium. As we discuss this reach, we will explore the network/affiliate relationship and how it, too, is changing. We encounter new television technologies and their promise to change the organization and operation of the television business as well as the interaction between medium and audience. We focus on the remote control, cable, VCR, DVD, personal video recorders, satellite television, and digital television, including HDTV.

Technological and Economic Roots

Before we delve deeper into the nature of this powerful medium, we must examine how it developed as it did. This is not a straightforward task, because the history of television is "fraught with contradictions and unsubstantiated claims and [is] spiced here and there with commercial politics" (Fisher & Fisher, 1996, p. xv). Among the things we do know for certain, though, is that by 1923 Charles Francis Jenkins had applied for the first U.S. television patent and a Frenchman, Édouard Belin, was actually giving demonstrations of television. Belin was quoted in the December 9, 1922, *London Times* saying, "I cannot make you see the Statue of Liberty at New York, but I can transmit the image of a point of light by wireless, and thus show you the principle on which, when the proper apparatus is developed, you will be able to see the Statue of Liberty while sitting in a room in London or Paris. . . . If a single point of light can be transmitted by wireless, so can a million" (as cited in Fisher & Fisher, 1996, p. 51).

It was an Englishman, John Logie Baird, who would eventually build that "proper apparatus." But like Jenkins and Belin, his reliance on **mechanical scanning** would render him an important, but minor, character in television's development rather than one of its true creators.

Mechanical and Electronic Scanning

In 1884 Paul Nipkow, a Russian scientist living in Berlin, developed the first workable device for generating electrical signals suitable for the transmission of a scene that people could see. His **Nipkow disc** consisted of a rotating scanning disc spinning in front of a photoelectric cell. It produced 4,000 picture dots (**pixels**) per second, making a picture composed of 18 parallel lines. Although his mechanical system proved limiting, Nipkow had demonstrated the possibility of using a scanning system to divide a scene into an orderly pattern of transmittable picture elements that could be recomposed as a visual image. (This is still the operational basis for contemporary television as well as for the new digital television sets.) By 1925 Baird was able to transmit moving images using a mechanical disc, and in 1928 he successfully sent a television picture from London to Hartsdale, New York.

Electronic scanning, dividing a picture electronically into transmittable elements—television as it exists today—came either from another Russian or from an American farm boy; this is just one of the "facts" in television's history about which there is dispute. Vladimir Zworykin, an immigrant living near Pittsburgh and working for Westinghouse, demonstrated his **iconoscope tube,** the first practical television camera tube, in 1923. In 1929 David Sarnoff lured him to RCA to head its electronics research lab, and it was there that Zworykin developed the **kinescope,** an improved picture tube. At the same time young Philo Farnsworth had moved from Idaho to San Francisco to perfect an electronic television system, the design for which he had shown his high school science teacher when he was 15 years old. In 1927, at the age of 20, he made his first public demonstration—film clips of a prize fight, scenes from a Mary Pickford movie, and other images. In that same year the Federal Radio Commission granted its first television station license—to Jenkins's mechanically based W3XK.

But as we've seen, mechanical scanning would be eclipsed by electronic scanning. Farnsworth and RCA spent the next decade fighting fierce patent battles in court to determine who would control this technology. In 1939 RCA capitulated, agreeing to pay Farnsworth royalties for the use of his patents. (For this reason many broadcast historians credit Farnsworth with the invention of electronic scanning, though dispute does still exist.)

RCA made the first true public demonstration of television in the form of regularly scheduled 2-hour NBC broadcasts from the World's Fair in New York on April 20, 1939. These black and white broadcasts consisted of cooking demonstrations, singers, jugglers, comedians, and puppets, just about any-

Philo Farnsworth and Vladimir Zworykin, pioneers in the development of television.

thing that could fit in a hot, brightly lit studio and demonstrate motion. "Now," said RCA's David Sarnoff, "we add radio sight to sound." People could buy sets at the RCA Pavilion at prices ranging from $200 for the 5-inch screen to $600 for the deluxe 12-inch model.

Hoping to avoiding some of the technical problems that marked radio's earliest days, the FCC held hearings in 1941 that led to the establishment of the National Television System Committee. The committee's charge was to establish common standards for the new medium. Those standards, in use in the United States from 1941 until television's full digitization in 2006, are called **NTSC** standards. (Other countries established different standards, Europe's PAL, for example.) They mandate a picture composed of 525 lines, scanned completely 30 times per second in a process called **interlaced scanning;** that is, the electron beams sweep the picture tube twice, creating half the image's lines on the first pass and then filling in the gaps in the second.

The picture portion of the signal relies on amplitude modulation (AM) and the sound portion on frequency modulation (FM). With these technical standards in place, the FCC granted construction permits to the first two commercial stations in 1941. It would grant only a few more because the country was soon to be dragged into the Second World War.

The Freeze

When World War II ended in 1945 there were nine commercial stations authorized by the FCC, six of which were on the air. But with war's end and growing national prosperity, the FCC was soon flooded with hundreds of applications for television licenses. The commission quickly realized the inability of its prewar technical and regulatory standards to meet that demand. In 1948, therefore, it declared the **Television Freeze,** a 6-month period in which it would authorize no new stations while it solved four pressing problems: (1) developing an industry standard for color, (2) reducing interference between stations in nearby cities operating on the same channel, (3) finding spectrum space for additional channels and assigning those channels, and (4) reserving channels for educational television. These problems proved more difficult than the commission had envisioned, and the 6-month freeze lasted 4 years, ending in 1952 with the publication of the **Sixth Report and Order,** the fundamental blueprint for the technical operation of television.

Color CBS and RCA had been competing since the earliest days of television's technical development to be the first to produce a practical system of color television. At the time of the freeze, the already available CBS system was mechanical—it employed rotating colored discs that separated colors and sent them to a single tube. The still-developing RCA system was electronic—it used electronic filters to divide a scene into colors and send them to three tubes: one red, one green, and one blue. In 1947 the FCC chose the RCA system as the nation's standard because its more sophisticated engineering promised greater improvement in the future and it was compatible with existing NTSC receivers. RCA continued to produce black and white sets but promised to have color available very soon. During the freeze, however, the FCC tired of waiting for the promised RCA system and ruled that the CBS color television should be the new standard. But there was no demand for the new color sets. Manufacturers would not make them because they were small and clumsy, those that were produced were too expensive, and there were very few programs broadcast in color. In 1953, after the freeze, the commission rescinded its decision and remade the RCA system the standard, allowing color television to begin its fuller development.

Interference The FCC imposed limits on how close to one another two television stations operating on the same channel could be (170, 190, or 220 miles, depending on the part of the country). No stations were put off the air, but several had to change positions on the dial, for example, from Channel 6 to Channel 4.

Spectrum Space and Allocation The fact that stations operating on the same channel now had to have so much distance between them meant that more channels would have to be made available. The commission accomplished this through its alteration of the television portion of the **broadcast spectrum,** authorizing the use of scores of new channels for UHF transmission. (Different forms of electronic communication employ different portions of the electromagnetic spectrum for the transmission and reception of their signals. The broadcast spectrum in the United States operates from 300 to 3,000 kHz, where AM radio operates, to 300 to 3,000 MHz, where UHF television operates.) The FCC refused, however, to move *all* television from the existing VHF band to **UHF** (ultra high frequency), although many people advocated doing so as a way to allow all stations—affiliated and independents alike—to enter the television age on equal footing (Reel, 1979). This had the effect of institutionalizing the power of the national television networks from the medium's earliest days, because they had already secured affiliate relationships with virtually all the existing **VHF** (very high frequency) stations. The networks ignored the new UHF stations, and as a result, so did the viewers. UHF, too, was plagued by poorer signal quality and reach when compared to VHF, and no television sets then being made could even receive UHF signals without a specially purchased antenna.

The FCC then created an assignment table, allocating channels throughout the United States on a city-by-city basis. All parts of the country would have television service from at least two sources. Cities of reasonable size would have at least one but preferably two channels. UHF and VHF frequencies would be intermixed.

Educational Television Recalling the failure of the 1934 Communications Act to provide educational radio with usable AM frequencies, the FCC reserved channels for educational television on the basis of one educational license for every four commercial assignments. We will see more on educational television in Chapter 4.

The Fifties

In 1952, 108 stations were broadcasting to 17 million television homes. By decade's end 559 stations were received by nearly 90% of U.S. households. In the 1950s there were more television sets sold in the United States (70 million) than there were children born (40.5 million) (Kuralt, 1977). The technical standards were fixed, stations proliferated and flourished, the public tuned in, and advertisers were enthusiastic. The medium's content and character were established in this decade, as can be seen from these trends:

- Carried over from the radio networks, the new medium's genres were variety shows, situation comedies, dramas (including westerns and cop shows), soap operas, and quiz shows.

- Two new formats appeared, feature films (ABC premiered *Hollywood Film Theater* in 1957, although movies would not become successful on television until NBC's *Saturday Night at the Movies,* beginning in 1961) and talk celebrity shows. The latter were instrumental in introducing radio personalities to the television audience, who, for the first time, could see their favorites.

- Television news and documentary remade broadcast journalism as a powerful force in its own right, led by CBS's Edward R. Murrow and his *See It Now,* premiering in 1951, and NBC's David Brinkley and Chet Huntley. Huntley and Brinkley's 1956 coverage of the major political conventions gave audiences an early glimpse of the medium's power to cover news and history in the making.

- AT&T completed its national **coaxial cable** and **microwave relay system** in the summer of 1951, which enabled distribution of television programming nationwide. The entire United States was now within the reach of the major television networks, and they came to dominate the medium.

Who would own the growing number of television stations became an issue in the 1950s. In the name of diversity of opinion and expression, the FCC placed limits on station ownership, restricting a single person or company to financial interest in only one station in a market, and no more than seven nationally. The FCC suspended its one-to-a-market rule in 1999, allowing single ownership of two stations as long as at least eight other independently owned stations exist in that market. As with radio, ownership of more than one station in a market is called duopoly. National ownership limits have also been erased by the FCC. Now a single entity can own as many television stations as it wishes, as long as its total audience reach does not exceed 35% of the nation's television homes. Some companies already exceed the 35% cap—Paramount's 35 stations, for example, reach 40% of U.S. television homes; Paxon Communications' 71 stations reach 35.03%. As a result, they and others in the broadcasting industry are currently lobbying the commission to raise the ceiling to 50%. Still in dispute is the future of **local marketing agreements (LMAs),** deals by which one station operates another without formally owning it.

Four other events from the 1950s would permanently shape the medium's operation: the quiz show scandal, the appearance of *I Love Lucy,* McCarthyism, and the establishment of the ratings system.

The Quiz Show Scandal As with radio, the television networks served primarily as time brokers, offering airtime and distribution (their affiliates) and accepting payment for access to both. Except for their own news and sports coverage, the networks relied on outside agencies to provide programs. An advertising agency, for example, would hire a production company to produce a program for its client. That client would then be the show's sponsor, for example, *The Kraft Television Theatre* or *Westinghouse Studio One.* The agency would then pay a network to air the program over its national collection of stations. This system had enriched the networks during radio's heyday, and they saw no reason to change.

But in 1959 the quiz show scandal, enveloping independently produced, single advertiser–sponsored programs, changed the way the networks did business. Popular shows like *The $64,000 Question* and *Twenty-One* had been rigged by advertisers and producers to ensure that favored contestants would defeat less popular ones and to artificially build tension where a mismatch was anticipated. The audience was shocked that winners had been provided with the questions and answers before the broadcasts. Congress held hearings. The reputation of the new medium was being tarnished. The networks used this "embarrassment" as their excuse to eliminate the advertisers and ad agencies from the production/distribution process. They argued that because audiences held them responsible for what appeared under their name they were obligated to control their schedules. The networks themselves began commissioning or buying the entertainment fare that filled their prime time schedules. This has been the case ever since, but most historians now believe the networks simply used the scandal as an excuse to remove the advertisers from program production. The networks' goal was to exert more control over their schedules, building audiences over their broadcast day (Boddy, 1992). Because of the success of the morning and late night shows they themselves owned, they also recognized that greater profit could be made from **spot commercial sales** (selling individual 60-second spots on a given program to a wide variety of advertisers) than from selling blocks of time to individual advertisers.

Some critics argue that this change to spot sales put an end to the "Golden Age of Television." When sponsors agreed to attach their names to programs, *Alcoa Presents* or the *Texaco Star Theater,* for example, they had an incentive to demand high-quality programming. Spot sales, with network salespeople offering small bits of time to a number of different advertisers, reduced the demand for quality. Individual advertisers were no longer identified with a given show, and they had no stake in how well it was made—only in how many viewers it attracted. Spot sales also reduced the networks' willingness to try innovative or different types of content, because content familiarity and predictability attracted more viewers and, therefore, advertisers. There is a counterargument, however. Once the networks' own financial well-being became dependent on the programming they aired, they themselves became more concerned with its quality, lifting television from its dull infancy, remembered now as the "Golden Age" only by those small, early audiences committed to serious character-driven televised drama. Different historians and critics offer arguments for both views.

I Love Lucy In 1951 CBS asked Lucille Ball to move her hit radio program, *My Favorite Husband,* to television. Lucy was willing but wanted her real-life husband, Desi Arnaz, to play the part of her video spouse. The network refused. Some historians say the network objected to the prime time presentation of an interracial marriage, but CBS denies this. But Lucy made additional demands. Television at the time was live. Images were typically captured by three large television cameras, with a director in a booth choosing between

I Love Lucy was significant for far more than its comedy. Thanks to Lucille Ball's shrewd business sense, it became the foundation for the huge syndicated television industry.

the three available images. Lucy wanted her program produced in the same manner—in front of a live audience and three simultaneously running cameras—but these cameras would be *film* cameras. Editors could then review the three sets of film and edit them together to give the best combination of action and reaction shots. Lucy also wanted the production to take place in Hollywood, the nation's film capital, instead of New York, the country's television center at the time.

CBS, uncertain about this departure from business as usual, refused these requests as well. So Lucy and Desi borrowed the necessary money and produced *I Love Lucy* on their own, selling the broadcast rights to CBS. In doing so, the woman now best remembered as "that zany red head" transformed the business and look of television.

• Filmed reruns were now possible, something that had been impossible with live television. Not only did this create the **off-network** syndication industry (selling network shows for rebroadcast, see Chapter 10), but it demonstrated to programmers that viewers would watch the same episode of a television show more than once.

• The television production industry moved from New York, with its staged drama orientation, to Hollywood, with its entertainment film mindset. More action and more visually interesting material came to the screen because in theater people talk, whereas in film people do things, most often in an outdoor setting.

• Weekly series could now be produced relatively quickly and inexpensively. Because shows were now produced on film, the production of a 39-week series could be completed in 20 or 24 weeks. This meant less expensive production costs because actors, crew, equipment, and facilities did not have to be assembled and maintained for the full 39-week run of a series, as was required by live television. In addition, the same stock shots, for example, certain exterior views, could be used in different episodes.

• Lucy's success demonstrated to network programmers that television's attraction resided not in people's ability to see *live* activity over the air but in *content,* live or otherwise, that attracted and held viewers (Boddy, 1992).

McCarthyism In 1950 three former FBI agents operating a company called American Business Consultants published *Red Channels: The Report of Communist Influence in Radio and Television.* Its 200 pages detailed the alleged pro-Communist sympathies of 151 broadcast personalities including actor/director Orson Welles and newsman Howard K. Smith. Advertisers were encouraged to avoid buying time from broadcasters who employed these "Red sympathizers." The television industry caved in quickly, and the networks employed security checkers to look into people's backgrounds, refused to hire "suspect" talent, and demanded loyalty oaths from performers. In its infancy television had taken the safe path. Many gifted artists were denied not only a paycheck but also the opportunity to shape the medium's content.

Ironically, it was this same Red scare that allowed television to demonstrate its enormous power as a vehicle of democracy and freedom. Joseph McCarthy, the Republican junior senator from Wisconsin whose tactics gave this era its name, was seen by millions of viewers as his investigation of "Reds" in the U.S. Army was broadcast by all the networks for 36 days in 1954. Daytime ratings increased 50% (Sterling & Kittross, 1990). At the same time, Edward R. Murrow used his *See It Now* to expose the senator's lies and hypocrisy. Between the two broadcasts, McCarthy was ruined, censured by his Senate colleagues, and later died the death of a lonely alcoholic. Television had given the people eyes and ears—and therefore power—whereas before they had had little. The Army-McCarthy hearings and Murrow's challenge to McCarthyism are still regarded as two of the medium's finest moments.

The Nielsen Ratings The A. C. Nielsen Company began in 1923 as a product testing company, soon branching into market research. In 1936 Nielsen started reporting radio ratings, and it was doing the same for television by 1950. Very quickly, the Nielsen ratings became central to almost all aspects of the business of television and television production. Today, Nielsen offers a variety of services, as you'll see in some detail in Chapter 8.

Television and Its Audiences

The 1960s saw some refinement in television's technical structure, which influenced its organization and audience. Until 1962, television sets in the United States were unable to receive UHF channels unless special converters capable of detecting signals in the UHF spectrum were purchased and attached to them. But in that year, Congress passed **all-channel legislation,** requiring all sets imported into or manufactured in the country to be equipped with both VHF and UHF receivers. This had little immediate impact, because U.S. viewers were now hooked on the three remaining national networks and their VHF affiliates. (Dumont, one of the original television networks, closed shop in 1955.) Still, the UHF independents received a little audience attention, and now that the educational stations were attracting a few more viewers, they began to look less educational in the strictest sense of the word and began programming more entertaining cultural fare. The Public Broadcasting Act of 1967 united these two groups in an important network, the Public Broadcasting Service (PBS), which today has 350 member stations.

The 1960s also witnessed the immense social and political power of the new medium, for example, the influential Nixon-Kennedy presidential campaign debates of 1960, the broadcasts of the aftermath of President John F. Kennedy's assassination and his funeral of 1963, the 1969 transmission of Neil Armstrong's walk on the moon, and the use of the medium at decade's end by Civil Rights and anti–Vietnam War leaders to force profound alterations in the country's consciousness and behavior.

The 1960s also gave rise to a descriptive expression often used today when discussing television. Speaking to the 1961 convention of the National Association of Broadcasters, President Kennedy's new FCC Chair, Newton Minow, invited broadcasters to

> sit down in front of your television set when your station goes on the air and stay there without a book, magazine, newspaper, profit and loss sheet, or ratings book to distract you, and keep your eyes glued to that set until the station signs off. I can assure you that you will observe a **vast wasteland.**

Whether we agree or not with Mr. Minow's assessment of television, then or now, there is no doubt that we continue to watch it:

The worldwide, live broadcast in 1969 of Neil Armstrong's walk on the moon demonstrated television's immense reach and cultural power.

- There are 100.8 million television households in the United States: 99% have color, 76% have more than one set.
- In the average American home, the television is on 7 hours and 26 minutes a day.
- Sixty-nine percent of the American public turns to television as the source of most of its news about national and international issues.
- A top-rated program (*E.R.*, for example) will typically draw more than 30 million viewers per episode.

Neither can there be doubt that television is successful as an advertising medium:

- Total 1998 billings were $41.2 billion: $17.3 billion in national network and national syndication sales, $11.0 billion in national non-network sales, and $12.9 billion in local sales (Advertising Outlook, 1999).
- The average 30-second prime time network television spot cost $120,000 (30 seconds on *E.R.* costs $565,000 and on *Frasier* $475,000).
- The average 30-second spot on the 2000 Rams/Titans Super Bowl broadcast cost $2.2 million ($73.3 thousand per second). Emerging

Critical Issues *Women in Network Television*

Among the more subtle effects of cable on the television industry is that cable has helped crack network television's glass ceiling. Long denied top spots on the Big Three commercial networks, a number of women proved their worth at what were considered "lower echelon" cable networks. Eventually, the major networks, painfully aware of the success of the cable networks, decided that there was no reason women could not be successful in their operations. Now women occupy significant managerial positions in all forms of television—cable and network—joining ABC's head of programming Jamie Tarses who, having risen to that position in 1995, became a network television pioneer (Adalain, 1998, p. 1).

It has become "a good day for women," according to Margaret Loesch (as quoted in Rathbun, 1998, p. 14). What the newly installed president of the Odyssey Channel was referring to was the wave of high-profile female hires in the second half of 1998. In addition to Loesch's own appointment, the last 5 months of that year saw Susanne Daniels named head of entertainment programming for the WB network, Nancy Tellem became president of CBS Entertainment, Anne Sweeny was promoted to the presidency of Disney/ABC Cable Networks, Fran Shea became acting president of E! Entertainment Television, Laurie Younger became ABC's senior vice president and chief financial officer, and Angela Shapiro was named head of ABC Daytime. But the promotion many women in television considered the most significant was that of Pat Fili-Krushel who, on July 31, 1998, assumed the presidency of the ABC Television Network. In doing so the 45-year-old, one-time ABC secretary became the first female president of a major television network. Moreover, her duties were expanded from those of the man she replaced. In addition to responsibility for ABC's business operations and affiliate relations, she was handed control of all of the network's entertainment programming, news, and sports.

Why the sudden rush to hire women into leading positions? Although women "for a long time [were not] automatically considered the first choice" when important positions opened up, they have built "a body of experience that has to be reckoned with," said Loesch. "You can't justify overlooking them." Anne Sweeny concurred, "There isn't anything mysterious about it. We did the work. Now it's time" (as quoted in Rathbun, 1998, p. 14).

Why do you think women were ignored in the past when top network jobs became available? Do you think it matters if a man or a woman heads a network? Why or why not? What changes, if any, do you anticipate in the kinds of shows we'll see on channels with women in the top spots? Even though women now occupy a significant number of

cable network Oxygen paid $2.5 million for 30 seconds, and another unidentified advertiser paid $3 million.

- A 30-second local spot can fetch up to $20,000 on a top-rated special in a major market. (*Broadcasting & Cable Yearbook 2000*; various issues of *Broadcasting & Cable*)

Network Television

As it has been from the beginning, the business of television is dominated today by a few centralized production, distribution, and decision-making organizations. These **networks** link affiliates for the purpose of delivering and selling viewers to advertisers. The large majority of the 1,243 commercial sta-

In 1995, ABC's Jamie Tarses became network television's first female head of programming. Three years later, ABC made its one-time secretary Pat Fili-Krushel the first female president of a major television network.

important positions at the national network level, they are still woefully underrepresented as station owners, general managers, and program producers at the local television level. Why do you think this discrepancy exists? What will or should happen to rectify this imbalance?

tions in the United States are affiliated with a national broadcasting network: ABC has 224 affiliates, NBC 212, CBS more than 200, and Fox has 172. Scores more are affiliated with so-called **weblets** (newer national networks striving to develop complete schedules along the lines of the more established competitors), which include UPN (United Paramount), the WB (Warner Brothers), and Pax. Although the networks' share of audience is declining and cable has introduced us to dozens of new "cable networks" such as ESPN, MTV, Comedy Central, and A&E, most programs that come to mind when we think of television were either conceived, approved, funded, produced, or distributed by the broadcast networks. It should be no surprise, then, that despite their declining audience share the Big Three commercial networks were the only broadcast or cable operations to have more than $3 billion in revenues in 1999. ABC had $3.5 billion, NBC $3.4 billion, and CBS $3.3 billion ("Top 25,"

1999). The involvement of female executives in this financial success is discussed in the box titled Women in Network Television.

The Affiliates

Local affiliates carry the network's programs and the commercials they house (they are said to **clear time**) for a number of reasons.

1. Networks allow affiliates to insert locally sold commercials in a certain number of specified spots in network programming. The affiliates are allowed to keep all the money they make from these local spots.

2. Networks make direct payments to local affiliates for airing their programs (and more to the point, the commercials they house). For example, NBC pays **compensation** to its affiliates for airing *Frasier.* Compensation, however, has diminished dramatically in recent years as the networks have been losing audience to cable and other video alternatives. In fact, Fox pays no compensation to any of its affiliates. Many affiliates of the other networks now receive no compensation at all for some or most of the shows they carry, and the networks have even begun to ask affiliates to help them pay for high-cost programming such as professional football and some hit series. In February 2000, NBC, which had been paying San Francisco affiliate KRON $7.5 million a year in compensation, even went as far as to decline any payments at all to the station and demand a $10 million payment *from* KRON's new owners as a condition of continued affiliation. Unhappy with this $17.5 million revenue turnaround, Young Broadcasting, the new owners, declined, ending a 50-year-long affiliation between the station and the network (Sinton & Carman, 2000). However, NBC soon found a station willing to pay it for affiliation; San Jose's Channel 11 agreed to pay the network $40 million a year in **reverse compensation** for 10 years.

3. Financial risk resides with the network, not the affiliate. The networks pay for the programming, and failure cost them much more than it does the affiliates, which, at worst, suffer only diminished ad revenues until a more successful alternative is scheduled (Chapter 10).

4. Affiliates enjoy their network's prestige, using this to their financial advantage. Not only do affiliates charge higher advertising rates for local programming that borders a network offering (for example, *Eyewitness News at 6* that immediately precedes *CBS Evening News With Dan Rather*), but they typically charge more for local spots at other times of the day than can independent stations in the same market.

5. Affiliates get network-quality programming. Few local stations can match the promotional efforts of the networks; few locally produced programs can equal the budget, the glamour, and the audience appeal of network programming.

Career opportunities in local commercial television are the focus of the box titled Working in Commercial Television.

Careers *Working in Commercial Television*

For all intents and purposes, the administrative and managerial structures that characterize local radio stations (management, programming, technical, and sales) characterize local television stations. There are, however, some differences.

In radio, market size is an important factor in determining how many duties a given person might have to perform. That is, in larger markets there is more division between positions; in smaller ones, there is less. A small market GM, for example, may also be involved in sales. But television is somewhat immune from this influence, because the scope of a television operation is much larger than that of radio. Specificity in expertise and task, therefore, is typically the case. So, where a person in radio (especially small market radio) may learn all aspects of the business as her or his career unfolds, in television there is less mobility across jobs.

But the greatest difference in radio and television station operation is in programming. Both radio and television stations have program directors (although in television the position is more likely to be called program or programming manager), but in television this is a bigger job. Making locally produced television content (e.g., news, public affairs, sports, and children's shows) is more complicated than making local radio content. There are usually many more people answering to a television program manager than there are to a radio PD. Of course, this means that there are also more jobs in local television production than in local radio production.

Television programmers have infinitely more program choices to consider than do radio PDs. They must deal with the network, if the station is an affiliate, but there are literally hundreds of syndicators offering a wide array of content (Chapter 10). Yes, there are many radio networks and syndicators as well, but because radio is format-based, a radio PD will, in actuality, deal with only a very few—those offering content in the station's format. In contrast, a television station will typically offer different types of content at different times of different days, and it is the programming chief's duty to decide what goes when and where.

Among the people answering to a television programming manager are a production manager, who monitors all aired content (local, syndicated, and network), and the production staff, including directors, producers, video editors, camera people, and technical personnel such as set and lighting people and floor managers, that is, all those involved in studio production. Most stations will also have a film and videotape director, who reports to the programming manager through the production manager, and whose job it is to receive and ship films and videotapes, screen them to ensure not only their physical condition but their appropriateness for the station's program and advertising standards, and mark them for inclusion of commercials, public service and promotional announcements, and other breaks.

Also reporting through the production manager will be the art director, who designs, constructs, and maintains sets and produces graphics for the station's various on-air presentations such as public service announcements, commercials, and promotional announcements and its off-air efforts such as public appearances by station personalities. Especially at larger stations, the news staff also typically reports to programming.

The Networks and Program Content

Networks provide what appears on the vast majority of local television stations, but they also influence what appears outside of network time, that is, when affiliates, independent stations, and cable channels program their own content. This "non-network" material not only tends to look much like network programming, but more often than not it is programming that originally aired on the networks themselves.

Why do network and network-type content dominate television? *Availability* is one factor. There is 50 years' worth of already successful network content available for airing on local stations. A second is that the *production and distribution* mechanisms that have long served the broadcast networks are well established and serve the newer outlets just as well as they did NBC, CBS, and ABC. The final reason is us, the audience. The formats we have become comfortable with, our television tastes and expectations, have been and continue to be developed on the networks. We will take a detailed look at network and other television programming in Chapter 10.

New and Converging Television Technologies

As we saw at the beginning of this chapter, the networks' domination of the television industry and audience is in steady decline. New and converging television technologies have altered almost every aspect of television from the simpler time of the Big Three networks (Figure 3.1). Let's take a closer look at some of these changes.

Remote Control

One in-home technology whose impact is being felt by the television industry in ways you might not expect is the remote control, currently in more than 90% of U.S. homes. Viewers increasingly **zap** commercials, jumping to another channel with a mere flick of a finger. Two-thirds of all viewers zap commercials. The remote control also facilitates **grazing** (watching several programs simultaneously) and **channel surfing** or **cruising** (traveling through the television offerings focusing neither on specific programs nor the commercials they house). The advertising industry estimates that at least half the audience misses television commercials because of these maneuvers. As a result, advertisers are increasingly moving their dollars to other channels and media that promise more committed, loyal, and likely-to-stay-put consumers; that is, outlets such as demographically specific cable channels and narrow interest magazines.

Cable

The greatest damage to the control of television by the traditional broadcast channels has been inflicted by cable. We'll study cable television and the industry that supports it in detail in Chapter 6. Here, though, we'll look at the medium in terms of its impact on television in general.

In 1948 television salesman John Walson erected a tower to bring signals from Philadelphia to his town of Mahanoy City, Pennsylvania. He could not sell sets if people could not find channels. Some historians place cable's genesis in Palm Springs, California, others in Astoria, Oregon, still others say Walson's fellow Pennsylvanian Robert Tarlton is cable television's father. Credit

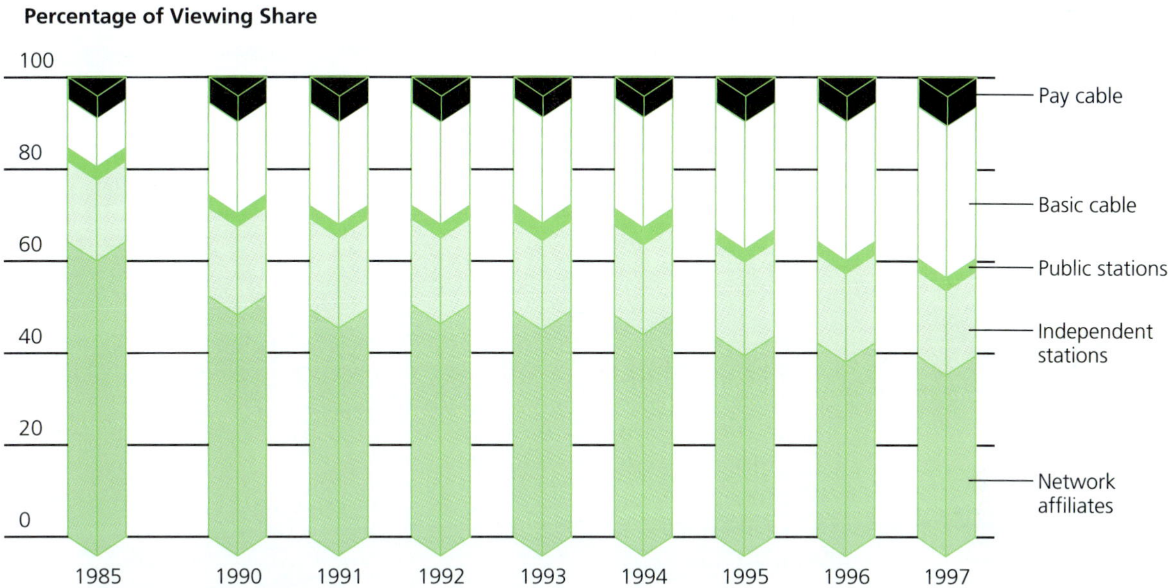

Figure 3.1 **Viewing Shares of Different Television Channels**
The decline of network affiliates' share of the total audience and the parallel rise in the proportion of that audience that has turned to cable is evident in this table. Note, too, how the audiences for independent and public television stations have stayed fairly constant, meaning that the affiliates' loss has primarily benefited cable. These data represent all television viewing, in all television households, 7 days a week, 24 hours a day. *Source:* National Cable Television Association, 1998.

for developing the first cable system is another of those historical disputes. Nonetheless, within 2 years there were 14 such community antenna television companies in the United States, all designed to improve reception by importing television signals. Today more than 68% of all U.S. homes—67 million cable households, or 200 million viewers in 34,000 communities—are wired for cable reception. The **pass-by rate,** the percentage of homes that have access to one of the country's 11,800 cable systems, is 93%. Cable's **penetration,** those homes that choose to take it, is more than 90% of all passed-by households. The cable industry generated $41 billion in gross revenue in 1999 (*Broadcasting & Cable Yearbook 2000*). And where the precable television audience had very few choices (three commercial networks, public television, and an independent station or two sitting way out on the end of the UHF dial), today's cable audience has 100 or more channel options. These numerous new outlets often provide channels for innovative, first-run series like HBO's *The Sopranos* and *Oz* and Showtime's *Fast Track* and *The Hunger.*

In addition, with the increased diffusion of **fiber optic** cable, which uses signals carried by light beams over glass fibers, 500-channel cable systems are becoming technologically feasible. By its mere existence, cable offers alternatives that were not otherwise available.

The Sopranos has become emblematic of cable's ability to present controversial and high-quality fare. This HBO production was passed on by all the major over-the-air networks, only to become the most highly honored show of both the 1999 and 2000 seasons.

Empowering the Independents Cable has had another effect on the networks that is more subtle but every bit as powerful. It has helped equalize the size of the audience for independent and affiliated stations. The fact that NBC affiliate channel 4 might appear as cable channel 11 and sit next to independent channel 44 (cable channel 12) erases the historical audience bias for VHF. The affiliate is now only one of many options, which further diminishes the distinction between it and its independent neighbors. In addition, cable provides a solution to UHF's reach and reception inadequacies. During the 1980s, the newly powerful independents also enabled creation of a viable fourth television network, Fox, and the flowering of the first-run syndication business (more on this in Chapter 10).

Fox Only ABC, CBS, and NBC survived the 1950s as successful television networks. Ever since, speculators and dreamers have planned a "fourth network," but its development was hampered by the lack of available VHF channels. As cable eradicated the distinction between VHF and UHF, media magnate Rupert Murdoch was quick to move, uniting his own chain of stations with powerful independents in important markets. His Fox Television Network won viewers away from its competitors with innovative and popular programming including *The Simpsons, In Living Color, Melrose Place,* and *The X-Files.* The WB, UPN, and Pax, although not yet reaching Fox's level of success, draw even more viewers away from the traditional broadcast networks.

Videocassette Recorders

Introduced commercially in 1976, videocassette recorders (VCRs) now sit in 90% of U.S. homes. In some places, Flagstaff, Arizona, for example, they are in more than 97% of all homes. There are approximately 30,000 video rental stores in the United States, and annual revenues for videocassette sales and rentals exceed $16 billion. Naturally, viewing rented and purchased videos further erodes the audience for traditional over-the-air television. The good news for the television industry is that VCRs allow **time-shifting,** or taping a show for later viewing, and 65% of taping from television is from network affiliated stations (*Broadcasting & Cable Yearbook 1997*). Therefore, content (and commercials) that might otherwise have been missed can still be viewed. The bad news is that VCRs also permit **zipping,** fast-forwarding through taped commercials, and much of what is taped is never actually watched.

DVD

In March 1996 **DVD** went on sale in U.S. stores. DVD might stand for digital video disc or digital versatile disk or, according to its developers, it might stand for nothing at all, being merely a series of letters that sound "high tech" (Apar, 1997). The 4.7-inch DVD discs look exactly like audio CDs and deliver high-quality digital images and sound. Discs can be rented or purchased. A DVD player looks much like a VCR machine, and viewers can stop images with no loss of fidelity, can watch a movie subtitled in a variety of languages, can search for specific scenes from an on-screen picture menu, can opt for viewing in wide-screen format, and can access information tracks that give background on the movie, its production, and its personnel. In addition, scenes and music that may not have been used in a movie's theatrical release are often included on the disc, as are alternative endings.

The diffusion of DVD is affected by a number of factors. Cost is one. Most people already have VCR machines and tapes, and many do not consider the DVD's technical strengths and improved sound sufficient to warrant an expensive changeover. The lack of content is another. Fox is the only major studio yet to make its recent movies available in the new format, and there were only 2,300 titles available at the end of 1998. Until more titles are available to consumers, they will be slow to buy players. Until there are more players, studios will be slow to make more titles available. A third factor affecting the

A typical DVD player.

diffusion of DVD is that for the first 4 years of the technology's life the discs were not recordable, rendering them inferior to VCR for many users. (Recordable DVD became available to buyers in summer 2000, but their cost—$2,000—made them unattractive to most consumers.) Still, 4 million machines were sold in 1999, a 300% increase over the previous year, bringing the total to 5.4 million machines and 130 million discs in people's homes. Industry analysts predict that DVD's penetration will be 10% of all U.S. homes by the end of 2000 (Graser, 2000).

Direct Broadcast Satellite

The technology for direct delivery of television signals from satellites to ground locations has been available for some time. (NASA launched the first direct broadcast satellite in 1976 to further global cooperation in medicine.) But its diffusion among home viewers has been hampered by the easy availability of cable, the high cost and large size of receiving dishes, and its inability to deliver local stations. These problems largely have been overcome. With the development of a small, affordable receiving dish, DirecTV began **direct broadcast satellite (DBS)** service in 1994, offering 150 channels. It was soon joined by several other satellite program providers, including Rupert Murdoch's American Sky Broadcasting (ASkyB). Then, in late 1999, Congress passed legislation removing restrictions on importing local content to local markets, allowing DBS providers to deliver these hometown channels to their subscribers.

Today, due to mergers and failures, only two DBS providers operate in the United States. DirectTV, owned by General Motors' subsidiary Hughes Electronics, had 8 million subscribers in early 2000. EchoStar/Dish Network, of which Fox parent company News Corporation holds a significant stake, had 3.1 million subscribers (Higgins, 2000).

According to the FCC, at the start of 2000 12.5% of all multichannel subscribers were DBS viewers and two thirds of new subscribers to a home television service choose satellite over cable (McConnell, 2000). Cost seems to be the determining factor. Although overall cable rates are lower than satellite

As the size of DBS receivers has shrunk, their popularity has grown.

rates, the per-channel cost of satellite delivery is lower than cable's. Viewers who want a lot of channels for their money may choose satellite, as will those who don't mind paying overall higher costs for specific content not available (or available for additional cost) on cable, for example, a complete menu of collegiate and professional football. Cable hopes to thwart the advance of DBS with increased digitalization, allowing it to bundle even greater amounts of programming and information for delivery to people's homes. In that way not only can its per-channel cost be lowered to a level below that of DBS, but it can provide services, specifically those based on interactivity, not available to DBS subscribers.

The box titled You Make the Call—Predicting Successful Technologies presents one way of determining for yourself which of these technologies is most likely to survive in our current technology-rich environment.

Digital Video Recorders

New technology writer John Quain (1999) enthused, "The term *convergence* has been buzzing around the computer business for years. For the technology-bound digerati, it has represented the Holy Grail of acceptance: access to the mass-market audience for television. Despite many attempts, including

The Real World *You Make the Call—Predicting Successful Technologies*

One of the first mass communication textbooks, *The Process and Effects of Mass Communication,* offered an interesting way to judge your own media choices. Author Wilbur Schramm (1954) asked the question, "What determines which offerings of mass communication will be selected by a given individual?" (p. 19). The answer he offered is called the **fraction of selection,** and it looks like this:

$$\frac{\text{Expectation of Reward}}{\text{Effort Required}}$$

His point was that people weigh the level of reward they expect from a given medium or message against how much effort they must make to secure that reward.

Today, millions and millions of dollars are being gambled on your choice between DBS and cable, VCR and DVD, and VCR and DVR. Who wins—and who loses—will be decided by viewers like you. To enter this high-stakes business for yourself, subject these three sets of competing technologies to your own personal fraction of selection.

For DVD, for example, effort required might include the two trips to the rental store; expectation of reward might include low rental cost. Expectation of reward for DBS could include its large number of channels, and effort required might include the cost of installing the receiving dish and its unsightliness on the side of your home.

Compare these competing technologies—DVD versus VCR; DVR versus VCR; and DBS versus cable—and determine your fraction of selection by following these seven steps.

1. Make a list of the expected rewards for each of the technologies in question.
2. Make a list of the effort required for each of the technologies in question.
3. Assign a value of 1 to 10 for each reward based on your estimation of how important that reward is to you (more expectation of reward gets a higher score).
4. Assign a value of 1 to 10 for each effort based on your estimation of how "costly" to you the effort really is (greater effort required gets a higher score).
5. Total all the promise of reward scores.
6. Total all the effort required scores.
7. Divide effort into reward and you'll arrive at your personal score for each technology.

In this way you can determine which technologies you think will survive (at least as represented by your selection of one over the other).

To make this exercise more meaningful (and possibly more fun), go beyond the advantages and disadvantages discussed in the text by experiencing these technologies firsthand before you make your final decisions.

Internet-on-your-TV boxes, video phones, and 'smart' cable converters, the marriage of computers and television has never been consummated—perhaps until now" (p. 64). This union produced the **digital video recorder (DVR),** which puts a significant amount of control over content into viewers' hands. In March 1999 Philips Electronics unveiled its version, TiVo, and soon after, Replay Networks introduced its ReplayTV.

What can viewers do with DVR? They can "rewind" and play back portions of a program while they are watching and recording it without losing any of that show. People can digitally record programs by simply telling the system their titles. By indicating what their favorite shows are, viewers can

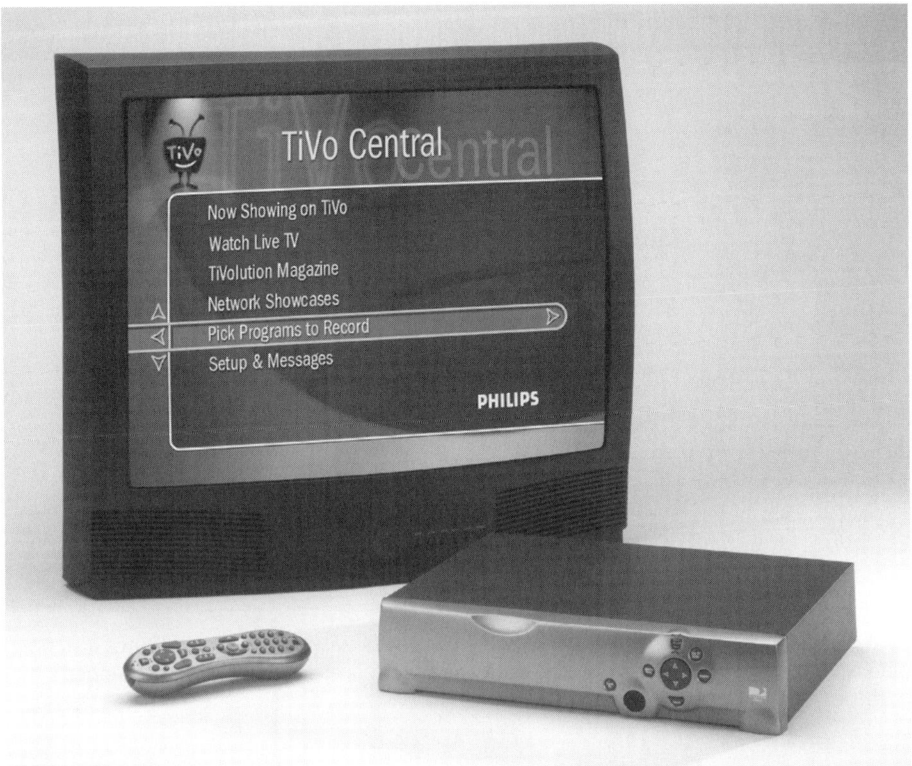

Digital video recorders such as ReplayTV and TiVo (pictured here) promise to change how we watch television.

instruct DVR to automatically record and deliver not only those programs but all similar content over a specified period of time. This application can even be used with the name of a favorite actor. Punch in Adam Sandler, and DVR will automatically record all programming in which he appears. The system accomplishes this through an automatically placed, nightly toll-free phone call to the service provider. The call lasts no more than 3 minutes, and if a viewer should use the phone while this call is being placed, DVR hangs up and makes the necessary call later.

DVR does not deliver programming as do broadcasters, cablecasters, and DBS systems. Rather, it is used *in addition* to these content providers. To use either ReplayTV or TiVo, viewers must buy a special receiver and pay a monthly service charge. It is too early in the life of this technology to predict its success or failure. There are questions about its cost. The devices themselves are expensive, and TiVo requires a monthly fee. Both systems require access to a telephone jack, often a problem in rooms where a television typically might sit such as a den or living room, and the technology faces a legal copyright

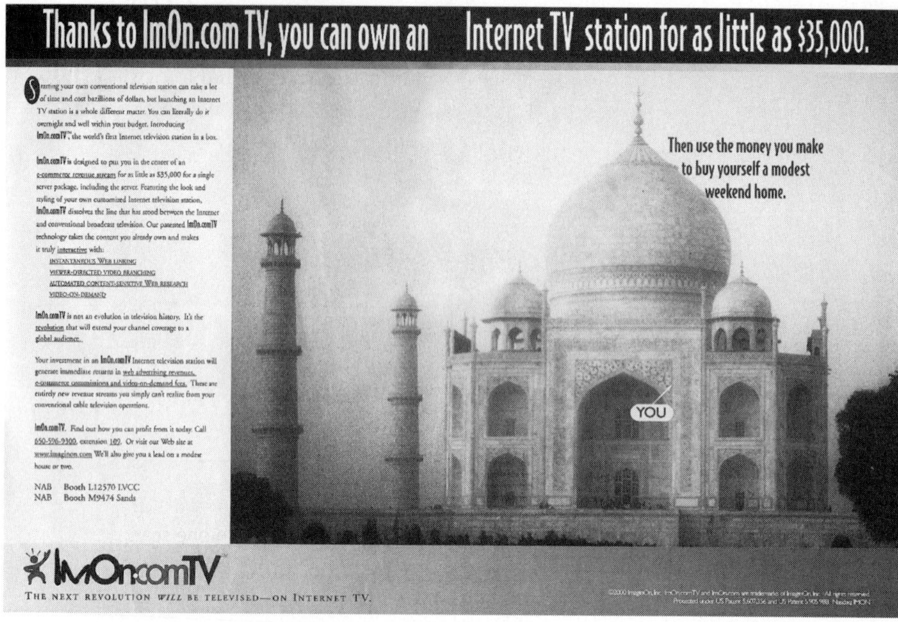

As this ad from ImOn.comTV makes clear, the rules of television—what constitutes a station, who can own and operate one, the necessity of an FCC license—have already changed because of the Internet.

challenge from the broadcast and cable industries who fear its commercial-skipping potential. But, as with DVD, its mere presence is additional evidence that television viewing is changing dramatically in the 21st century.

The Internet

Perhaps the greatest change in the business of television and the medium's relationship with its audience sits just on the horizon in the form of the Internet. Chapter 7 details several different forms of Internet video or TV over the Internet. Here we will limit ourselves to a brief comment on the Internet's impact on the amount of television people watch. In June 1999 household penetration of personal computers surpassed 50%. People used their at-home computers about 35 hours a week, 15 hours of which were dedicated to the Internet and the World Wide Web. Where did these hours come from? Seventy-eight percent of the households were online during prime time, taking them away from the television set. Nielsen reports that homes with Internet access watch 15% less television than do people without (McChesney, 2000).

Digital Television

Digitalization of video signals reduces their size; therefore, more information can be carried over telephone wires (either a cable or phone company's) and

Tech Talk *Comparing NTSC Television and HDTV*

The differences in picture and sound quality between conventional television sets and HDTV are immediately apparent when viewing. The reasons can be seen in this comparison of the two.

	NTSC	*HDTV*
Aspect ratio (width versus height)	4:3	5.33:3
Maximum number of pixels	307,200	2,073,600
Number of lines per picture	maximum of 525	720–1,080
Technological basis	analog	digital
Audio	stereo	multichannel
Scanning	interlaced	interlaced and/or progressive

Interlaced: Electron beams sweep across picture tube, creating half of the lines, then fill in the gaps on the second pass.

Progressive: The entire picture is built line by line in one scan.

First scan

Second scan

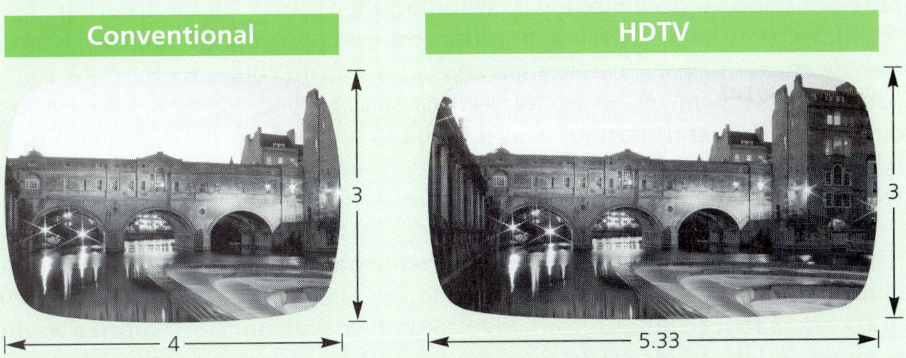

stored. We saw in Chapter 1 that this convergence of telephone, video, and data transmission fuels the rush of mergers that characterizes the contemporary telecommunications environment. Traditional television broadcasters also see digitalization of television signals as their salvation, because broadcasters would be able to carry multiple forms of content on the spectrum space currently used to carry their one broadcast signal. But digitalization for the purpose of transmitting multiple signals conflicts with the use of their spectrum space to transmit high-definition digital television (**HDTV**). This puts broadcasters in something of a bind. Viewers want beautiful, clear, widescreen high-definition images, but they also want a lot of channels of video and other data. If broadcasters devote their entire spectrum space to the transmission of high-definition images, they will lose audience to cable, the Internet, and DBS, which all offer multiple channels of programming and data. If broadcasters use digitization to divide their channels (called **multiplexing**) and, in effect, become mini cable companies themselves—offering over-the-air services such as clearer digital video images (although not of HDTV quality), Internet access, paging services, and constant data flow of information like stocks and sports scores—they will lose HDTV set owners who want the even crisper images of DVD. This conflict is slowing the diffusion of digital television, and it has opened the door to increased regulatory scrutiny of television.

When Congress passed the Telecommunications Act of 1996, it considered broadcasters' request for additional spectrum space with which they could experiment with HDTV. The Japanese were making great strides in the area, and U.S. broadcasters argued that it was a matter of great national concern that they, rather than the Japanese, set the technological standard for what was clearly the television of the future. You can study NTSC and HDTV specifications in the box titled Comparing NTSC Television and HDTV.

But because the airwaves belong to the people (Chapter 2), many in Congress felt that broadcasters should pay for the additional spectrum space. Proponents of the sale of the spectrum predicted that $70 billion could be collected through the auction of the necessary frequencies. This money could help reduce the federal deficit, fund new children's and educational television program services, and wire every school in the country to the Internet. The broadcasters argued, however, that the cost of buying the frequencies would have been an unfair burden in their competition with computer network companies. These companies were also entering the video business, with online delivery to personal computers and **WebTV,** which features high-definition television sets connected to the Internet that serve as "family-sized" monitors (see Chapter 7).

Declaring its intent to revisit the sale issue later, Congress gave the spectrum space to the broadcasters free in what Senator Bob Dole called "a multibillion dollar giveaway" and the "worst form of corporate welfare" (as quoted in Hickey, 1996, p. 40). Conservative columnist William Safire called the gift a "ripoff . . . on a scale vaster than dreamed of by yesteryear's robber barons" (p. 40).

Then, in the summer of 1997, several broadcasters, including ABC, announced that they would give up their HDTV plans and use the additional spectrum space to multiplex a number of pay services, for example, channels that deliver stock quotes. Angry Washington politicians called hearings on the "double-cross," the *New York Times* on August 17, 1998, editorialized against "welfare for broadcasters," and the White House established what became

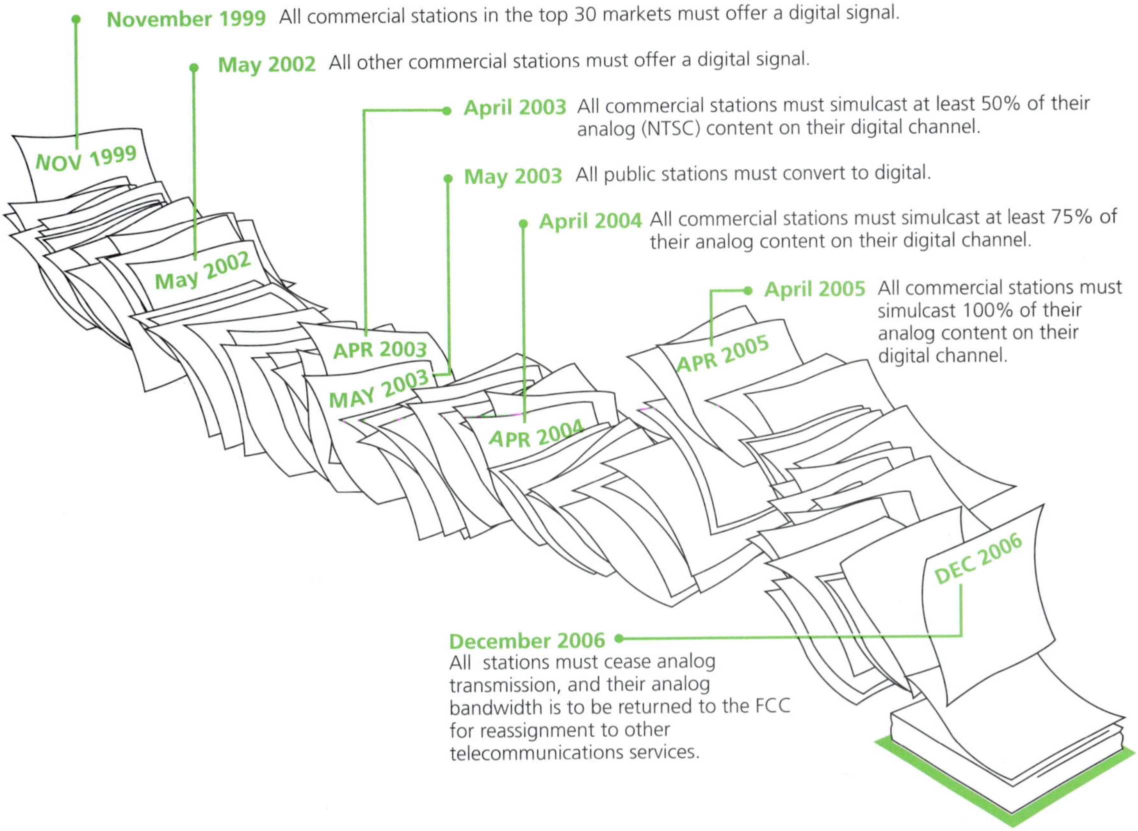

Figure 3.2 Digital Television Rollout Timeline

The FCC has mandated a schedule for the introduction of digital television. Forty-two stations in the 10 largest markets voluntarily began offering digital signals in November 1998. This was 6 months ahead of the commission's May 1999 requirement that all commercial stations in these markets transmit digital signals. However, the Congressional Budget Office released a report in September 1999 announcing it believed neither the broadcasters nor their audiences would be ready for the move to full digital television by the 2006 deadline (Stern, 1999a). Nonetheless, this is the FCC's official schedule and all stations must meet it. You might find it interesting to monitor broadcasters' ability to comply over the next few years.

known as the Gore Commission. Composed of 24 industry, regulatory, and political people, this commission explored broadcasters' use of digital television and their public service responsibilities (Figure 3.2). We'll revisit this issue in Chapter 12, but for now the digital television revolution is progressing slowly for a number of reasons unrelated to government scrutiny.

The first is cost, both to broadcasters and to viewers. Digital receivers now cost from $5,000 to $10,000 for basic sets, and with all programming still available for traditional analog receivers, relatively few people are willing to incur this expense. Broadcasters, too, must invest in new equipment—transmission

Tech Talk *Digital Television Formats*

The analog television that we've known for years (at least in the United States) employs a picture composed of 525 lines. Until a single digital television standard can be agreed on by set manufacturers and broadcasters, several formats will exist side by side. Here are the specifications of four of the most common.

Format	*Specifications*
480 I	480 horizontal lines, interlaced. Provides picture detail equivalent to today's best analog sets.
480 P	480 lines, progressively scanned. Offers twice as many lines at any given moment than today's sets. Images are sharper than with 480 I.
720 P	720 lines, progressively scanned. Permits wide-screen (movie aspect) images. Considered HDTV; offers a much more detailed image than either of the preceding formats.
1080 I	1,080 horizontal lines, interlaced. Permits wide-screen (movie aspect) images. Considered HDTV. Image quality is equivalent to 720 P format.

Source: Knight Ridder News Service, 1999.

towers, cameras, and editing and other facilities. It's estimated that the changeover to digital will cost each and every television station in the United States $10 million (Gomery, 1998). Some critics feel that the broadcasters are overstating the costs they will incur in the changeover to ensure that the FCC does not start charging them more for spectrum space. The reality, they claim, is that most stations would be replacing their technology anyway as part of routine equipment purchases. Further, buying new equipment, digital or otherwise, is written off against taxes, meaning that the public is in fact bearing much of the cost of broadcasters' transition to digital.

A second problem has to do with technical standards. Similar to television's earliest days, this dispute is also over scanning. The telcos and computer companies want **progressive scanning,** that is, the format compatible with today's personal computers. In progressive scanning the entire picture is built line by line in one scan of the television's electronic beam. Sets equipped to receive progressive signals can also receive interlaced signals. Many broadcasters want the standard to be interlaced scanning, the format compatible with their own current standards for manufacturing and broadcasting. (This format cannot accommodate progressive scanning.) After intense lobbying by both sides, the FCC refused to set a standard, leaving the telecommunications industry to set a standard for itself. CBS chose interlaced, ABC and Fox chose progressive, and NBC uses one signal for some of its programming and the other for the rest. But in mid-1998 Sony (an interlaced proponent) and Microsoft (a progressive proponent) announced that they would collaborate to ensure development of fully compatible receivers (Hundt & Levin, 1998). The specifications of four of the leading digital television formats are presented in the box titled Digital Television Formats.

A third problem in the diffusion of digital television is that cable operators must also be willing to change over to digital. It does broadcasters who have made the expensive conversion to digital no good if cable operators cannot or will not devote valuable channels to digital signals. Some small steps have been made to resolve this difficulty. Although the over-the-air broadcasters have long lobbied the FCC to pass federal **must-carry rules** requiring cable operators to carry all digital as well as analog channels, some cable operators are making the move voluntarily. Time Warner, for example, has announced an agreement with CBS and its affiliates to carry its digital broadcasts.

These many obstacles notwithstanding, the FCC requires that all television stations convert completely to digital transmission by 2006. At that time they must return their analog licenses to the government. The commission has also mandated that network affiliates in the 10 largest markets offer digital service by May 1999, with the whole country having access to digital television by 2002 (noncommercial stations have until 2003).

Notwithstanding these obstacles, the changeover has begun. CBS made its digital television debut on October 29, 1998, with its broadcast of the launch of the space shuttle *Discovery* (three days in advance of the FCC's November 1 target date for digital). CBS also produces its NFL football games in both digital and analog formats. A few days after CBS's launch, ABC offered an HDTV broadcast of *101 Dalmatians* and Fox did the same with its airing of *The Lost World.* NBC offers Jay Leno in HDTV. Other program providers are also making the move to digital. DBS provider DirectTV began offering two HDTV channels in the spring of 1999, and cable's HBO and Discovery have also committed to digital programming. In all, by January 2000 there were 109 stations transmitting digital signals in markets totaling 60% of all U.S. television households.

Chapter Review

Mechanical methods of television transmission were developed as early as 1884 by Paul Nipkow, but it was the electronic scanning developed by Zworykin and Farnsworth that moved the medium into its modern age. World War II delayed the diffusion of television but hastened its technical development. At war's end, a mature medium, complete with network structure, formats, stars, and an economic base, was in place. So explosive was television's growth that in 1948 the FCC froze new station permits until a plan could be put in place for the orderly and efficient growth of this new technology. In this period a standard for color television was adopted, interference problems were solved, new spectrum space was opened, and channels were set aside for noncommercial broadcasting. By 1960, 559 stations were being watched in 90% of all homes in the United States.

During this time, the quiz show scandal, Lucille Ball's production innovations, McCarthyism, and the onset of television ratings served to define the new medium's character.

The business of television is still dominated by the networks, although their share of the audience is in decline. New and converging television technologies promise great change in the business of television and in the relationship between this medium and the audience. Cable has eroded the networks' audience by providing numerous alternatives for viewers. In addition, cable has empowered the independents, allowed for the creation of a powerful fourth network, and given new life to the first-run syndication business. The VCR has further eroded the audience of the networks and allows viewers to time-shift and zip. DVD promises to bite into the audience for over-the-air television even more, and DVR is placing greater control over content in viewers' hands. The remote control permits zapping, grazing, and channel surfing. And DBS, hampered at first by cable's success and the size and cost of receiving dishes, is now a well-entrenched technology.

But the greatest technological revolution in television comes from digital television, both in the form of HDTV and multiplexing. The diffusion of digital television is being slowed by governmental scrutiny, the cost of the changeover to both broadcasters and viewers, conflicting technical standards, and the unwillingness of cable operators to make channels available for the new digital broadcasts. Despite these problems, broadcasters are moving into the digital age, banking on convergence to make them successful.

Questions for Review

1. What is the importance of each of these individuals to the history of television? Paul Nipkow, John Logie Baird, Vladimir Zworykin, Philo Farnsworth, and Newton Minow.
2. What is the distinction between mechanical and electronic scanning?
3. What is NTSC?
4. What are interlaced and progressive scanning? What are the advantages and disadvantages of each according to their proponents?
5. Why did the FCC call for the 1948 Television Freeze? What issues did it hope to resolve?
6. What limits on television station ownership exist today?
7. What was the impact on television of the quiz show scandal, *I Love Lucy,* McCarthyism, and the Nielsen ratings?
8. What is all-channel legislation? How has it affected the television industry and what we see on our screens?
9. What is the nature of the network/affiliate relationship? What does each provide for the other?
10. What is compensation, and how and why is it changing?
11. What is meant by cable television's pass-by rate and penetration?
12. How did cable empower independent stations and influence the development of the Fox network?
13. How have cable, VCR, remote control, DVD, DBS, and DVR affected the networks?
14. What factors initially impeded the diffusion of DBS? Do they still operate?
15. What is multiplexing? Why is it attractive to broadcasters?
16. Why would broadcasters want to transmit HDTV? Why would they not want to transmit HDTV?

Questions for Critical Thinking

1. Do you use any of these new video technologies—DVD, DVR, or DBS? If yes, how has each changed your viewing (specifically, how much you watch, what you watch, and when you watch it)?

2. Do you agree with Newton Minow's assessment of television? If so, what can be done to improve the medium's performance?

3. There is no question that digitalization will change television as we know it, but how much? Aren't people's television habits just too ingrained, too set, for there to be major alterations in who delivers our programming and what that programming will look like?

4. How should Congress distribute spectrum space, through sale or through free distribution? What difference does this make to viewers?

5. When digital television is truly here in 2006, will you be willing to watch television on your personal computer? Will you be willing to access information on your living room television set? Just how sophisticated a new television viewer do you think you'll be?

Engaging Technology

1. Television's history is far more complex and interesting than can be detailed in a text such as this. Access <http://web.mediahistory.com/teevee.html> on the Web. From there you can read about TV's history and link to scores of other TV history sites. Determine for yourself what the "truth" is of the historical disputes that are mentioned in this chapter.

2. There are no more visible players in the changing television environment than the big commercial networks. They have for so long defined the medium that their welfare has come to symbolize the medium's current state and its future. Visit their Web sites to determine how they plan to face the challenge of the digital age. ABC is at <http://www.abc.com>, NBC is at <http://www.nbc.com>, CBS is at <http://www.cbs.com>, and Fox can be found at <http://www.fox.com/>. Then search for the sites of the weblets: the WB, UPN, and Pax. Visit these sites and determine how they plan to keep the pressure on the four majors.

3. Users, manufacturers, and distributors of DVD are engaged in an aggressive effort to ensure the technology's survival. Use your search engine to find as many DVD sites as you can. How many are hardware sales sites? How many sell discs? Can you find any that discuss or debate the merits of DVD? What are the primary claims for and against this technology?

Chapter Four

Noncommercial Broadcasting

September 16, 1962, was to have been an historic night of television, real must-see-TV. The hosts for the evening were FCC Chair Newton Minow of "vast wasteland" fame and Edward R. Murrow, director of the United States Information Agency and broadcast legend. It was the debut of New York City's Channel 13, WNDT-TV, marking "the transition of 'educational television' from a straightforward instructional medium for college extension courses in secondary broadcast markets to what would become, by the end of the decade, the multifaceted PBS network we know today" (Jarvik, 1997, p. 1). The lineup of shows would not be out of place on today's public television: *Television and the World,* a 30-minute documentary from England's BBC; *Around My Way,* a 10-minute documentary about children's paintings; *Once Upon a Day,* a local news program; and an installment of *Russian for Beginners.* Murrow and Minow also previewed upcoming broadcasts of *Master Class,* featuring cellist Pablo Casals, and a French movie, *A la Recherche d'Albert Camus.*

And therein lies the problem for public broadcasting. This type of content is unlikely to appear on commercial television, so noncommercial broadcasting is an appropriate outlet for it. But to how much of the public does it really appeal? And isn't much similar fare now available on numerous cable channels? Today's noncommercial broadcasting faces this and other dilemmas, some rooted in its development, some a function of its means of economic support, and still others brought on by the rapidly evolving telecommunications technologies.

Preview

In studying **noncommercial broadcasting** we investigate radio's earliest educational uses and how the nation's experience with radio shaped the debate surrounding the development of educational television. We trace the transformation of educational television to public television and examine in detail the nature, structure, programming, and audience of television's Public Broadcasting Service and radio's National Public Radio. We also look at the relationship between these networks and other entities in the world of noncommercial broadcasting, particularly the Corporation for Public Broadcasting and content providers such as Public Radio International. To do this, we must look at how these organizations plan to survive and prosper in the face of converging technologies and fragmenting audiences.

The Roots of Educational Broadcasting

The first educational broadcast operation was experimental station 9XM at the University of Wisconsin in Madison. The station began broadcasting in 1917 and instituted a regular schedule of programming in 1919 as station WHA. Two years later, Brigham Young University began noncommercial, educational broadcasting in Utah. Within one year 74 noncommercial stations were on the air, and by 1936 a total of 202 radio licenses were granted to schools, colleges, and universities. These stations were educational in the strictest sense of the word. Their schedules were composed of agricultural information for farmers, household information for homemakers, lectures and other "extension" lessons for the general public, and over-the-air courses for college credit. But running a radio station was expensive, and without advertising support, many of these early noncommercial stations could not survive, eventually selling their frequencies to commercial broadcasters. In 1928 alone, 33 educational stations sold out to commercial operations.

Disturbed by this trend, two powerful philanthropic groups, the Carnegie Corporation and the Payne Study and Experimental Fund, financed the Advisory Committee on Education by Radio (ACER) in 1929. The committee united noncommercial stations and others interested in championing educational radio and issued an influential report in 1930 calling for these changes:

- Reservation of frequencies for noncommercial broadcasting (called **reserved channels**)
- Establishment of an Office of Education for Radio in the Department of the Interior, headed by a permanent advisory committee composed of educators, members of the public, and commercial broadcasters

- Establishment of secure and regular funding for educational programming
- Informing the President of the United States about educational radio's importance

In 1931 the Payne Study established another group, the National Committee on Education by Radio, whose goal was to fight for reserved channels in opposition to the recommendation of a Rockefeller Foundation–backed advisory committee advocating the idea that educators' radio needs could be met by commercial broadcasters. As we saw in Chapter 2, the commercial broadcasters, unwilling to cede valuable frequencies to noncommercial licensees, prevailed in the debates leading to the passage of the Communications Act of 1934. No new reserved channels were allocated. Commercial broadcasters promised to serve educators and other civic groups, but they still coveted the licenses held by the dwindling number of educational stations. As a result, and encouraged by the hard financial times of the Great Depression, 164 noncommercial stations either allowed their licenses to lapse or sold out to commercial entities by 1937 (Jarvik, 1997, p. 6).

Educational radio was given a reprieve with the development of FM (Chapter 2). In its allocation plan for FM radio in 1945, the FCC reserved the 20 discrete FM channels between 88.1 and 91.9 MHz for educational use. Eventually, some 700 noncommercial FM stations would join the 20 noncommercial AMs still on the air.

Educational Television

Even though television was considered a commercial medium from its inception, the military was making educational telecasts as early as 1945. And recalling the failure of the 1934 Communications Act to provide educational radio with usable AM frequencies, prodded mercilessly by Commissioner Frieda Hennock, and lobbied by the Ford Foundation–funded Joint Committee on Educational Television, in its 1952 Sixth Report and Order the FCC reserved channels for educational television on the basis of one educational license for every four commercial assignments. Of the 242 stations authorized, 80 were VHF and 162 were UHF. The problem for the majority of educational television broadcasters was that very few people had UHF receivers.

This handicap was overcome somewhat by two factors, one economic, one regulatory. In television's early days, the commercial broadcasters, not yet attracting large amounts of advertising revenue, made significant amounts of airtime on their channels available to noncommercial groups, which provided programming that was either very inexpensive or free. So educational programming, when it did come to the air in greater amounts later, was not totally alien to viewers. Then, in 1962, Congress passed all-channel legislation to take effect over 2 years, requiring all sets imported into or manufactured in the United States to be equipped with both VHF and UHF receivers. As we saw in Chapter 3, this had little immediate impact on educational stations, but they finally were able to attract a bit more of an audience. Soon educa-

FCC Commissioner Frieda Hennock

tional stations began to look less educational in the strictest sense of the word and began programming more entertaining cultural fare. Combined with the allocation of additional channels for noncommercial television (eventually bringing the number of reserved allocations to 600, of which about 240 remain vacant), this led to calls for a more robust system of noncommercial stations and their interconnection. A first step toward that goal was the formation of National Educational Television (NET). Here again, the Ford Foundation was instrumental, "pouring money into educational television to the tune of $100 million a year, sponsoring NET and instructional projects around the country, and influencing the development of flagship stations for the network" (Jarvik, 1997, pp. 18–19).

Beginning in 1959, scores of noncommercial television stations were part of NET's content-sharing cooperative. But NET was plagued by two problems. The first was that it did not produce national-quality fare; it simply served as a clearing house for the distribution of programming produced by its **member stations.** Second, it delivered the programs to its members by mail and other forms of physical shipment, a method called **bicycling.** Clearly, if noncommercial television, and radio for that matter, were to become more meaningful alternatives to the commercial broadcasters, something grander was required.

From Educational to Public Broadcasting

It was the Great Society programs of President Lyndon Johnson and the efforts of the philanthropic Carnegie Commission that provided a grand vision. In 1964 Johnson ordered his Office of Education to hold a conference dedicated to developing a long-range financing plan for educational television. What resulted was establishment the next year of the Carnegie Commission on Educational Television. This group released its *Public Television: A Program for Action* in January 1967. The decision to stop talking about "educational" television was intentional, and "public" television was substituted. The goal was twofold: to shed the dull, stodgy connotation of "education" and to declare that this would be the "public's" medium, as opposed to the advertisers' or government's medium.

This report called for creation of a Corporation for Public Television, federal financing of public stations, establishment of production centers, and true network interconnection. So popular was the proposal that with little change this scheme became law 11 months later with passage of the Public Broadcasting Act of 1967. Among the few alterations was inclusion of a ban on editorializing, a requirement that public broadcasters be objective and balanced in their programming, and most significant, the inclusion of radio in the plan. Hence, the Corporation for Public "Broadcasting" (CPB) came into existence.

Public Broadcasting Today

The Corporation for Public Broadcasting remains the central organizing institution of public broadcasting in the United States. The creation of this body, today composed of 10 board members appointed by the president and confirmed by the Senate, was based on two goals. The first was to create an influential, national body that could provide vision and leadership. A second goal was to insulate public stations from government control. (There was a fear that if the government paid the bills it might exert undue pressure.) To accomplish this, federal money went first to the CPB and then was distributed to public stations.

Because the CPB was not allowed to own or operate the networking structure uniting its public stations, it established a television network, the Public Broadcasting Service (PBS), in 1969 and a radio network, National Public Radio, in 1970. This is the structure that remains today, although it is important to note that PBS is not as much a network as it is an *interconnection agency.* That is, PBS charges its member stations dues, and then offers those stations content produced by a range of program suppliers, paid for with pooled station money, corporate underwriting, funds from CPB, and other sources. NBC, in contrast, produces content itself and then offers its affiliates financial incentives to carry that content (see Chapter 3). PBS member stations are also free to schedule and even rerun content when they wish, although PBS strongly encourages its stations to air its most popular programming, for example, *The Newshour with Jim Lehrer* and *Nightly Business Report,* at the same time to facilitate their national promotion.

The vision for public broadcasting that was encouraged and supported by the CPB was, and is, noble. Public broadcasting was to be an alternative to commercially oriented radio and television; it was to be a strong "second force" in American broadcasting. That is, it was to be "a system devoted to excellence and to the exploration of the full range of programming possibilities, without bondage to the marketplace" (Brown, 1979, p. 29). In addition, the *public* in public broadcasting "was meant to be synonymous with 'people'" (p. 29). That is, public broadcasting was intended to be responsive to the people rather than to advertisers.

Criticisms of Public Broadcasting

Not everybody was delighted with public broadcasting. From both the political right and left came complaints about government financing. Conservatives were philosophically opposed to government support. If there was a need or desire for this type of broadcasting, they argued, the market would provide it. Liberals feared not government funding but *annual* government funding. They worried that Congressional appropriations, even if made every 5 or 10 years, put far too much pressure on public broadcasters to satisfy their federal benefactors.

A number of other complaints about public broadcasting were raised in its early days and remain today. It is important to note that these criticisms typically come from the conservative side of the political spectrum, whose antipathy toward public broadcasting is not shared by the majority of Americans. This fact was made quite clear in the huge public outcry (even from citizens classifying themselves as politically conservative) against the Republican Congress's efforts to "zero out" CPB funding in the mid-1990s. Here are some of the arguments put forth at that time.

- *Public broadcasting is elitist.* Its content is aimed at a very small portion of the American public, typically those with higher income and educational levels. If all the people pay for the service through their taxes, public broadcasters should work harder to serve all the people. This was among the central arguments of members of Congress who attempted to deny public broadcasting federal funding in the 1990s. CPB counters this criticism with data that demonstrate that less than one half of its income comes from government sources, with a mere 15% coming from Congressional appropriations (Figure 4.1). Not only are tax dollars not public broadcasting's primary source of

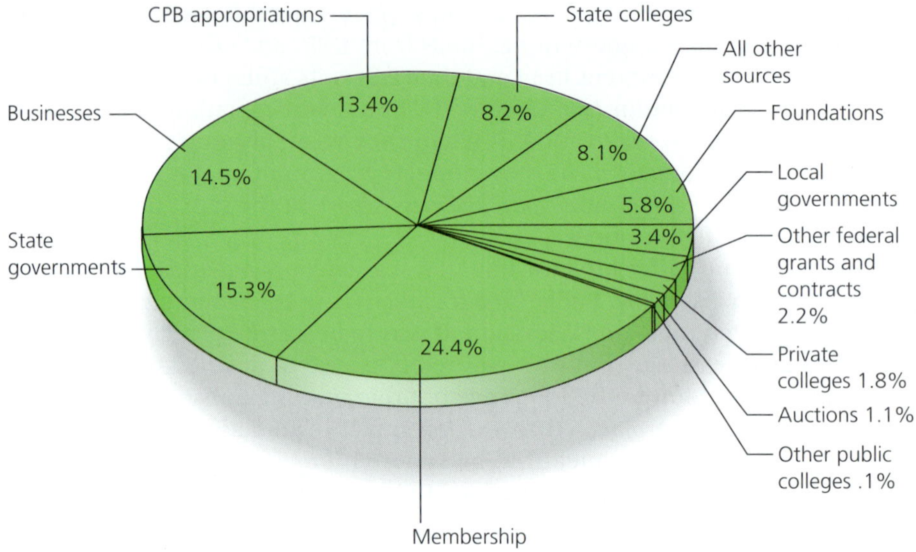

Figure 4.1 **Who Pays Public Broadcasting's Bills?**
Public stations had $1.94 billion in income in 1997. That money came from a variety of sources, with less than half coming from tax-based sources of local, state, and federal governments. *Source:* Corporation for Public Broadcasting. (2000, May 15). [On-line]. Available: <http://www.cpb.org/research/faq>.

income, but a large majority of the public supports public broadcasting, refuting the charge of elitism: 89% of the public agrees that public television "offers intelligent programming"; 88% feels that "regardless of how much people watch it, it's good to know it's there"; and 86% agrees that "it's an important alternative to network television." After military defense and police and law enforcement, the public rates public television and public radio as providing people with the best value for their tax dollars, making them "better buys" than even medical research, highways, and environmental protection (Corporation for Public Broadcasting, 1999).

• *Public broadcasting's reliance on corporate underwriting borders on commercialism.* **Underwriting** involves a contribution from a corporation or other source in exchange for the on-air announcement that that gift made a particular broadcast possible. Naturally, the benefactor's name is prominently mentioned. Critics contend that public broadcasting's contemporary, relaxed rules render the distinction between underwriting and advertising nearly invisible. Cable news channel MSNBC's John Hockenberry, for 12 years a popular and respected National Public Radio correspondent, said of public radio, "By the time I left NPR in 1992, it was an audience-driven, revenue-driven entity, not unlike corporate media outlets" (as quoted in Cox, 1999, p. 41).

Public television is often criticized for its commercialism. In fact, PBS is sometimes called the "Petroleum Broadcasting Service" because of the presence of so much oil company money financing its programming. Naturally, the concern is that corporate money will lead to some programs being made and aired—and not others.

The parallel concern is that corporate dollars can shape a program's content. For example, controversy erupted inside and outside PBS in 1999 over its broadcast of *John Glenn, American Hero.* "Most of the one-hour show was just what you'd expect of a documentary with such a title. But suddenly it veered into an upbeat discussion of the International Space Station that was barely related to the rest of the program and gave almost no hint of the financial and political problems that have dogged the massive undertaking" (Effron, 1999a, p. 44). The program had as its sole corporate sponsor the Boeing Company, holder of a $5.63 billion contract with NASA to design and build that very station. *John Glenn* producer Patrick Butler said that Boeing had "nothing to do with what was in or what was out" (p. 44). But critics say that underwriting "creates its own brand of influence, affecting the subject matter being chosen and the way stories are told. Nobody has to say, 'Give us the money and we'll be nice to you or your cause'" (p. 44). One PBS producer not associated with the documentary called the *John Glenn, American Hero* incident "content for underwriting," another called it simply "content corruption" (p. 44).

Another concern surrounding heavy underwriting is that as public broadcasters pursue corporate money they will begin to behave no differently than their commercial counterparts; that is, they will stress ratings and demographics. Given the complaint that public broadcasting is already elitist, what will happen to the service if it even further "refines" its programming to attract corporate underwriters/advertisers in search of viewers with up-scale demographics?

PBS strives to control the more commercial aspects of underwriting through issuance of program funding standards that include restrictions such as limiting any individual underwriter's announcement to 15 seconds, with no more than 60 seconds of announcements to be broadcast at one time, and a complete prohibition against any reference to tobacco products. In 1990, however, with the FCC's blessing the system began permitting *enhanced* underwriting announcements, which do permit limited depictions of corporate spokespeople, limited depictions of the underwriters' products in use, brand rather than corporate underwriting, and inclusion of underwriters' 800 telephone numbers and Web addresses. Although the guidelines do prohibit comparing an underwriter's product or service to that of a competitor and "calls to action" encouraging viewers to patronize an underwriter's business, for example, for a special sale, these rules remain controversial, as the definition of "limited" is open to much interpretation. Still, PBS reminds its affiliates "that the on-air appearance and overall effect of each credit must be in keeping with the noncommercial nature of public television" (Public Broadcasting Service, 1997, p. 1).

• *Public broadcasting is overly reliant on British content to fill its programming schedule.* This criticism is directed primarily at PBS. Television critic Les Brown (1979) says the network's "reliance on British television for most of its

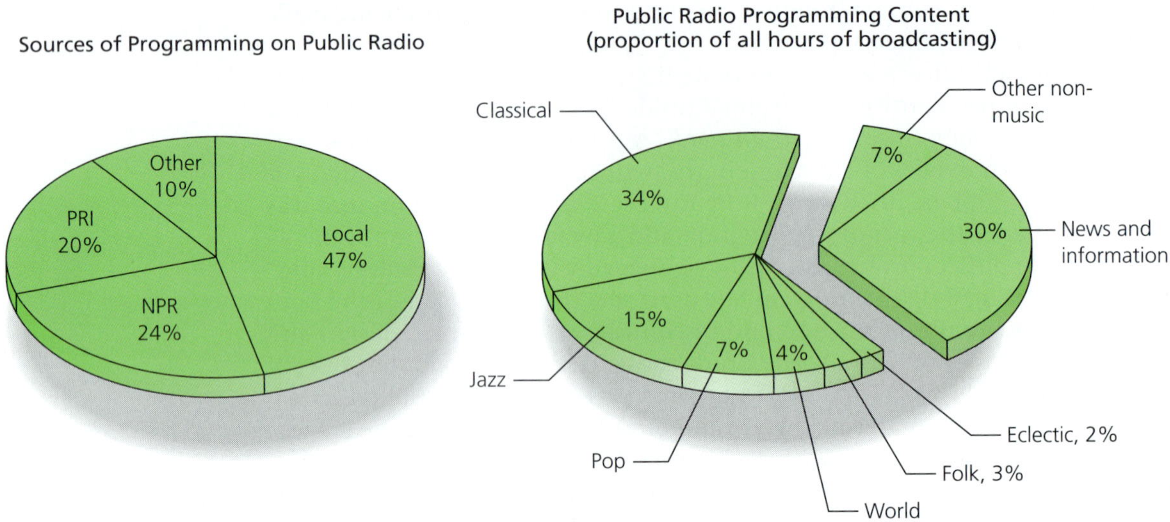

Figure 4.2 What's on Public Radio?

Free from commercial radio's strict format rules, public radio stations program a wide variety of music and information. The first table shows the sources of content for all public radio stations in terms of the percentage of all their broadcast time. The second indicates the proportion of that time devoted to different types of content. *Source:* Corporation for Public Broadcasting. (2000). [On-line]. Available: <http://www.cpb.org/>.

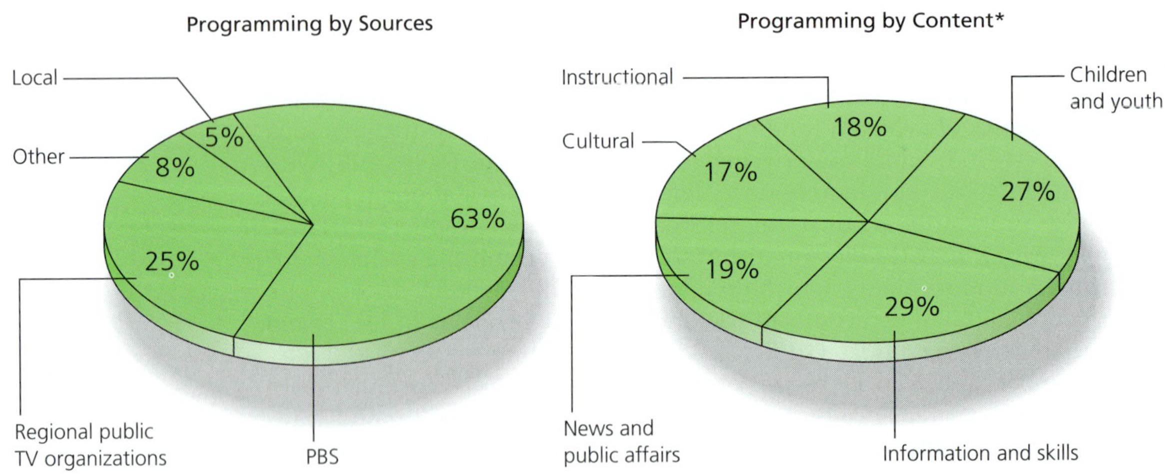

*Total exceeds 100% because some content fit more than one category

Figure 4.3 What's on Public Television?

Free from the demands of ratings and demographics, and with no advertisers to satisfy, public television stations devote nearly 30% of their broadcast hours to programming for children. Information and skills, public affairs, and news account for nearly half of all its broadcast hours. The first table shows the sources of content for all public television stations in terms of the percentage of all their broadcast time. The second indicates the proportion of that time devoted to different types of content. *Source:* Corporation for Public Broadcasting. (2000). [On-line]. Available: <http://www.cpb.org/>.

This scene is from *I, Claudius,* a classic presentation of *Masterpiece Theatre.* Despite the series' long history of airing high-quality drama, *Masterpiece Theatre* embodies two common complaints about public television—that there is too much reliance on British imports and that there is too little local programming.

important fare verges on scandal" (p. 6). Others simply say that PBS stands for "Primarily British Stuff." The network counters that, regardless of their source, British-provided series like *Masterpiece Theatre* and *Mystery!* are among its most popular year in and year out, and others such as *The Forsyte Saga, Civilization,* and *Upstairs, Downstairs* have "become almost synonymous with public broadcasting" in the United States (Jarvik, 1997, p. 259).

- *There is too little local programming on public broadcasting.* Only 4% of the programming that appears on all of the public television stations in the United States and 49% of that airing on all the nation's public radio stations is produced locally (see Figures 4.2 and 4.3). The criticism here is that public broadcast outlets do little to serve their local communities and, in this way, are no different from their commercial counterparts. In response, public broadcasters point to the fact that more people tune in (and at fund-raising times, more people contribute money) when national fare like NPR's *Car Talk* or PBS's *Mobil Masterpiece Theatre* airs.

The Real World *Judging Public Broadcasting's Promise Versus Its Performance*

The Corporation for Public Broadcasting offers its own answer to the question, "What is public broadcasting?" Its Web site (<http://www.cpb.org/research/faq/history/>) provides us, as viewers and listeners, an opportunity to judge the several criticisms of public broadcasting we've studied in this chapter against the standards public broadcasters have set for themselves. Spend a week listening to or watching a public broadcasting operation; then judge for yourself how well that operation meets the goals of public broadcasting and if the critics' complaints have any merit. Consider these ideals that the CPB is committed to.

Excellence

The best in radio and television programming every day . . . a dependable source of information in a fast-changing world . . . music and dance that make your spirits soar . . . or a well-guided tour of almost anything from a bioengineering laboratory to the ins-and-outs of your car.

Did the station you spent time with present programming you would call "the best"? Was it dependable? How did it make your spirits soar (if it did)?

Diversity

Public broadcasters bring to their communities programs that help Americans bridge the gaps between different racial and ethnic groups. From news and public affairs to cultural and performance programming, public broadcasting provides viewers and listeners insight into the values and cultural heritage of diverse groups.

How diverse was the content you encountered? Were there groups—ethnic, cultural, political, or social—that you think were unfairly excluded? Which and why?

Community

Public broadcasting mobilizes volunteers to serve their communities. Local stations reach beyond their world of spectrum, frequency, and programming to assist communities in addressing issues such as violence and literacy.

Did you hear or see information about station-led community action? If yes, what was it? Why do you think the station took on that particular cause?

- *Public broadcasting has become a weak imitation of commercial broadcasting.* This criticism is leveled primarily at public television. Little local programming, demographic-driven, heavily reliant on corporate sponsors—is this PBS or Fox? For example, many supporters of public television were saddened and some outraged in 1999 as commercial networks filled their schedules with inexpensive, home video–based "reality shows" like *When Good Pets Go Bad* and *World's Scariest Police Chases* and PBS went along, airing a 4-week series titled *Escape! Because Accidents Happen.*

The dilemma for public broadcasters is to find the balance between being a true alternative and being ignored. When public broadcasters schedule content that attracts large numbers of viewers and listeners, they are accused of acting like commercial broadcasters. When they air material that appeals to smaller (alternative) audiences, they lose public and financial support and are labeled "elitist." You can evaluate public broadcasting's performance for your-

Learning

General-audience programs, rich in culture and information, stimulate life-long learning. Many public stations offer special services to teachers and students at all levels of education, from *Sesame Street* to high school diploma–equivalency programs to college-credit courses.

Other than Sesame Street, *what general-audience programs and special educational services did you encounter? Did you find them potentially valuable for yourself? For others?*

Noncommercialism

Viewer and listener satisfaction is not for sale to anyone. Public broadcasters don't choose programs with an eye to selling time and attention.

Was the station you monitored completely noncommercial? Were there underwriting announcements that you thought were appropriate for a public station? Were there any that you thought crossed the line into advertising?

Independence

The industry is a loose confederation of noncommercial broadcasters who prize their editorial independence. American public broadcasting doesn't answer to a bureaucracy or to a government. Its counterparts overseas, however, are either centrally directed, the voice of the national government, or both.

How independent do you think the station you listened to or watched really is? Did you find evidence of bias or support for a particular position or group?

The Future

Public broadcasting provides more information, creates greater opportunities for learning, and conducts business with stations, producers and others better, cheaper, and quicker, using new media such as the Internet.

Did your station offer Internet-based services? If so, what were they? Did you take advantage of them? Describe that experience.

self using the information in the box titled Judging Public Broadcasting's Promise Versus Its Performance.

- *Public broadcasting, especially television, is an anachronism that is no longer necessary in the 500-channel, digital era of near limitless choice.* A noncommercial alternative to commercial broadcasting was once a necessity, but critics now ask why taxpayers should underwrite the production and distribution of documentaries and arts programming found in bountiful supply on cable. What these critics ignore, however, is that nearly 40% of all Americans are served neither by cable nor DBS (Chapter 3). These people, too, are taxpayers.

National Public Radio (NPR)

Public radio is more insulated from criticism than is its television counterpart because radio is expected to program for smaller, discrete audiences

Terry Gross of *Fresh Air.*

(Chapter 2). Radio in the United States today is in fact local. In providing large amounts of high-quality national news and information programming, NPR does serve as an alternative.

There are nearly 700 NPR member stations (the network, like PBS, does not use the term "affiliate"). These stations attract more than 20 million listeners a week with a mix of national and local programming (see Figure 4.2) not typically found on commercial stations. Still, it is fine national shows such as *Morning Edition, Fresh Air with Terry Gross,* and *All Things Considered* that not only define public radio for most of its listeners but glue them to their sets for an average of 8 hours a week. In fact, the audience for *Morning Edition* regularly exceeds that of television's most popular morning news program, NBC's *Today,* by more than 3 million people (Heilbrunn, 2000).

Public Radio International (PRI)

Nearly 20% of the programming that airs on public radio is provided by Public Radio International. Founded in 1983 as American Public Radio, its goal was to bring a wider array of diverse programming to noncommercial radio. PRI links more than 600 affiliates by satellite and has an average weekly listenership of 18.8 million people for the 400 hours of programming it distributes.

PRI acquires, develops, funds, and distributes radio programming in four comprehensive categories: news and information, comedy and variety, specials and documentaries, and music. Some of PRI's most popular programs are Garrison Keillor's *A Prairie Home Companion, Marketplace, The World, Beyond Computers,* and *To the Best of Our Knowledge.* Stations can be PRI affiliates and NPR member stations at the same time and, in fact, most are. NPR and PRI are also partners, along with the CPB, in America One, a 24-hour English language radio channel that serves all of Europe. This service is headquartered in Munich, Germany, and distributes its programming to listeners by DBS, cable television, and AM and FM radio. In addition, PRI distributes much of its programming worldwide on its satellite-based World Radio Network. PRI's mission, and how it attempts to fulfill it, are the subjects of the box titled Defining *Diversity* the PRI Way.

Public Broadcasting Service (PBS)

The Public Broadcasting Service is a private, nonprofit television program distribution company (in other words, a television network) that is owned and operated by its 350 member stations. The public knows PBS best from its award-winning and long-running children's programming (for example, *Where in the World Is Carmen San Diego, Puzzle Place,* and *Mister Rogers' Neighborhood*), news shows (*The NewsHour with Jim Lehrer*), documentaries (*Nova* and *Frontline*), and special live broadcasts of momentous events like the Watergate and Iran Contra Hearings and the Clarence Thomas confirmation and Bill Clinton impeachment proceedings.

The network produces none of its own programming, relying on outside producers for all its fare. For example, American Program Service (APS) pro-

Critical Issues *Defining* Diversity *the PRI Way*

Public broadcasting is sometimes criticized as being too elitist. Its audience tends to be middle class and well educated and, therefore, its programming reflects that audience's interests. PRI, however, was founded expressly to bring diversity to public radio. But what does *diversity* mean to these broadcasters, and how do they attempt to foster it? PRI answers the first of these questions in its mission statement (<http://www.pri.org>):

> The mission of Public Radio International is to engage listeners with distinctive radio programs that provide information, insights, and cultural experiences essential to understanding a diverse interdependent world.
>
> The central challenge before public radio is to help the scope of understanding and respect within American society for the experience of all nations, all ethnic and racial groups, and all classes. Radio's universal availability, intimacy, and evocative power can fire imagination and nourish the spirit, make connections between American life and cultures around the globe, and encourage the exchange of ideas and search for common principles fundamental to a civil society.

The answer to the second question—How does PRI attempt to foster diversity through its programming?—can be seen in the number and variety of programs it produces and distributes. As you read this small sample of its shows, can you identify any groups that are ignored? Do you think PRI goes too far in its pursuit of diversity? Does some of this programming seem too narrowly focused to you, or is it important that people from different groups learn about other groups, no matter how small? Can you think of any other broadcast or cable channel that provides a similar range of material? If not, why aren't others producing and distributing this kind of content?

The Writer's Alliance. A daily 5-minute show in which Garrison Keillor reads short poems and highlights of that day in history.

Common Knowledge: Science and Tradition in Native America. A series of programs investigating how science and traditional Alaskan culture accommodate one another in the areas of health, marine mammals, anthropology, and native scientists.

History Told by Those Who Lived It: Remembering Slavery. Two 1-hour shows featuring restored recordings and dramatic readings of interview transcripts from former slaves.

The Mississippi: River of Song. A series of programs highlighting the music born along that giant river, including Native American music from its headwaters in northern Minnesota, St. Louis blues, Arkansas rockabilly, farm belt bluegrass, and New Orleans jazz.

Afropop Worldwide, World Cafe, Jazz After Hours, and *Mountain Stage.* Several different regular series featuring music rarely found on commercial radio.

Music from Chautauqua, Spoleto Chamber Music Series, and *Pittsburgh Symphony.* Several different regular series featuring a different kind of music rarely found on commercial radio.

Which, if any, of these programs have you heard? Which interest you? Why?

---

duces *Nightly Business Report* and *Lord of the Dance,* member stations WGBH and KCET produce *This Old House* and *The Puzzle Place,* respectively, and the Children's Television Workshop provides the network with *Sesame Street.* Like its commercial counterparts, programming decisions are made by a programming vice president and other high-level network executives.

Regardless of its source, public television programming attracts a significant and fairly sizable audience. Each week, 96.2 million viewers—39% of

M-38 Figure Caption to Come

the total U.S. television audience—tune in to a PBS station, watching for a household average of 2 hours and 49 minutes. The network's average prime time rating is 2.1, making it higher in average evening viewership than any individual cable network.

Commitment to Children Public television exhibits a particular commitment to high-quality children's programming. Twenty-nine percent of all public television broadcast time is devoted to children's fare. Along with *The NewsHour with Jim Lehrer,* four of PBS's top-five-rated shows are *Barney & Friends, Sesame Street, Wishbone,* and *Kratts' Creatures.* All are gentle, commercial-free, and overtly encourage learning. A 1996 survey by the Roper Company found that, after watching PBS's children's programming, 29% of kid viewers would read a book about the subject they'd just seen, 29% would make something related to what they had viewed, 25% would do additional research, and 21% would conduct an experiment on or test what they'd viewed (Corporation for Public Broadcasting, 1999).

The network's commitment to children was evident in its January 1999 announcement of the PBS Kids Millennium Project. At the center of the $125 million project was the September 1, 1999, launch of the 24-hour interactive digital PBS Kids Channel. At that time only one third of the United States was reached by digital public stations, so analog public outlets were given access to the channel's content, but not its interactivity. Those public stations that are digital will offer an analog version of the channel until a greater proportion of the public itself has gone digital (Chapter 3). The Kids Channel carries perennial PBS favorites including *Barney & Friends* and *Wishbone* but also highlights numerous new programs such as the fully interactive *Zoboomafoo,* a preschool nature series enhanced with games, puzzles, and children's activities (Katz, 1999, p. 56).

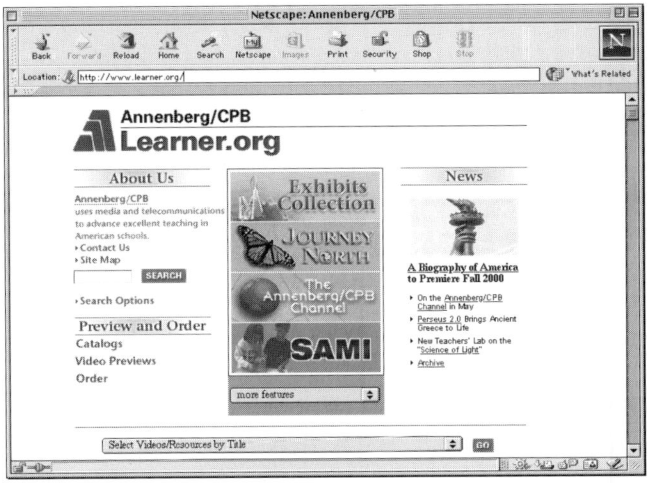

This home page from Learner.org, a joint Annenberg/CPB venture, represents the convergence of the Internet and public broadcasting. Learner.org serves as a portal, leading to CPB-produced information and education sites.

Still Educational After All These Years Public television never completely abandoned its educational roots. Instead, it has used its **branding** as a source of high-quality children's and informational programming to take a leading role in U.S. educational television efforts. For example, the PBS Adult Learning Service (ALS) was founded in 1981 to deliver college and university **telecourses.** ALS, the world's largest provider of over-the-air instruction, now enrolls more than 400,000 students at nearly 2,000 colleges and universities. Through the PBS Going the Distance program, 135 colleges and universities partner with 56 public stations to offer full degree programs entirely through **distance learning.**

Through the Ready to Learn Service, begun in 1994, 95 public stations are linked to schools and day care centers in an effort to ensure that every child in the United States will be able to read by the time he or she starts first grade. PBS provides free books in English and Spanish to those who cannot afford to buy them, conducts workshops on boosting kids' learning at home for teachers, caregivers, and parents, and publishes *PTV Families* and *PTV para la familia,* companion magazines to the Ready to Learn Service broadcasts.

PBS even runs an instructional video (as opposed to broadcast) service, the National Instructional Satellite Service. The service provides 1,000 hours of instructional programming a year to 30 million students and 2 million teachers at more than 63,000 preprimary and primary schools nationwide. High schools in 28 states use another public broadcasting service, the Satellite Educational Resources Consortium, to offer their students study in advanced science, math, and foreign languages. These courses are self-taught; that is, they do not require local teachers.

Finally, the CPB has converged public broadcasting and the Internet in two new projects. K–12 Internet Testbed brings together 15 television stations and schools, libraries, museums, and colleges in their areas to create cooperative Internet projects for grade school children; and the Annenberg/CPB Projects offer Internet-based information and education. At <http://www.learner.org>, for example, you can learn to speak French.

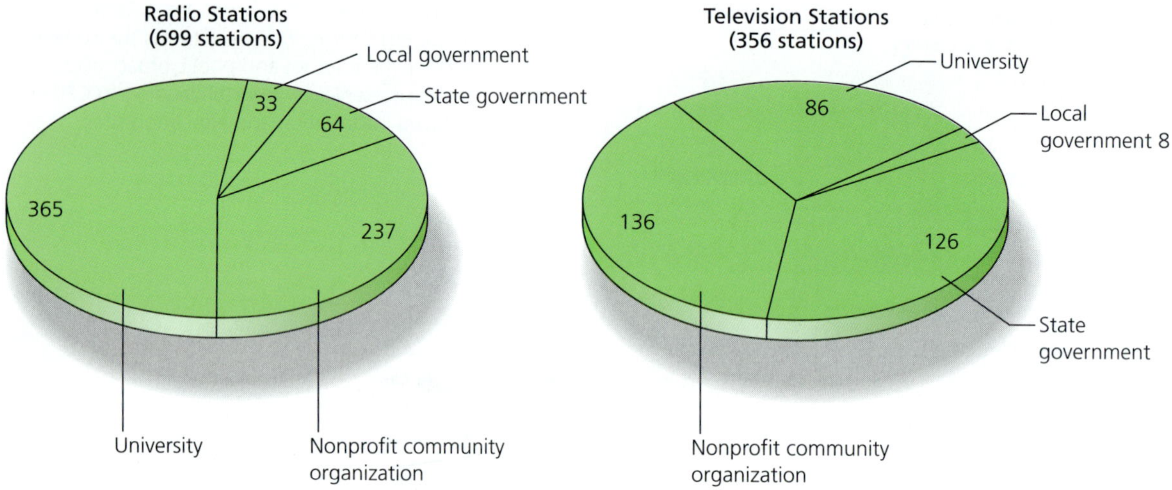

Figure 4.4 Who Runs Public Broadcasting Stations?
Public radio and television stations in 2000 are run by a variety of organizations. *Source:* Corporation for Public Broadcasting. (2000). [On-line]. Available: <http://www.cpb.org/>.

Other Noncommercial Broadcasters

All public stations are noncommercial, but not all noncommercial stations are public. Other common forms of noncommercial broadcast operations are college radio, community radio, and religious radio and television.

College Radio

Although more than half of all public radio stations are housed on college and university campuses (Figure 4.4), there are another 800 college and university radio stations that are not NPR members. In fact, the single most common form of noncommercial broadcast operation is college radio.

Some college stations are part of the academic curriculum; that is, students earn credit for courses associated with running the station and learning about its operation. Others are extracurricular and operate more or less like clubs, possibly with a faculty or staff advisor. In either case 90% of college stations air music primarily, with some public affairs, and both forms of content tend to be **alternative programming,** that is, content rarely heard on commercial stations. In addition, college stations avoid the strict formatting that characterizes their commercial cousins, most relying instead on **block programming,** wherein large segments of related radio content are scheduled at the same time each day or week, providing listeners with a consistent schedule but allowing stations to air several different formats. For example, for several years KSJS at San Jose State University aired *Radio Azlan* (a collection of music and public affairs shows geared toward Latinos) from 6 to 10 p.m. every weeknight, followed by *Da Underground* (a collection of music and public affairs

These materials from some of the 30 stations belonging to the Adventist Radio Network demonstrate just a bit of the variety inhernet in noncommercial religious broadcasting.

programs of interest primarily to African Americans) from 10 p.m. to 2 a.m., jazz on weekends, and progressive rock on weekdays.

Block programming allows college stations to fulfill two important goals. The first is that this variety of significant amounts of diverse content permits college radio to serve groups that are typically disenfranchised or ignored by commercial broadcasters. The greater the variety of blocks of time available, the greater the number of people who can be served. Most college stations serve as "laboratories" for training students in broadcasting in general and radio in particular. The second benefit of block programming, then, is that this diverse content encourages a wider variety of students to learn more of radio's aspects.

Community Radio

The least common form of noncommercial radio operation is **community radio,** stations underwritten by community groups, local businesses, and listener donations. Like college stations, community stations eschew strict formatting and rely on block programming. The most common operators of community stations are school boards, civic associations, and not-for-profit foundations. A well-known example of this latter form of governance are the stations of the Pacifica Foundation (Figure 4.5).

Community stations typically have one or all of these goals or characteristics:

- They are nonprofit and are subject to a board of directors drawn from the community.
- They are run by groups representing the overall community and work to serve the entire public rather than special interest or demographic groups.

wmua fm 91

CALL THE WMUA F

	MONDAY	TUESDAY	WEDNESD
6 am - 9 am AM ECLECTIC	**Acoustic Cafe** Best of the 90's singers, songwriters, folk, acoustic, blues, and folk rock Lee Larcheveque	**Your Weekly Fixx** Try it once, you'll love it. DJ Ezra	**Oblivion Express** Exploring musical horizons from the 20's to the 90's R&B, Americana and everything in betwee Fern
9 am - noon JAZZ	**The Jazz Garden** A broad spectrum of Jazz from 1950 to the present. Mark Gardener	**Transnational Jazz Conspiracy** Improvised music in and out of the jazz idiom. Michael Ehlers	**Java Jazz** A show featuring jazz and other improvisational Cheryl K.
noon - 2:30 pm WORLD	**!Kita** Native American programming for the valley! Indigenous music of the Americas. Native American issues and interviews. Monique Fordham	**Nuevo Mundo** Latin alternative from across the Americas, plus news and commentary. Ramon	**Mustt Mustt** World music with an emphasis on South Asian clas popular music. Has a pulse on topical issues from ir to campus politics. Satish Kolluri
2:30 pm - 5 pm BLUES	**13th floor** transmit the funky soul vibration bring the psychedelic rhythm shower drop the bomb on the planetary work machine brendon rule	**Mixed Bag** A mixture of blues and beyond from today and yesterday. Gene Phelps	**Nothin' But da Blues** Post World War II, Electric Urban, and Contemp Dave Hart
5 pm - 5:30 pm NEWS	WMUA N		
5:30 pm - 6:30 pm PUBLIC AFFAIRS	**Undercurrents** News not covered by the mainstream media. Views from the Left, news from between the lines. Holroyde, Traschen & Tremblay	**Making Contact** Making Contact is a half-hour syndicated show focusing on pressing social issues and democracy. **Upfront** Social justice and diversity issues from a worker perspective.	**Sportsline/Joanie O'Brien Sho** Sports talk call-in show. Women's Basketball head co O'Brien talks b-ball with Justin Rudd, David Rose, and
6:30 pm - 9 pm ROCK	**History is Bunk** Rhythm as an universal idiom in the music we are exposed to... hardcore, punk noise, industrial... a foolhearted dialectic. Sol Cotler	**File Under Easy Listening** I guarantee you will not hear Boston's "More Than a Feeling," but you may hear Sleater Kinney's version of the same song. Chris Lind	**Lost in the Discount Rack** Sweaters beware: attack of the turtlenecks Hardcore, punk, and indie. Craig Holland
9 pm - midnight URBAN CONTEMP	**100% Pure Energy** Playing the best in dance and hip-hop. Hai Nguyen	**Tha Hitt List** Three hours of urbanized solution in yo temples scopin gettin yo third eyes open. Big ANT	**Mission 101** DJ Mgosi
midnight - 3 am PM ECLECTIC	**3 Hours of Everything** If you want to hear it all... Adam De Young	**The Regan Show** The best of today's hits, yesterday's favorites, along with the Weird & Wild News of the Week, the 2 O'Clock Couples, and this day in history. Regan	**Uncommonly Smooth** Only the best in hip-hop, R&B, jazz, and slow jams. An mixture of old school music to the most recent ar Raydiance

This program schedule from WMUA (91.1FM; http://wmua.org) in Amherst demonstrates not only the freedom inherent in alternative and block booking, but their potential scope. The student-run University of Massachusetts station airs programming that can conservatively be called alternative—polka, gospel, Latino, death metal, and future-weapon music. And its commitment to block programming shows in its nine blocks—AM Eclectic, Jazz, World, Blues, News, Public Affairs, Rock, Urban Contemporary, and PM Eclectic—that stretch through five, and in some cases, seven days a week.

- They allow the public access to their facilities and airwaves and work to present a wide range of perspectives, especially those typically ignored by commercial media. (Adams & Massey, 1995, p. 188)

Religious Radio and Television Stations

A distinct form of community station that is becoming more common is the noncommercial religious broadcast operation. Religious broadcasters typically operate on both the AM and FM radio bands and on television's UHF chan-

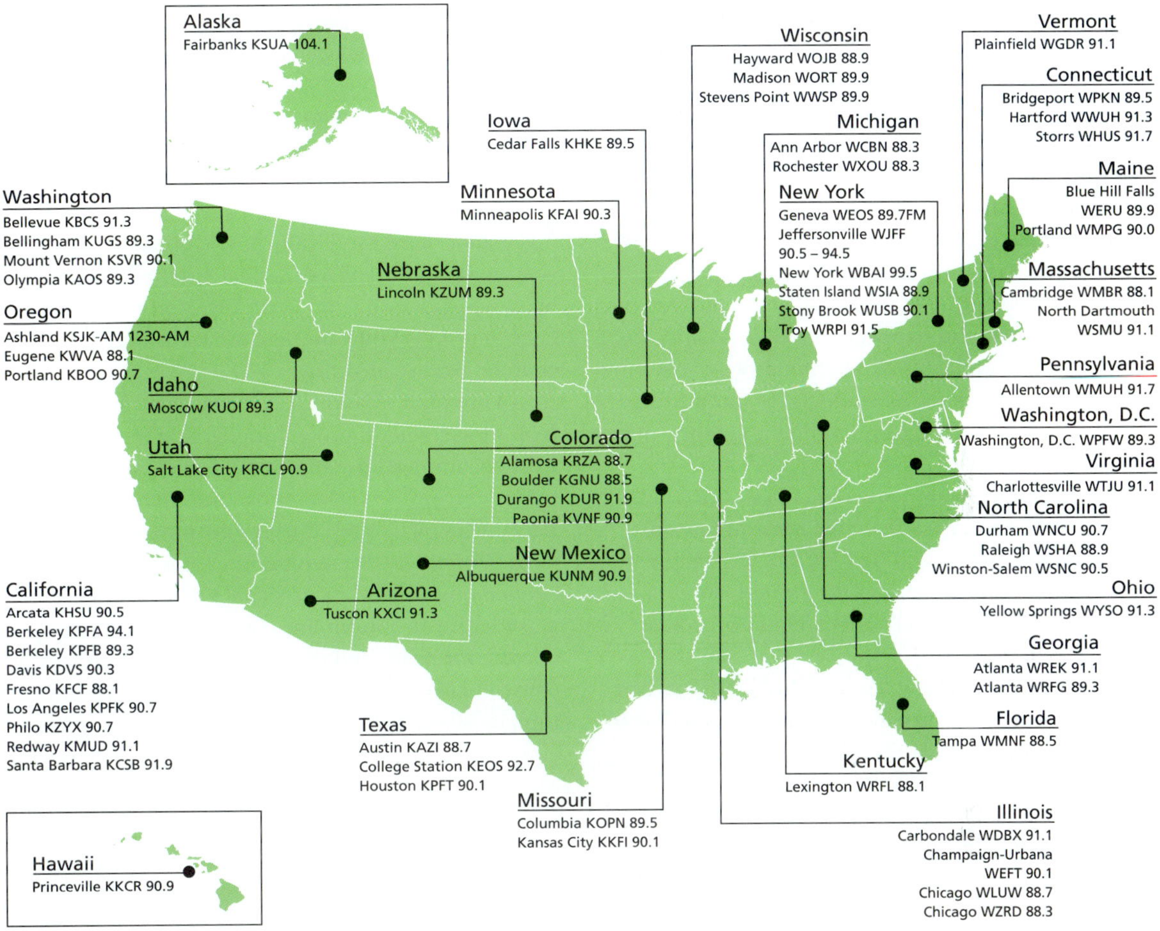

Figure 4.5 Pacifica Radio

Pacifica Radio is the nation's first listener-supported, community-based radio network. Its flagship stations are KPFA Berkeley (94.1), KPFK Los Angeles (90.7), WBAI New York (99.5), KPFT Houston (90.1), and WPFW Washington, DC (89.3). In addition, it has nearly 60 affiliates in 27 states. Its Web site states its mission and lists all the stations that belong to the Pacifica family.

The mission of Pacifica Radio:

- To promote cultural diversity and pluralistic community expression;
- To contribute to a lasting understanding between individuals of all nations, races, creeds and colors;
- To promote freedom of the press and serve as a forum for various viewpoints; and
- To maintain an independent funding base.

Use this list to find a Pacifica station near you. After listening to it for a day or two, determine for yourself if its programming is consistent with the Pacifica mission statement. If it isn't, why not? If it is, how? How does your chosen station sound different from commercial stations with which you are familiar? How is it different from or the same as other noncommercial stations with which you may be familiar? *Source:* Pacifica Radio. (2000, May 15). [On-line]. Available: <http://www.pacifica.org>.

Careers *Working in Noncommercial Broadcasting*

If your campus has a radio station, you may already be working in noncommercial broadcasting. College radio is an ideal place to "see the big picture." Along with community and religious stations, these small, not-for-profit operations typically allow people to perform a wide variety of on- and off-air jobs. Not only can you gain valuable experience in a wide array of broadcast skills, but you can learn what you like and don't like. The idea of being a DJ or other on-air talent may have been what brought you to the station, but community relations, with its high level of public contact, may strike you as a better long-term career goal.

Noncommercial broadcasting doesn't necessarily mean small broadcasting. Large noncommercial radio and television stations are in every major, and even some smaller, cities in the United States. Naturally, for someone wanting a career in broadcast sales, these might not be the best places to start a career, but fund-raising, community relations, and research—all necessary in a noncommercial operation—are important skills for people interest in sales. It might be difficult to break into broadcast sales in a large market commercial broadcast operation, but a big noncommercial station might be a good choice.

For this same reason and because of the difference in economic support, public broadcasting is a good starting point for people interested in a broadcasting career. Pay is generally lower, and hours are generally longer. But as with college and community radio, there is typically ample opportunity to wear many different hats, either simultaneously or by moving laterally and vertically. And the experience and skills you gain here are transferable to commercial broadcasting.

All broadcast stations must concern themselves with management, programming, and technical issues. Commercial stations have chief engineers; so do noncommercial stations. Commercial stations must keep logs; so too must noncommercial stations. Ultimately, there is another "major similarity between public and commercial broadcasting . . . and that is the bottom line. Commercial stations operate for profit and must be expense and revenue conscious. Public stations, while nonprofit in nature, are financially accountable and must conform expenses to anticipated revenues" (Pringle, Starr, & McCavitt, 1991, p. 268).

nels. These stations are often difficult to characterize. Some religious radio and television stations are licensed as commercial broadcasters and simply have a religious format. Other stations are licensed as noncommercial but sound remarkably similar to the commercial outlets because of their frequent calls for donations and the sale of religious items. In addition, some noncommercial religious stations are affiliated with one denomination, and others make their airwaves available to many different religions. Some noncommercial religious stations intentionally take on the look or sound of commercial stations; for example, a number of religious radio stations format Top 40–style "God Rock," others adopt an All-Talk sound. And just as Pacifica is a network of community stations, there are networks of religious stations serving commercial and noncommercial broadcasters alike, the Protestant Radio and Television Center being a well-respected example.

There can be no ignoring the high visibility and influence of national religious broadcasters and cablecasters such as Pat Robertson and Jerry Falwell. Televangelists such as these religious leaders, however, typically employ commercial rather than noncommercial outlets. Still, they have been a big part of the success of local religious broadcasting as they have turned viewer and listener attention to the presence and power of religiously oriented content.

The benefit of making a career in noncommercial broadcasting, in whatever form, is the subject of the box titled Working in Noncommercial Broadcasting.

Chapter Review

Although many educational stations were on the air in radio's early days, lack of adequate financing and the FCC's refusal to set aside AM frequencies for fuller involvement in broadcasting by noncommercial interests doomed educational radio to the margins until 1945 when the commission set aside space on the FM band. Near this time, the push for more robust educational television took hold, leading to the passage of the 1967 Public Broadcasting Act. This legislation created the Corporation for Public Broadcasting (CPB), which, in turn, created the television network Public Broadcasting Service (PBS) and the radio network National Public Radio (NPR). The establishment of a strong public broadcasting system is not without its critics. The complaints are that public broadcasting is elitist, too dependent on corporate underwriting, overly reliant on British programming, insufficiently local, a weak imitation of commercial broadcasting, and unnecessary in today's 500-channel universe. There are reasonable counterarguments to each of these criticisms.

Despite this criticism, public broadcasting enjoys widespread support. NPR's 700 member stations rely on their own local material and national-quality programming produced by the network and by independent producers such as Public Radio International (PRI). PBS's 350 member stations air an impressive array of children's, cultural, and public affairs content, all provided by individual member stations or independent producers.

Public broadcasting is not the only form of noncommercial broadcasting on the air. There are also 800 college and university radio stations, a variety of community radio stations, including those affiliated with Pacifica Radio, and numerous religious radio and television stations.

Questions for Review

1. What are reserved channels?
2. What accommodation did the Communications Act of 1934 make for noncommercial radio? What accommodation did the Sixth Report and Order make for noncommercial television?
3. What was NET? What was its function?
4. What is bicycling? How did it hamper the development of noncommercial television?
5. What are the arguments of those who see public broadcasting as elitist? What is public broadcasting's defense?

6. What are the arguments of those who see underwriting as damaging to public broadcasting? What is public broadcasting's defense?
7. What are the arguments of those who see public television as overreliant on British programming? What is public television's defense?
8. What are the arguments of those who see too little local programming on public broadcasting? What is public broadcasting's defense?
9. What are the arguments of those who see public broadcasting as a weak imitation of commercial broadcasting? What is public broadcasting's defense?
10. What are the arguments of those who see public broadcasting as unnecessary in today's multichannel universe? What is public broadcasting's defense?
11. What are the various roles of the CPB, PBS, NPR, and PRI in noncommercial broadcasting?
12. What differentiates the relationship between PBS and its member stations from that of a commercial network and its affiliates?
13. What characterizes college radio? Community radio? Religious broadcasting?

Questions for Critical Thinking

1. Can public broadcasting ever escape charges of elitism? Because it must program as an alternative to the commercial mass media, isn't some elitism inevitable?

2. What do you think of government funding of public broadcasting? Does it inevitably lead to safe, uncritical programming, or is it an important protection against even heavier corporate influence? Explain your answer.

3. What do you think of public broadcasting's growing reliance on corporate underwriting? Is it a good way to free public broadcasters from government handouts or an invitation to the presentation of increased amounts of business-friendly content? Explain your answer.

4. Are you a regular listener to noncommercial radio or a regular viewer of public television? If so, which content do you typically consume, and why do you do so? If not, why not?

5. The CPB was established, in part, as a buffer between government funders and public stations. There were fears that if the government was paying the bills it could exert undue influence over public station programming. This seems reasonable. But then doesn't it follow that advertisers, those who pay commercial broadcasters' bills, exert undue influence over commercial station programming? Are you more comfortable with advertiser influence over broadcast content than you are with government influence over content? Why or why not?

Engaging Technology

1. Because noncommercial radio and television operations serve the public rather than advertisers, it is no surprise that they have turned to the Internet to stay in touch with their listeners and viewers. Visit these sites to see how the different operations characterize themselves. Then contact the organization (all these sites are interactive) and ask any questions that come to mind from your examination of their Web sites. Share your experience with the class. National Public Radio is at <http://www.npr.org>. The World Radio Network is at <http://www.wrn.org/>. PRI is at <http://www.pri.org>. PBS is at <http://www.pbs.org>.

2. Pacifica Radio is one of the best known community radio networks, and it is also one of the most controversial. Its progressive, some would say radi-

cal, politics make it the darling of millions of listeners and the bane of many others. After you've listened to some of its programming, contact Pacifica through its Web site (<http://www.pacifica.org>) and register your opinion. Explain to the class your communication with the network and its response.

3. Do you have college or community radio experience? If so, describe to your classmates your motivations for involvement with this form of noncommercial broadcasting.

Chapter Five

Corporate and Industrial Telecommunications

The first few months of 2000 saw some big news in the telecommunications industry. Bob Ladrach, president of Digital Images Inc., based near Raleigh, North Carolina, was named one of the top 100 multimedia producers in the world by *AV Video and Multimedia Producer* magazine. Mr. Ladrach had received the award—based on excellence in projects, accomplishments, and progress—once before, and repeat winners are rare. However, a panel of distinguished peers considered Mr. Ladrach's accomplishments sufficiently impressive to honor him again on the 25th anniversary of the magazine.

In other news Logan Productions, Inc. (LPI) completed one of the first high-definition (HD) video programs in the midwestern United States for insurance company Northwestern Mutual Life. The program, *Leveraging Life,* is being shown at Northwestern Mutual's regional meetings across the country. LPI is among the first production companies in the Midwest to invest in high-definition video, and Northwestern Mutual was eager to see what the technology could do. Joanne Lemke, media specialist at Northwestern Mutual, said, "We chose HD because we were looking for a fraction of the price" ("Industry News," 2000).

The two companies worked closely together from creative concept to postproduction. "We enjoyed working with Logan Productions," Lemke said. "They were as dedicated to the project as we were, and the result is a professional, high-quality program we are excited to show to our agents around the country" ("Industry News," 2000). Jim Logan, President of LPI, said many of his clients were eager to integrate high-definition video into their productions. "It provides an image similar to film that is a gigantic step up for video, and a step ahead of what their competition is doing." He added, "The projected image is truly theatrical in quality" ("Industry News," 2000).

Also of note, in Atlanta, Georgia, Philips Consumer Electronics hired Harris Digital Media, Atlanta to create a looping trade show video featuring its new audio CD recorder. Transferred to DVD, the video is being played at technology and electronics exhibitions. Digital Media, Atlanta programmed the DVD to auto loop, allowing it to play repeatedly without a human attendant and without it experiencing the kind of wear and tear that render film and video loops unviewable by the hundredth playback.

Preview

Big news? In all probability you did not even hear about these events. In fact, you probably haven't even heard of the companies, never mind the people involved. Yet these reports, people, and companies come from the contemporary telecommunications industry that employs more audio and video production personnel than any other.

There are a finite number of radio and television stations and a relatively static number of cable channels. But there is no limit to the number of businesses, organizations, and government agencies that can and do contract for, produce, and distribute audio, video, and Web material to meet organizations' specific needs and interests. As a result, more money and effort are expended in the production of nonentertainment audio and video than are devoted to the more common entertainment fare that comes to mind when we think of radio and television ("Out of the Boardroom," 1991). Specific-use, nonentertainment telecommunications is called many things: nonbroadcast, business, private, **corporate, industrial,** or **organizational telecommunications** (or video or audio). This business produces and distributes material that

- Serves very specific instructional, economic, public relations, or individual needs of the producing (or contracting) organization.
- Is very narrowly targeted at a specific and known audience.
- Is distributed over channels not usually available to general audiences.

Throughout this chapter we will use the term *corporate and industrial telecommunications* to refer to this form of telecommunications as we study its inception, its current practice, and how this typically less glamorous but important use of electronic communication is changing largely as a result of emerging and converging technologies.

Roots of Corporate and Industrial Telecommunications

Before World War II, sound reproduction existed primarily in two forms. One form was recording sound to a metal master disc from which vinyl records could be made. The second was wire recording, literally magnetic recording on metal wire (Chapter 2). The first method was too expensive and inefficient for noncommercial use, and the second produced relatively poor fidelity. As a result, there was little corporate or industrial use of audio until after the Second World War when soldiers returned to the United States from Germany with the kind of plastic tape recordings with which we're now familiar. But by that time television had arrived, so the history of corporate and industrial telecommunications began and remained primarily the story of corporate and industrial video—until the arrival of digital technology.

To be sure, during the 1950s organizations communicated with their various audiences through mediated means. Films, slides, and the newly arrived audiotape were standard parts of most companies' internal and external communication strategies. In 1956 the Ampex Corporation began to market an inexpensive 1-inch-wide videotape. And this, along with development of smaller, lighter, less expensive video cameras, led to an expansion of the media operations of those companies already making use of existing technologies and opened up new possibilities for thousands of small- and medium-sized organizations. Then, in the 1970s, a newer, more portable, less expensive, color videotape—the ¾-inch U-Matic videocassette—was developed, attracting even more organizations to corporate and industrial video. Because the new videocassette allowed color recording, the content it carried seemed less "educational" to viewers now attuned to network television's high production values. Because the tape itself was housed in a cassette, not only was ease-of-use assured but duplication and distribution were simple and inexpensive. These were important factors for companies in need of a fast, inexpensive way to communicate with their rapidly expanding operations and to train their growing workforce.

A variety of technological, business, and social factors combined in the 1970s and 1980s to establish corporate and industrial video as a major telecommunications industry:

1. Introduction of even cheaper, easier-to-use, better quality ½-inch VHS videotape cassettes encouraged more organizations to take advantage of corporate and industrial video.

2. Introduction of the camcorder, in which both camera and tape recorder are housed in a single piece of equipment, did the same. The camcorder made production easier and cheaper, as one person could serve as camera person, sound person, director, and editor.

3. The world in which organizations had to operate became much more complicated during the '70s and '80s. As a result, employees had to be trained and retrained, not only as their jobs became more sophisticated (for example,

with the computerization of many formerly manual tasks) but as changing cultural mores required employers and employees alike to reconsider their racial, social, and gender assumptions. Well-produced video was an inexpensive and effective way to meet these expanded training requirements.

4. As companies grew through mergers, buyouts, and international expansion, so too did the size and complexity of their operations and their need to communicate effectively and efficiently with all their parts. Again, inexpensively produced and distributed video met this need.

5. As organizations expanded, travel to their various far-flung operations became increasingly inefficient, expensive, and time consuming. Why not let technology ease the burden?

Corporate and Industrial Telecommunications Today

In 1973 the International Industrial Television Association was founded to serve as a unifying professional organization for those working in the industry. A mere 5 years later it dropped "industrial" from its name to better reflect the wider range of uses being made of video technologies. Then in July 2000 it changed its name again. The International Television Association became the Media Communications Association (MCA). It has more than 10,000 members worldwide and remains the organizing body for an industry that does more than $50 billion in business a year (<http://www.itva.org>). And although it is correct to say that the corporate and industrial telecommunications industry of today is a direct descendant of the corporate and industrial video of the 1970s and '80s, there have been some significant changes, most of which are tied to changing and converging technologies.

Converging Technologies and Applications of Corporate and Industrial Telecommunications

The changing and converging technologies that are redefining other telecommunications sectors are also reshaping corporate and industrial telecommunications. One effect of the emergence of these relatively inexpensive but sophisticated production and distribution technologies has been the shift by organizations away from in-house production to a heavier reliance on independent producers. Even the biggest companies now farm out much of their telecommunications production work to a growing number of small, specialized producers, for example, freelance shops specializing in CD-ROM production, business-to-business and instructional Web creation, production houses with strengths in video and special effects, and operations with expertise in **video news releases,** which are preproduced reports typically funded by a public relations company or an organization's in-house public relations operation

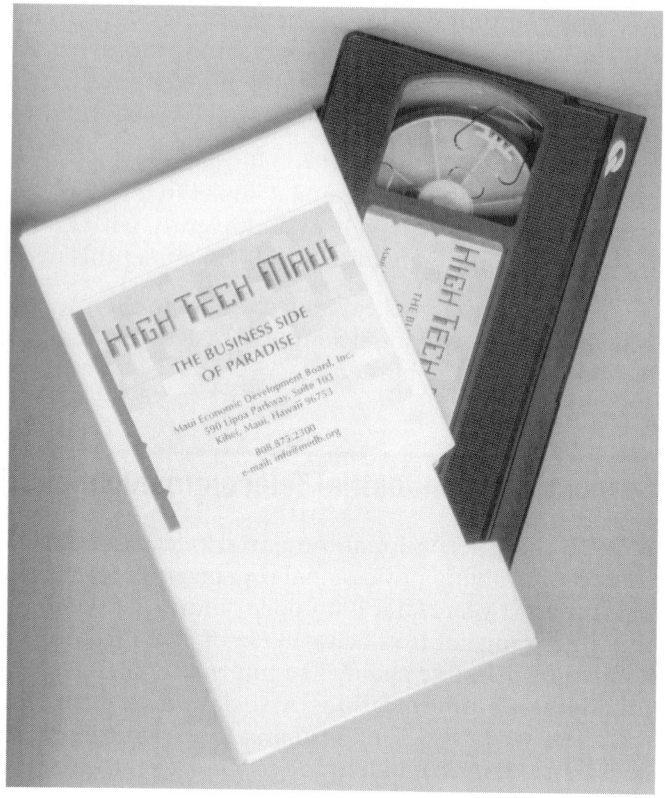

The vast majority of video production conducted in the United States is not for broadcast. It is for corporate and industrial use, for example, for the production of promotional videos such as this one from the Maui Economic Development Board, which can be sent to companies seeking to locate in Hawaii or to broadcast stations for use as a video news release.

and distributed free of charge to television stations and cable channels on videocassette. This latter form of production is not without its critics, as can be seen in the box titled Video News Releases.

Satellites and Videoconferencing

The proliferation of government and private satellites has led to sharp increases in the use of **videoconferencing,** the meeting of different people in different locations via satellite-distributed audio and video that permits full and instantaneous interactivity. Sales of videoconferencing equipment have grown at an annual rate of 40% to 50% throughout the 1990s, reaching $1.4 billion in 1997 (Lawlor, 1998).

A number of factors other than the availability of satellites have fueled this growth. One is cost. Digital technology has converged with audio and video production and transmission technologies to bring the price of videoconferencing within reach of even small companies. **Group systems,** designed for videoconferencing meetings held in large rooms (corporate boardrooms,

Critical Issues *Video News Releases*

Hundreds of production houses and thousands of people work to produce video news releases. These short video reports, provided to virtually every commercial television station in the United States, typically have these features:

- They look exactly like genuine news reports, employing the visual and aural conventions viewers typically associate with television news.
- They are narrated by a speaker whose voice, intonation, and delivery match those of bona fide television news reporters.
- They carry the voice-over on a separate audio channel so the station can delete the original narration and have its own anchor or reporter narrate to give the appearance that the report originated locally.
- They are accompanied by a script in the event the local station wants its own reporter or anchor to do the narration but needs help in writing it.
- They come free of titles or other graphics, allowing local stations to use their own logos and video character typefaces.

Video news releases can be used in their entirety or in part, and most providers consider even a 5-second excerpt (a long time in television news) to be a success.

As television news operations scale back their staffs because of the demands for greater profit brought about by conglomeration and other market forces (Chapter 3), defenders of the video news release argue that this is one of the few ways local news can acquire visuals and details about subjects of interest to the public. Critics, however, point to the ethical problem inherent in the production of a piece of public relations designed expressly to look like something it isn't—original broadcast journalism. These provided-by-outsiders reports can never be true or honest, say the critics, because they simply are not "news."

Put yourself in the position of the owner of a growing corporate video production house. You accept a contract from the makers of a new sports energy drink. You work with them on the script for the video, and you are certain the product works as well as its makers say and that the script and subsequent video report present the truth. Or perhaps you are certain that the video represents the truth as the product's manufacturer sees it. Do you have any misgivings in producing the video, knowing that all or part of it will be aired on hundreds of television news shows?

Now, put yourself in the position of a television station news director. You decide to use 10 seconds of the video news release in a station report on your town's upcoming 10K Charity Fun Run. At the end of your news program, as the credits roll, you list the manufacturer and production house as the source of the report. Do you have any misgivings about using material provided by an obviously interested party as part of what most, if not all, of your viewers assume is an objective news report produced by your station?

Finally, imagine yourself as a television news viewer. Are you aware that television industry observers estimate that from 50% to 70% of all video aired on local news is provided by outside sources? Does this bother you? How much do you value objectivity in television news? As long as the material is identified as being provided by an outside agency, isn't that sufficient?

Defend your answers from each of these three role positions. Then, after reading the discussion of ethics in Chapter 13, revisit the issue of video news releases to see if you would still respond in the same way. Explain any changes.

Picturetel is among several companies producing PC-based desktop videoconferencing systems. These low cost, easy-to-use systems put videoconferencing within reach of even the smallest organizations.

for example), dropped from $60,000 to $6,000 during the 1990s. Several companies, 3M and Picturetel most notably, even manufacture smaller, less expensive PC-based desktop videoconferencing systems. Digitalization has helped in another way as well. Now that audio and video signals can easily be digitally compressed, they are less expensive to transmit because they use less satellite time and channel space.

Nontechnology-based factors are also spurring the expansion of videoconferencing. Organizations, both commercial and public, are globalizing, and travel is an increasingly expensive and inconvenient enterprise. In addition, especially in the digital age, companies want to rush their products to market at an ever faster pace. Teleconferencing is quicker than shipping real people to real locations for meetings and shipping them back again.

Satellites and Video Networks

Most big organizations have utilized video networks since the beginning of television. The U.S. military, for example, made early use of over-the-air instructional television to link its far-flung bases. Other organizations, as we've

Few consumer industries make as wide and as varied use of corporate and industrial telecommunications as the health and fitness industry. In this brochure we can see audio, video, and the Internet integrated into a workout routine.

just seen, have relied on the distribution of videotapes to their different locations. Microwave relay systems, in which super-high frequency signals composed of very short waves are sent from land-based point to land-based point, made this bicycling obsolete. Companies were now able to link as many sites as they wished to one or more central production and distribution centers. Microwave technology also makes interactivity possible; that is, receiving sites can send as well as receive microwave signals. But it is satellites that are making the biggest impact on corporate and industrial video networks because, although still frequently used, microwave has an important limitation. Its signals are **line of sight;** that is, they can only be received by sites that are in a direct line with one another. Put differently, they cannot negotiate the curve of the earth. So, even when placed on the highest buildings or towers, microwave signals can travel only about 30 miles. To solve this problem, several relay sites must be used. Each site in this network picks up the signal, amplifies it, and sends it along to the next, which does the same. Satellite renders this transmission and retransmission process unnecessary, and in some situations satellite distribution works where microwave simply cannot, for example, when an organization wants to communicate with a site that is across an ocean.

A microwave relay station.

Video networks are used primarily for training and communicating with clients and customers. Their use for training is obvious—a large employer can train countless people in numerous locations through its networks. The Seventh-Day Adventists, IBM, Intel, and Fireman's Fund are only a few organizations that train, retrain, and orient their people via video networks. You can find another use for video networks at department stores that are part of a national chain; Macy's TV and Bloomingdale's TV are examples. Monitors are set up near specified merchandise, for example, near junior miss girls clothing. And on those screens are satellite-delivered, MTV-like videos showing off the latest fashions, all of which, of course, are available just to your left and right. This is called **point-of-purchase video promotion.**

Satellite-Delivered Media Tours

Corporate and industrial telecommunications are frequently used for public relations purposes to communicate with the various publics with which organizations and businesses interact. Considering the number and variety of these publics—organizations' own employees, stockholders, the communities in which they operate, government agencies, various media outlets, the

From the largest department store to the hippest specialty chain to the smallest mom and pop music store, retailers make good use of point-of-purchase video promotions.

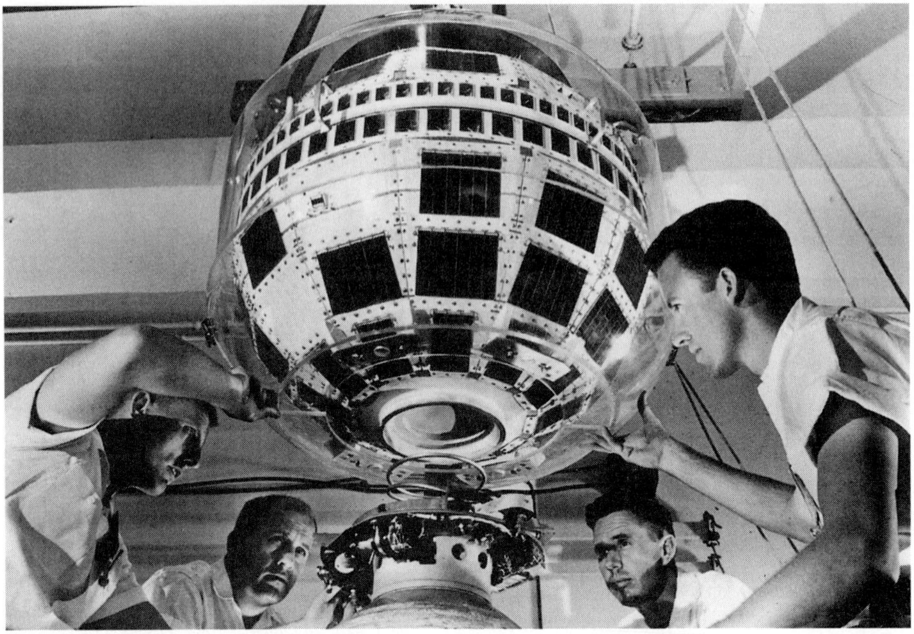

Among the first, and certainly the most famous of all communications satellites is Telstar, launched in 1962. Chapter 11 tells more of this early bird's story.

investment community, customers, and clients—it is no surprise that satellites are often used. One example is the **satellite-delivered media tour,** where spokespeople can be simultaneously interviewed by a worldwide audience of media or business professionals connected to the on-screen interviewee via telephone.

Telecommunications Satellites

As we've seen, the proliferation of telecommunication satellites has proven to be a boon to corporate and industrial telecommunications, just as it has been to virtually all the other telecommunications industries.

There are more satellites in our skies than ever before, all employed for a variety of telecommunications functions. At the start of 1999, for example, there were 773 working communications satellites in orbit around the globe, with an additional 1,400 scheduled for deployment by the year 2008 (Lo, 1998). Each of these satellites has many **transponders,** which receive and retransmit specific telecommunications services. These services are leased to telecommunications service providers—commercial, noncommercial, and corporate and industrial.

How these satellites are used is determined by the height and nature of their orbits. There are three types of satellites:

- **geosynchronous Earth orbit (GEO) satellites:** used to link cable television networks to operators' headend facilities (Chapter 6) for DBS and for other television signal distribution. They also serve important meteorology, maritime, and intelligence operation functions. These satellites orbit the Earth at a height of 22,300 miles directly above the equator. At that altitude they spin with the Earth, remaining in a fixed location above the planet.
- **middle Earth orbit (MEO) satellites:** used primarily for global positioning systems like those on boats and cars and for voice and data transmission. These satellites travel around the Earth at altitudes ranging from 1,000 to 22,300 miles.
- **low Earth orbit (LEO) satellites:** used for spying, remote sensing, and data communication. Flying at a distance of from 100 to 1,000 miles above the ground, these satellites, because of their proximity to the Earth, are used for e-mail and paging, mobile telephones, videoconferencing and video networking, and high-speed Internet access.

CD-ROM, Videodiscs, and Multimedia

One of the earliest uses of corporate and industrial media was instruction, training, and orientation. This is still the case, but now much of this work is conducted using digitalized, individual-user technologies. Instead of scheduling large groups of people for training sessions and setting aside space for group training, individual employees can now learn at their desks or stations—or even at home. In addition, these technologies allow for self-paced learning and immediate, corrective feedback.

CD-ROM, read-only discs—exactly like the CDs you play at home or in your car—permit storage of a large amount of easily accessible, indexed information. **Videodiscs** are also similar to common CDs, but they have **multimedia** capabilities and can present instruction in sophisticated audio and video. Videodiscs are interactive and can be programmed to demonstrate the outcome of an employee's choices. For example, a trainee pilot may choose to land her plane with the wind, as opposed to against it, and the videodisc can show the possible outcome of this and other choices. With the recently introduced recordable discs, employees' responses can even be captured for later assessment and evaluation by supervisors.

A CD-ROM.

The Internet and the World Wide Web

The Internet, with its e-mail and computer programs used to run e-mail mailing lists, bulletin boards, and discussion groups (see Chapter 7), has become integral to the internal telecommunications needs of countless businesses and organizations. The World Wide Web, carried on the Internet, has become just as important, but with its audio and video capabilities and its hyperinteractivity, it is used more for communicating with outside publics than with in-house respondents. Business-to-business sales, public relations, promotions,

Careers *Working in Corporate and Industrial Telecommunications*

A wider array of telecommunications jobs are available with businesses and other organizations than with the more "glamorous" broadcast and cable industries. These positions are classified primarily in terms of the functions they serve for the organization, that is, whether they provide internal or external communication. Within these two broad categories are numerous job opportunities.

Internal Communication

Communication With Employees

- Communication about organization benefits
- Communication from headquarters to other sites
- Communication from sites to headquarters
- Management conferences
- Marketing developments
- Orientation of new employees
- Periodic news
- Policy announcements
- Stockholder meetings

Manufacturing

- Documentation
- Improvement of methods
- Quality control
- Testing

Training

- Customer training
- Dealer training
- Improvement of operations
- Introduce new procedures
- Management development
- Role playing
- Safety training
- Sales training
- Technical training
- Upgrading proficiency

Research and Development

- Communication with marketing
- Recording and documenting live tests

and education are primary functions of corporate and industrial Web sites. Perhaps the most important contribution of the Internet to corporate and industrial telecommunications resides in its ability to facilitate videoconferencing and distance learning at costs much lower than can be had using satellites. But high-quality, full-screen, interactive Internet video will come only with the fuller diffusion of significantly more sophisticated modems and access to high-capacity networks than most people and businesses now have (see Chapter 7).

The box titled Working in Corporate and Industrial Telecommunications enumerates some of the many career opportunities available in this field.

Converging Functions

At the start of this chapter we saw that corporate and industrial telecommunications is characterized by its three most common functions: (1) it serves specific instructional, economic, public relations, or individual needs; (2) it is narrowly targeted at a specific and known audience; and (3) it is distributed

External Communication

Communication With External Publics

Audio or video facility tours
Community relations
Consumer relations
Convention exhibits
General public relations
Government relations
Orientation programs for visitors
Press conferences
Press relations
Recruitment of new employees
Video news conferences
Video news releases

Sales and Promotion

Explanation of services
Introduction of new products
Product demonstrations
Reports of sales trips
Reports on competition
Sales meetings

Marketing

Ad promotion and testing
Communication with research and development
Point-of-purchase promotions
Reports on new applications
Reports on new and emerging markets
Trade shows

People hoping to make their careers in corporate and industrial telecommunications typically must have good skills in several areas, especially organization, audio and video production, using (if not creating for) the Web, and writing. Unlike broadcast and cable operations, corporate and industrial audio and video departments tend to be small enterprises requiring that a relatively small staff meet a variety of organizational and institutional communication needs. But this is no reason to avoid a career in this realm of telecommunications. In fact, it is a fine way to develop and hone a variety of skills that can be profitably employed in a wide number of other telecommunication industries.

Source: Adapted from E. Marlow, 1989.

over channels not usually available to general audiences. But convergence is erasing the distinction between this and other forms of telecommunications. Because of the ease with which much content can now be created, distributed, and exhibited, traditional mass communicators are increasingly involving themselves in corporate and industrial applications, and traditional corporate and industrial operations are increasingly distributing their work to larger, more general audiences.

Here are just a few examples of this blurring of distinctions between corporate and industrial and other better known forms of telecommunications:

- CNN, CBS, and other cable and broadcast giants regularly produce 30-minute programs for very narrowly targeted, specific viewing, such as during plane flights and ship cruises. United Airlines, for example, offers passengers *NBC News Magazine in the Air,* produced by the network's news division and targeted to the travel and business demographics.
- Marketing agencies increasingly use the Internet to help businesses interact. Its instantaneousness and interactivity make the Net well

The Real World *A Typical Corporate Telecom Operation*

Viking Freight is a division of Federal Express. It specializes in next day delivery of less-than-truckload freight throughout the western United States, from West Texas to Hawaii and Alaska, and also provides service to the rest of the country, Canada, and Mexico through arrangements with other carriers.

This 35-year-old company has 6,500 employees and is a general commodities motor carrier. General commodities include retail goods, electronics, automotive parts, clothing, and industrial equipment. Viking's 65 strategically located service centers allow Viking's customers to enjoy an industry-leading 99% on-time service level.

Its Skills Development/Employee Communication Department has three employees: a manager, a production specialist, and a training coordinator. So, what do these three people produce in the way of corporate and industrial telecommunications for this successful, mid-sized company?

The production specialist is responsible for graphics, scriptwriting, sound effects, preproduction, production, postproduction, and final delivery of all audio and video material. Typical audio productions include "Blue Line Fever," aimed at the company's line drivers, "Sounds of Safety," for all Viking operations employees, "Sales Talk," hosted by the vice president of sales and used to support the sales force, and "Leader's Perspective," hosted by the division manager, which includes as guests the company president and vice presidents. The target audience is all Viking supervisors and managers, and the goal is to provide them with corporate and industry news and information, policy changes, and skill development tips and to keep them up to date on company policy and activity. The production specialist records all audio sessions and edits the master versions of the tapes.

All production work is done in-house, and the finished product is then sent out for duplication. The Skills Development/Employee Communication Department staff then takes responsibility for distributing the material, either on CD or audiocassettes. Each audio series can have as few as 250 and as many as 2,000 copies, depending, naturally, on the number of employees targeted.

Most of the department's video productions provide training. Some programs are aimed at a very select audience; others are aimed at larger groups. Titles include "Telephone Skills," "Better Customer Service," "Sexual Harassment Awareness," "Employee Benefits," "Professional Driving," "Proper Freight Handling," and "Safety."

Another interesting use of in-house instructional videos is Employee Meeting Videos, quarterly highlight shows that are very popular with Viking's

suited for business-to-business communication, for example, in connecting distributors and suppliers.

- Virtually every major broadcast and cable network, as well as the major film studios, make video and film clips available for distribution over the Internet (Chapter 7). This material can and is accessed and downloaded for inclusion in all sorts of corporate and industrial productions, including instructional CD-ROM and training and orientation videos.
- Video news releases and satellite-delivered media tours, discussed earlier in this chapter, are typically produced and distributed by corporate and industrial production operations (serving some organizations' specific instructional, economic, public relations, or individual

workforce. They provide employees with company news and financial performance as well as new company and industry developments. Every 3 months the production specialist travels to one of the service centers to shoot tape of the action at the center, for example, people working on the dock and drivers making pretrip inspections.

All shooting is on location, but the narration is provided by the company's president, who reads off a teleprompter. This work is done at Viking's in-house production studio in San Jose, California. Video dubs are made in-house, and as many as 100 copies may be distributed to all Viking locations.

Viking's customers, too, can request the production of special interest videos. For example, if a customer has freight that requires special handling, she can contact the Skills Development/Employee Communication Department with the information to be covered in the special video. Once preproduction planning is complete, the production specialist travels to the customer's site and shoots the necessary footage. Postproduction (editing, effects, titles) takes place at Viking's studios, and copies are then distributed to all involved service centers.

The Skills Development/Employee Communication Department also assists the company's Marketing Department. It shoots still photographs to accompany trade magazine articles, produces cover designs and logos for company CD-ROMs, and assists in setting up projection computers and slide shows for a variety of marketing needs.

This division of Viking also publishes *Independent Times,* a newsletter distributed to all Viking employees. Its six or seven issues a year, like most company newsletters, offer brief articles on the most recent winners of the President's Safety Awards, lists of employees receiving Service Awards for years in service with Viking, credit union information, employees' personal stories, and birth announcements.

And the department does not ignore the Web. Using the company's intranet, EZInfo, the production specialist produces Web sites for employee use. A virtual tour of the company's administration is a typical self-built site, as are the Education Reimbursement Program Enrollment Form and a virtual Employee Handbook.

What kind of person do you think succeeds in this environment? Could you? What technical skills would you have to develop or improve? What personal skills? What work habits?

Can you gain access to a similar operation near you? If so, visit it (or otherwise communicate with the professionals involved) to see how similar or different it is from Viking's operation. Why do you think the similarities and differences you do find occur?

needs). But these productions are exhibited over channels quite readily available to general audiences as well as the narrowly targeted, specific, known audiences.

Internet and satellite transmission; the availability of low-cost production, distribution, and exhibition technologies such as desktop computers, videocameras, videotape, videodisc, and CD-ROM; and organizations' and people's willingness and facility using converged technologies for a variety of personal and professional needs increasingly renders the distinctions between corporate and industrial and other types of telecommunications industries almost meaningless.

The box titled A Typical Corporate Telecom Operation provides a peek at a successful west coast operation and the kinds of content it produces.

Chapter Review

An important use of telecommunications technologies resides in their corporate and industrial applications. This nonbroadcast audio and video typically serves the specific and individual needs of the producing organization, is narrowly targeted, and is distributed over channels not usually available to the general public. Advances in technology, new ways of doing business, and even social change combined in the 1970s and 1980s to turn corporate and industrial telecommunications into a large and prosperous industry. The development and convergence of even more sophisticated technologies continue today, and corporate and industrial telecommunications often finds expression in satellite use for videoconferencing, video networks, and media tours. Satellites make possible a number of corporate and industrial telecommunications activities, and they circle the Earth in increasing numbers at different orbit heights.

New digital technologies such as CD-ROM, videodiscs, and multimedia, especially when combined with interactive capabilities, are also reshaping the industry. The Internet, already in use for e-mail, LISTSERV, and other corporate functions, will occupy an even more pronounced part of corporate and industrial telecommunications with the wider diffusion of faster modems and high-capacity networks.

Questions for Review

1. What three aspects characterize corporate and industrial telecommunications?
2. Why is the early history of corporate and industrial telecommunications largely the story of corporate and industrial "video"?
3. What was the impact of first 1-inch, then ¾-inch, and then ½-inch videotape on corporate and industrial telecommunications?
4. What five factors combined in the 1970s and 1980s to expand the corporate and industrial telecommunications industry?
5. What is the primary unifying professional organization of corporate and industrial telecommunications?
6. What are video news releases?
7. What are videoconferencing and group systems?
8. How have satellites affected corporate and industrial telecommunications?
9. What is a microwave relay system? What is its primary drawback?
10. What is point-of-purchase video promotion?
11. What is a satellite-delivered media tour?
12. What are the different categories of satellites, and what is the primary use of each?
13. What are CD-ROM and videodiscs? What are their advantages as training technologies?
14. How is the Internet currently serving corporate and industrial telecommunications? What limits its greater use?

Questions for Critical Thinking

1. Have you ever considered a career in corporate and industrial telecommunications? Why or why not? What advantages might a career in corporate and industrial telecommunications have over one in commercial broadcasting? What disadvantages might there be?

2. Have you ever been trained or oriented using technology media rather than by a human being standing before you? Describe that experience. How were your questions answered? Would you have preferred a human instructor? Why or why not?

3. Have you ever been part of a video conference? Describe that experience. Did you feel that you could accurately "read" your electronic conference partners? What technical (and nontechnical) aids were provided to help boost understanding?

4. Do you use e-mail bulletin boards? If you do, or have, describe its strengths and weaknesses as a means of communicating with people sharing a common interest.

5. Do video news releases trouble you at all? Why or why not? How adept are you at identifying them when they are part of the local news you watch on television?

Engaging Technology

1. The primary professional organization for corporate and industrial media producers is the MCA. Visit its Web site at <http://www.itva.org> to see what the organization is about. If you know someone who is a member, ask her or him to log on with you. Examine its job listings to see the range and number of career opportunities available in this lucrative corner of the telecommunications business.

2. *AV Video and Multimedia Producer* is a well-respected magazine among corporate and industrial media practitioners. It is especially strong in its presentation and evaluation of new and emerging technologies and their application to corporate and industrial media production. Check its Web site at <http://www.kipinet.com/av_mmp/> for the latest update. Share with the class your evaluation of where corporate telecommunications are headed.

3. Defining yourself either by the profession you wish to pursue when you graduate or by the one you may be engaged in while in school, find and join a relevant e-mail group. Describe the information available and the nature of the communication exchange you find there.

Chapter Six

Cable

John Walson was a lineman for the Pennsylvania Power & Light Company. He was also the owner of an appliance store. That this resident of tiny Mahanoy City, Pennsylvania, had these two careers is a coincidence that would create one industry and change many others.

Walson could not convince his neighbors to buy the televisions he was stocking in his store. The problem wasn't that the townsfolk were afraid of the new technology—in fact, they were hungry for it. Walson's sales were hampered instead by geography and topography: the Pocono Mountains sat smack between his town of 10,000 people and Philadelphia's three new television stations, some 85 miles away. In 1948 the enterprising Walson conceived of a plan. He erected a 70-foot-tall tower on New Boston Mountain, connected several television sets to it, and ferried potential television buyers to the location to impress them with the new medium. Once impressed, Walson deduced, these customers would rush down the mountain to buy televisions from his store. The plan had two deficiencies, one obvious, one not so obvious.

The primary flaw in Walson's strategy was location. If people were impressed enough with what they saw on the mountain to want to own a set, they would also have to be impressed enough to build a house on New Boston Mountain, because that was the only place near Mahanoy City where television could be watched. The second flaw resided in Walson's sense of decorum. He was embarrassed and uncomfortable "according to a newspaper account, taking prospective lady customers up on the mountain after dark" (Davis, 1998, p. 10). The New Boston Mountain experiment was a failure.

Preview

John Walson's initial failure led to a second idea, what we now call cable television. In this chapter we study how cable spread from the Poconos to become not only a major industry but a telecommunications technology at the center of much of the convergence we have been and will be reading about in this book. We see how cable developed technologically and economically and how confusing and contradictory government regulation stunted and shaped its growth. We learn about the operation of contemporary cable systems, the different types of cable programming, the various pay cable options, and public access cable television.

Advances in cable television technology—fiber optics, digital cable, interactive cable, multiplexing, compression, and video-on-demand—are examined both in terms of their functioning and how they may shape the cable we receive in the near future. Concentration of ownership, a factor in all telecommunications, is an issue in cable as well. We study its various forms—MSOs, vertical integration, and conglomeration—and the arguments for and against them. The interaction between the telcos and the cable industry promises even more convergence as numerous telecommunications services are bundled into cable, and we will also see how satellites are a part of this trend.

Technological and Economic Roots

To understand cable's roots, we have to return to Mahanoy City and John Walson. Recognizing the shortcomings of his initial endeavor, the entrepreneurial Walson then ran a wire from his tower to his store. He used Pennsylvania Power & Light Company utility poles (obtaining permission from his employers only after he connected two of his superiors to his system), and as the existence of that system became known, he began wiring the homes of those who bought his television sets—for a $100 installation fee and a monthly $2 service charge. By June he had 727 subscriber homes for his **community antenna television (CATV)** system (Chin, 1978). Although CATV is now an outmoded name for Walson's brainchild, cable television was born. It was two other Pennsylvanians who would help bring the baby to childhood.

Walson's cable was a twin-lead wire, somewhat akin to the cord that connects a lamp to an electric outlet. To attract even more viewers to his system, he had to offer improved picture quality. He accomplished this by using coaxial cable and self-manufactured boosters (or amplifiers). Coaxial cable—a copper-clad aluminum wire encased in plastic foam insulation, covered by an aluminum outer conductor, and then sheathed in plastic—had a wider **bandwidth** than did twin-lead wire. As a result, it allowed more of the signal to pass, and even permitted Walson to carry a greater number of channels.

As Walson was solving his problem, Milton Jerrold Shapp, who would eventually become the governor of his state, noticed thousands of antennas cluttering the roofs of department stores and apartment buildings. He borrowed Walson's idea, setting up master antennae and connecting all the sets in these buildings to them. This became known as **master antenna television (MATV).** Shapp's advance was the manufacture of boosters for commercial sale. Then, in Lansford, another set salesman, Robert Tarlton, decided to combine the best of Walson's system (coaxial cable) with the powerful, commercially obtained Jerrold boosters. Two years after John Walson began taking people up New Boston Mountain there were 70 cable systems serving 14,000 subscribers nationwide.

With expanded bandwidth and powerful boosters, all three of these cable pioneers began experimenting with the **importation of distant signals,** using cable not only to provide improved reception but to offer a wider variety of programming as well. They used microwave relays (Chapter 5) to deliver independent stations from as far away as New York to fill their then-amazing 7 to 10 channels. By 1962, 800 systems were providing cable television to more than 850,000 homes.

During this period, many over-the-air broadcasters saw cable as something of a friend. It extended their reach, boosting their audience numbers and increasing their profits. Then, in November 1972 a company called Sterling Manhattan Cable launched a new channel, Home Box Office, or HBO. Only a few hundred homes caught HBO's debut, but the broadcasters' mild concern at this development turned into outright antagonism toward cable in 1975 when new HBO owner Time, Inc. began distributing the movie channel by satellite, a process made possible by the FCC's 1972 authorization of domestic satellite relays. Now **pay cable** was eating into the broadcasters' audience by offering high-quality, nationally produced and distributed content. With the public's enthusiastic acceptance of pay cable, the medium reached maturity.

Cable Reception and Distribution

The way cable systems receive and distribute programming has changed little since the industry's earliest days. The process of getting a picture to a home screen begins at a receiving antenna, which includes microwave and satellite receiving equipment. The gathered signals are collected from the receivers at the **headend.** From there, they are sent over a **super trunk cable** to a **hub,** typically the cable system operation itself, where they are processed and boosted for distribution. The cable that leads from the hub into the community is the **trunk cable,** which divides into **feeder cables** to access individual neighborhoods or areas. The line that runs from the feeder to our homes is a **drop cable.** The layout of a typical cable operation is presented in the box titled The Cable Signal—From Source to Home.

Tech Talk *The Cable Signal—From Source to Home*

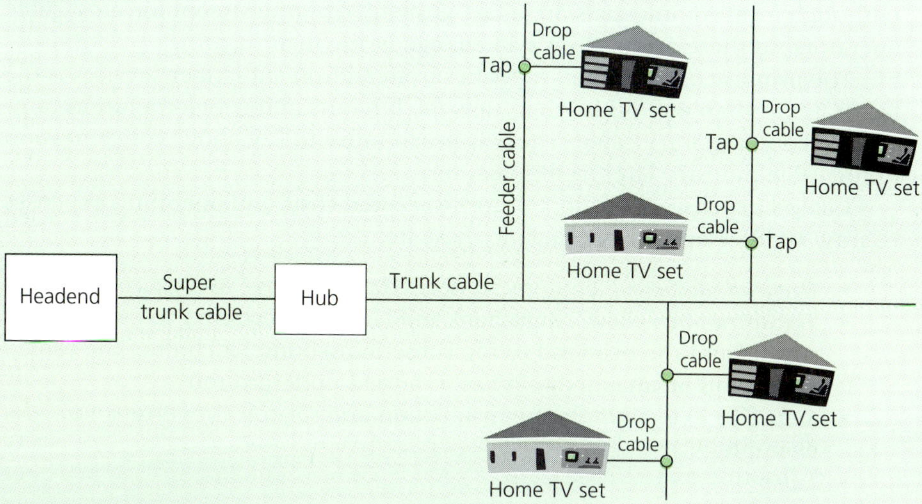

A typical cable operation collects and distributes content in this manner from headend to homes.

Cable's Economic Structure

The economics of the cable industry were established once cable became a medium to which people turned not for improved reception but primarily for expanded viewing options. At the outset, potential cable operators must make a substantial investment in the construction of their systems, with little hope of a quick return on those expenditures (Garay, 1988). New **franchise** operators (those with authority to offer cable service) spend heavily for such things as facilities construction, receiving and distribution equipment, and wiring neighborhoods and homes—all before a single subscriber pays a dime. Operators can gauge their likelihood of success by weighing a number of factors. The first is the pass-by rate—that is, the number of homes passed by or with the potential to take cable. Closely related is **density,** or the number of households per mile of cable. A mile of cable costs the same whether it passes 10 homes or 2,500 homes. Therefore, greater density offers the potential for greater penetration—the number of homes passed by the cable that actually subscribe. Increased penetration not only means more income from **basic cable** (the "free" channels provided automatically by virtue of subscribing) but also offers the promise of added income from pay cable, **pay-per-view,** and add-on services such as home security, various forms of data delivery, Internet access, and local and long-distance telephone service. In a reversal of the typical economic structure in broadcasting (television stations earn 90% of their income through the sale of advertising), the typical cable operation earns only 10% of its revenue that way, with 90% coming from subscribers (although local commercials are a fast-growing revenue stream for most cable operators, so these proportions will change in the near future).

The promise of additional revenues from add-on subscriber services is the factor, more than any other, that fuels the buyouts and conglomeration rampant in contemporary cable. There is much more to say about concentration in the cable industry later in this chapter.

Cable's expenses come in two forms: capital costs and operating expenses (Garay, 1988). Capital costs include these items:

- *Headend equipment:* equipment necessary to receive and process a signal
- *The distribution plant:* equipment and facilities necessary to get the signal from the headend to the subscribing home, including the cable, which can be attached to poles or buried underground
- *Subscriber equipment:* home installation and set-top converters, for example
- *Labor*
- *Maintenance*

Operating expenses include these items:

- Franchise fees paid to the municipality in which the system operates
- Copyright fees for imported material
- Utility pole attachment fees
- Distant signal transmission costs
- Production expenses for locally originated or access programming
- Fees paid to many (but not all) basic cable networks; for example, the Weather Channel and MTV are provided free whereas BET and ESPN charge a fee of a few cents per subscriber per month

Early Regulation of Cable

It is impossible to fully understand the development of cable television without understanding the ups and downs of its regulatory history. Cable industry's size, shape, and the content it offers have been variously limited and encouraged by frequently conflicting and shifting rules. As Joseph Fogarty and Marcia Spielholz (1985) wrote in the *Federal Communications Law Journal,*

> The history of FCC cable regulation is a complex interweaving of FCC opinions, court decisions, and technological advances, characterized by numerous shifts in opinion concerning both the source and scope of FCC cable jurisdiction and the value of new technology. At the heart of the development of FCC cable regulation lies the fact that until recently, the Communications Act of 1934 made no explicit provision for the regulation of cable television. (p. 113)

This means that the FCC had to make the rules as it went along.

In the beginning the FCC ignored cable, seeing it as simply an aid to over-the-air broadcasting; that is, it was a benign technology that did little more than extend the reach of licensed broadcasters. But when cable operators began to import distant signals from outside an operator's service area, the commission—intent on fostering television's diversity and localness—decided it was time to bring cable under its regulatory control. The FCC did this in 1963 with its ruling in a dispute between a microwave relay company, Carter Mountain Transmission Corporation, and a Wyoming television station, KWRB. The station objected to Carter Mountain picking up and delivering its signals to cable systems around that state. The FCC, concerned that cable would damage broadcasters' profitability, argued that "when the impact of economic injury is such as to adversely affect the public interest . . . it is our duty to determine the ultimate effect . . . and act in a manner most advantageous to the public" (Roman, 1983, p. 12).

Answering broadcasters' pleas for even more protection from the growing cable industry—there were now 1,325 systems operating nationally—the FCC expanded its regulatory control over cable. New rules promulgated in 1965 expanded FCC jurisdiction to all cable systems (not just those using FCC-licensed microwave relays, the legal basis for its intervention in the Carter Mountain dispute). This new level of regulatory control was based largely on two realities of cable at the time. First, the FCC had a stake in ensuring the operation of the nation's broadcast system, and a technology as significant as cable surely would have an impact on broadcasting. Second, cable systems enjoyed virtual monopoly positions in the areas where they operated and, like other monopolies (public utilities, for example), were subject to official oversight.

Among the restrictions that most angered the cablecasters were the **local carriage rules** (also called must-carry rules), which require cable systems to carry the signal of every television station within a 60-mile radius. Even more contentious were the **syndicated exclusivity rules,** which prohibited cable operators from importing programming that might duplicate those aired by a local station for a 30-day period. If a local broadcaster aired *The Adventures of Rin Tin Tin* on May 1, for example, the cable company could not deliver that program to people's homes for 15 days before that broadcast nor for 15 days after. The commission even began a concerted effort to limit cable's expansion into the top 100 markets by passing restrictive rules on the use of microwave relays to bring distant signals to cable operators wishing to serve those communities. In 1966, however, the FCC softened some of its restrictions, limiting syndicated exclusivity to one day and must-carry only to those television stations licensed in the cable operator's market. Still, the effect, if not the FCC's intention, of this rule-making was to slow the growth of cable.

But the FCC was not done with the upstart medium. In 1969 it began requiring that operators make available not only channels but equipment and studios for the production of locally originated programming. Many operators programmed little more than time and weather, but others began what we now know as **public access channels,** cable channels reserved on a first-come, first-served, nondiscriminatory basis for use by groups or individuals

The access channel's postproduction set up.

Community Television of Santa Cruz County's logo.

Much public access cable content is unsophisticated. But much, such as that produced by Community Television of Santa Cruz County, provides an important service to viewers. Here we see preparations for a cablecast of a community roundtable.

who maintain editorial control of their programming. This imposed a financial burden on many operators already struggling to make a profit under the commission's restrictive rules. Even among those operators who attempted to meet the spirit of the FCC local origination rules there was resentment that the federal government was telling them how to use their facilities.

Finally, in 1972, recognizing the inevitability of cable television, the FCC produced what it called its "definitive" rules for cable, hoping to shape rather than stop the medium's development (Hollins, 1984). The intent of the 1972 Cable Television Report and Order was to limit cable television to a secondary role, providing a supplement to over-the-air television. To that end, the following rules were implemented:

- All network and syndicated programming from imported distant television stations must be deleted.
- Systems serving the 100 largest markets must offer at least 20 channels.
- Operators of these systems must develop interactivity.
- Four channels must be set aside for public, educational, local government, and leased access.
- Systems with more than 3,500 subscribers must begin locally produced programming.
- Systems must submit to municipal franchising authority control over their rate structure.
- Telephone companies, local broadcasters, and the major television networks are prohibited from owning cable franchises.
- Cable pay channels could show only one feature film per week, and these films must be more than 2 and less than 10 years old. (This effectively banned premium programming.)

In exchange for accepting these restrictions, the FCC eliminated its rules designed to discourage cable's entry into the 100 largest markets, a significant development in cable's up-and-down regulatory battle with the commission and the catalyst for the industry's full flowering.

The cable industry fought mightily against the restrictions that limited its growth and profitability and, slowly the FCC began to rescind or soften many of the more onerous ones. When local governments began to step up *their* demands on operators, the cable industry (ironically) turned to the FCC for relief. The commission obliged with the Cable Communications Policy Act of 1984. The stated goal of this legislation was to provide regulatory stability for the once again growing cable industry and to fix some of the problems created by the 1972 rules. The Cable Act of 1984 made these modifications:

- Allowed local franchising authorities to require cable operators to provide local access channels as a condition of enfranchisement
- Prohibited cross-ownership of broadcast and cable operations in a single market
- Retained municipal franchising authority but limited its control over rates and programming

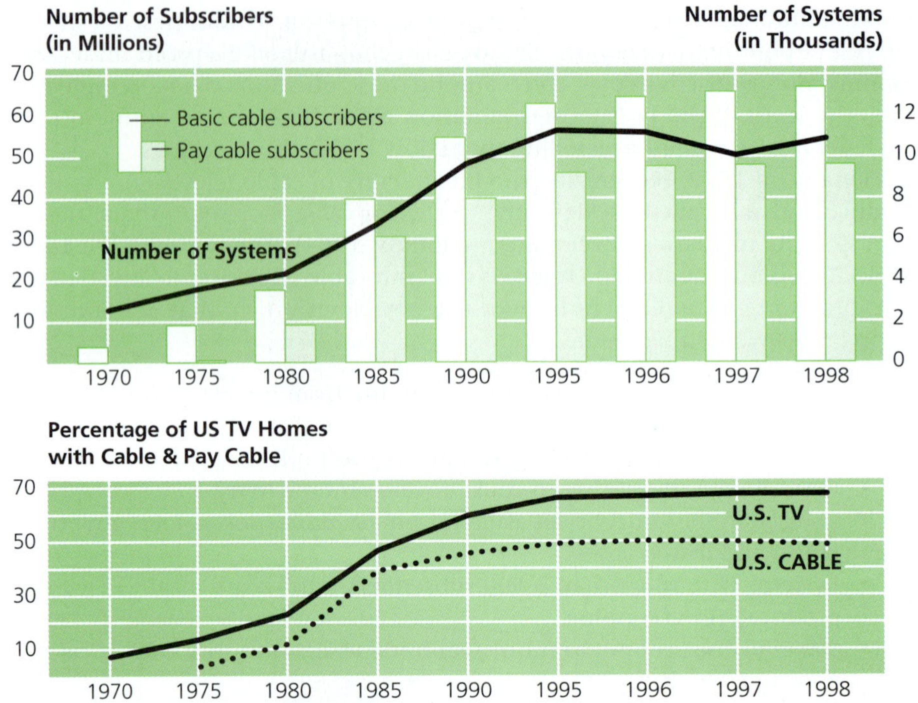

Figure 6.1 Cable Systems and Subscribers

Sources: National Cable Television Association, 1999; Paul Kagan Associates; *Marketing News Media,* March 16, 1998; *Cable TV Investor,* April 14, 1998; *Broadcasting & Cable Yearbook, 1999; Broadcasting & Cable Yearbook, 2000.*

- Provided increased protection for cable operators at franchise renewal time
- Protected subscribers' privacy by requiring cable operators to inform subscribers of any use they might make of their subscriber-related information
- Made it illegal to steal cable signals

Cable professionals celebrated these new rules, but the industry still faced many regulatory skirmishes (see Chapter 12). Nonetheless, with the regulatory relief provided by the Cable Act of 1984, cable joined broadcasting and telephony as a telecommunications giant in its own right.

Cable Today

Today that giant is composed of 11,800 individual cable systems serving 64,800,000 homes subscribing to basic cable (66% of all television households). More than 67% of these cable households receive pay cable (Figure 6.1). The

New England Cable News is one of the growing number of regional cable networks.

industry employs nearly 128,000 people and generates $41 billion in annual revenues (*Broadcasting & Cable Yearbook, 2000;* National Cable Television Association, 2000).

Programming

As we've seen, cable operators attract viewers through a combination of basic and pay channels, as well as with some local origination programming. There are more than 170 national cable networks and a growing number of regional networks: NorthWest Cable News serves Washington, Oregon, Idaho, Montana, northern California, and parts of Alaska; New England Cable News serves the states that give it its name; and several regional sports-oriented channels including Prime Sports Network Midwest serve the midwest. The financial support and targeted audiences for these program providers differ, as does their place on a cable system's **tiers,** which are groupings of channels made available to subscribers at varying prices.

Basic Cable Programming In recognition of the growing dependence of the public on cable delivery of broadcast service as cable penetration increased, Congress passed the Cable Television Consumer Protection and Competition Act of 1992. This act requires operators to offer a truly basic cable service composed of the broadcast stations in their area and their access channels. Cable operations also offer another form of basic service, **expanded basic cable,** composed primarily of local broadcast stations and services with broad appeal such as TBS, TNT, the USA Network, and the Family Channel. These networks program a wide array of programming not unlike that found on the traditional, over-the-air broadcast networks. There is news, sports, situation comedy, action adventure, movies, talk and game shows, soap operas, and commercials. An operator will often include more demographically targeted fare in the basic tier, such as kids-oriented Nickelodeon and the upscale A&E, as an inducement to get people to sign up. The Sci-Fi Channel, Weather Channel, and American Movie Classics (AMC) are often used this way, as are Black Entertainment Television (BET) and Spanish-language Galavision. The goal is not only to garner higher monthly fees and attract new subscribers but to make the "distance" between the basic and pay options smaller, encouraging viewers to take that last step. For example, basic cable might cost a subscriber $17. Expanded basic, which might include the Discovery Channel and the Disney Channel ("Why not, they're good for the kids") and even a few more interesting options like Comedy Central and E!, might cost only $10 more. The subscriber is now paying $27. The operator can then offer a **premium tier** that includes all programming from the lower tiers, plus a pay channel like HBO, for $37. To the subscriber that's "only $10 more," a seeming bargain.

Basic cable networks earn income either by charging operators a per subscriber fee (e.g., Bravo), selling advertising (e.g., VH1), or a combination of both (e.g., Lifetime). Operators choose to carry specific basic channels for a variety of reasons:

- A channel like sports-oriented ESPN, for which operators must pay, is essential to attracting subscribers. In the world of cable, ESPN is a necessity at the lowest tier.
- A channel like general-interest TNT, for which operators must also pay a small per subscriber fee, offers system operators **local avails,** that is, time that the operator can sell to local advertisers, keeping any income.
- Mind Extension University is attractive to operators because it is provided free, as are channels like C-SPAN.
- Home shopping channels like HSN pay local operators rebates of up to 5% of the sales they make from the operator's service area. Shopping channel QVC simply pays operators to carry it.

Because of concentration, operators are increasingly choosing to carry a specific basic channel because their owners (who have a financial stake in that channel) insist that they do. **Multiple system operators (MSO)** are companies that own several cable franchises. Time Warner, Liberty, and Cablevision own Court TV. AT&T Broadband has an interest in various prime channels. BET has powerful AT&T and Time Warner as part owners. Naturally, these networks are more likely to be carried by systems controlled by the MSOs that own them and less likely to be carried by other systems (Crandall & Furchtgott-Roth, 1996). This pattern also holds true for MSO-owned pay channels such as HBO and Showtime.

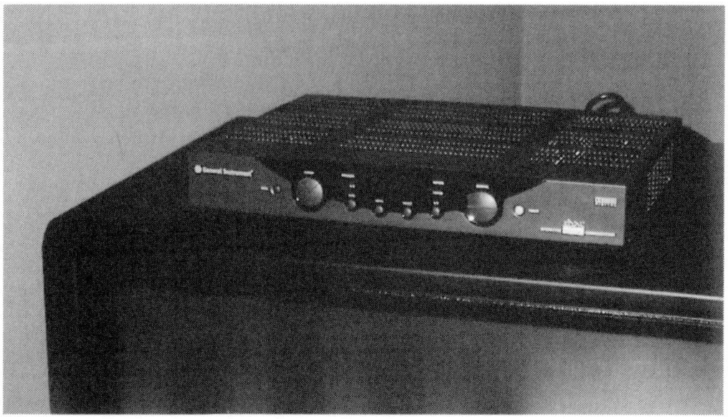

A typical set-top converter.

Pay Cable As the FCC lifted its restrictions on cable's freedom to import distant signals and to show current movies, HBO grew and was joined by a host of other satellite-delivered pay networks. By 1979 there were 10 premium or pay networks on the wire.

In addition to freedom from regulatory restraint, two important programming discoveries ensured the success of the new pay channels. After television's early experiments with over-the-air **subscription TV** failed, many experts believed people simply would not pay for television. So the first crucial discovery was that viewers would indeed pay for packages of contemporary, premium movies. These movie bundles could be sold more inexpensively than could films bought one at a time, and viewers were willing to be billed on a monthly basis for the whole package rather than pay for each viewing.

The second realization boosting the fortunes of the pay networks was the discovery that viewers not only did not mind repeats (as many did with over-the-air television) but welcomed them as a benefit of paying for the provider's package of films. Pay channel owners were delighted. Replaying content reduced their programming costs and solved their problem of how to fill all those hours of operation. Today, the most familiar and popular pay cable networks are HBO, Showtime, the Spice Channel, the Sundance Channel, and Cinemax.

An important factor in whether a cable operator carries a particular pay network is the availability of channel space—there must be room for the service on the operator's system. To encourage operators to find that room, pay networks offer them financial incentives in the form of revenue sharing. The money the local operator takes in for a premium network is split with that network: 30% to 40% goes to the program provider and 60% to 70% stays with the local operator. The pay channel often sweetens the deal by providing advertising and promotional support to the local operator in the form of preproduced commercials and announcements.

Pay services come in two forms: movie channels (HBO, Starz!, and Encore, for example) that offer packages of new and old movies along with big sports events and special events—all available for one monthly fee—and pay-per-view channels, where viewers choose from a menu of offerings (almost always of very new movies and very big sporting events) and pay a fee for the chosen viewing. In either case the subscriber must have a set-top converter to receive the paid-for channel, although most cable systems offer **addressable technology,** which enable pay services to be switched on and off at the hub. Many cable services are also experimenting with **interdiction technology,** which descrambles the signal outside the viewer's home, doing away with the set-top converter and enabling operators to sell pay services to homes that are not already basic cable subscribers.

Once in the home, people enjoy pay channels for their ability to present unedited and uninterrupted movies and other content not usually found on broadcast television, for example, adult fare and championship boxing and wrestling. Movies come to the pay channels in two ways. A film distributor may license a movie for a negotiated per-subscriber fee for a specified number of showings in a given period of time, or a movie channel may pay a flat fee to the distributor for a given movie. This is also how these channels acquire special events like sports and other original programming.

Technology and Cable Programming

An ongoing problem for cable is keeping subscribers once they have them. This is especially important as the industry faces mounting competition from DVD, VCR, and DBS. **Churn,** or turnover in subscribership, whereby new subscriptions in a given period of time are offset by cancellations, is damaging to a system's financial well-being, and it is pervasive. "Loyal cable subscribers may be an oxymoron," says cable industry consultant Dave Shepard (as quoted in Colman, 1998, p. 58). To keep customers, a variety of pricing strategies are promoted. For example, a cable system may offer certain highly attractive basic channels from a higher tier to lower tier subscribers on an à la carte basis. Others provide FM radio service for free with basic cable, piping local stations into subscribers' homes. Other strategies include free subscriptions to monthly cable guides and access to on-screen programming guides that offer constantly scrolling program and channel information and samples of the content that can be seen on those channels. Still others offer specialty pay channels such as DMX (talk- and commercial-free audio channels; see Chapter 2) and highly specialized text-based data and information channels.

Like all other telecommunications industries, the cable industry seeks out advances in technology to help it meet customer and audience demand. One existing technology, fiber optics, has already changed the industry, allowing systems to offer many more channels than they could with traditional coaxial cable. Fiber optic cable is made of thin strands (less than one hundredth of an inch in diameter) of very pure glass fiber over which signals are carried by light beams. Because fiber optic wire offers a very wide bandwidth that permits the passage of much more information, it can carry up to 600 times as much audio, video, or data information as an equivalently sized coaxial cable. A conventional 3-inch-thick cable contains 1,200 pairs of copper wire and can handle 14,400 phone conversations, but a ½-inch thick fiber optic cable

Tech Talk *Fiber Optics*

The operation of a fiber optic cable.

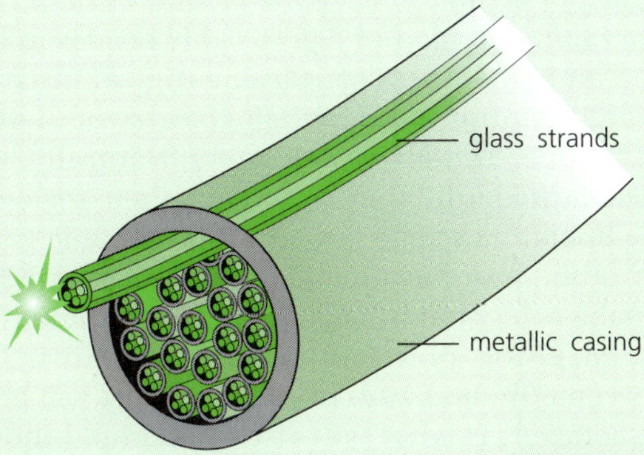

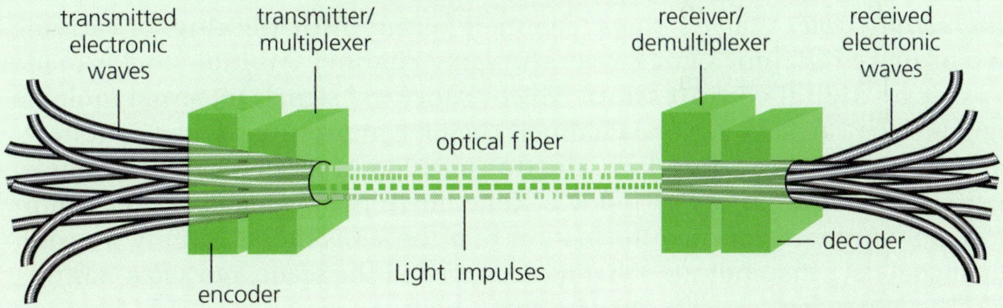

with 72 pairs of glass fibers can carry 3.5 million conversations. In May 1999 Nortel Networks announced the development of improved fiber optics technology that will "move 1.6 trillion pieces of data on a single fiber optic strand with each tick of the clock" (Healey, 1999b). Fiber optics wire has other advantages as well. It is cheaper and more flexible than metal-based wire, and signals suffer less degradation as they travel than they do over copper and aluminum cable.

What is sent over fiber optics is pulses of laser light. Those pulses are the equivalent of a digital signal's binary on/off structure (see Chapter 7), making them quite amenable to carrying digital signals. Fiber optics, shown in the box titled Fiber Optics, are at the heart of other digital technologies of interest to the cable industry.

One such advance is **digital cable television,** the delivery of digital images to subscribers' homes. At present digital cable has more to do with the services the system can offer than with the picture subscribers receive. This is because relatively few people have installed the converters necessary to receive digital cable, and as we saw in Chapter 3, even fewer have purchased digital sets requiring no converter. In fact, as of early 1999, "only about 60% of the 67 million cable homes are on systems rebuilt sufficiently to carry more than 54 channels. And only 3% actually have the (receiving) equipment to enjoy the cornucopia of digital cable" (Petrozzello, 1999, p. 19). Another impediment to more rapid diffusion of digital cable resides in cablecasters' dispute with over-the-air broadcasters concerning **digital must-carry,** potential FCC rules that would require cable systems to carry both digital and analog channels offered by over-the-air broadcasters. Naturally, the broadcasters want both channels carried until they shut down their analog operations. Just as naturally, cable operators see this as blocking valuable channel space that could be used for other, more profitable digital programming. The FCC has promised it will make a decision in 2000. Still, there is optimism surrounding digital cable. Kate McEnroe, president of American Movie Classics, predicts that by 2003, between 60% and 70% of all subscribers will buy digital service. Interactivity, she believes, is what will attract and hold those users (Katz & Peers, 1998).

Interactive Cable Cable's digital channels permit **multiplexing,** or carrying two or more different signals over the same channel. Multiplexing is made possible by **digital compression,** which "squeezes" signals to permit multiple signals to be carried over one channel. Digital compression works by removing redundant information from the transmission of the signal. For example, the set behind two actors in a movie scene might not change for several minutes. So why transmit the information that the set is there? Simply transmit the digital data that indicate what has changed in the scene, not what has not.

The expanded carrying capacity resulting from compression and fiber optics makes possible **interactive cable,** that is, the ability of subscribers to talk back to the system operator (one of the extra channels is used for this back talk). And *this* permits **video-on-demand,** the ability of viewers to call up any piece of content they want whenever they want to and have it delivered immediately. This technology is not yet available. But the cable television industry envisions that this technology will function through operators' maintenance of huge amounts of programming—old and new movies, old and new television shows, instructional videos, archival news footage, every football game your college ever played, video games—all residing in the form of digitally compressed data housed in immense databases. Once selected by the subscriber, the data would be sent to a set-top receiver/converter.

Public Access Cable Television Digitalization and compression have another effect on cable programming; they have renewed the debate over public access television.

We've already seen that public access channels were once required by the FCC, then un-required, only to become negotiable between the operator and the local franchising authority. As media historian William Boddy (1994) wrote, "As the cable industry underwent a rapid consolidation from small owners to highly capitalized multiple system operators in the late 1970s, the competition for large urban franchises became fierce. With channel capacity exceeding available programming services, cable operators looked at public access as an inexpensive bargaining chip with franchise officials" (p. 356). But that was then. Today, "the continuing consolidation of the cable industry into a handful of giant multiple system operators, and their moves into ownership positions in cable programming firms, has created a growing hostility toward public access from vertically-integrated cable operators dizzy at the prospect of increasingly lucrative commercial cable program services" (p. 357).

What digitalization and compression have added to the discord is the question of what to do with all those "extra" channels. Local franchising authorities see growing channel capacity as the perfect argument to maintain, or even expand, public access service. Many cable operators see it as a way to meet a wide array of subscriber needs and interests, fortifying them in their battle for profits (and survival) against competing media. Confounding the debate is the operators' belief that no one is watching public access television anyway. They point to the 1984 Cable Franchise Policy and Communications Act, which permits system operators to reclaim "underused" public access channels.

The future of local access, then, comes down to whose perception will prevail. As the number of channels grows, franchising authorities argue that "underused" is defined downward; that is, a local channel that serves a public service function but draws a relatively small number of viewers is a channel that is well used in a 500 channel environment. Many operators argue that the best use of these additional channels is to provide more sophisticated services to information- and programming-hungry subscribers. The box titled *Paper Tiger Television* details the work of one access champion. (We take a longer look at cable television programming in Chapter 10.)

Concentration

Gone are the days when systems were local mom-and-pop enterprises. Changes in the nature of cable system ownership parallel those in other media we've examined. As cable pioneer and current Time Warner executive Ted Turner explained, "We do have just a few people controlling all the cable companies in this country" (as quoted in "All Together," 1997, p. 14).

Concentration came to the cable industry initially in the form of multiple system operators. As cable experienced its greatest period of growth in the late 1970s, only the biggest and richest corporations could afford to build, buy, and improve operations in advance of the income they promised to generate. Today, more than half of all cable homes in the United States are served by one of three companies (Time Warner, AT&T Broadband, and Charter), and the five largest MSOs (the three just mentioned plus Cox Communications and Comcast) reach nearly 80% of all cable subscribers in the United States. Many MSOs own more than 100 systems each. Figure 6.2 lists the 10 largest MSOs and the size of their subscribership.

The Real World *Paper Tiger Television*

Public access television has a bad reputation. Even the most committed local access devotee must admit that too much of what fills the nation's access channels is self-indulgent, infantile, or silly. There are too many amateurish skits, bad take-offs on talk shows, and sophomoric *Star Trek* parodies.

Access also suffers from controversy. Its first-come first-served nature, absence of censorship (other than restrictions on obscenity and libel), and nearly full producer control leave public access television open to abuse by racists and other haters who have few other public outlets for their ideologies.

But isn't this what public access is supposed to be about? Isn't it supposed to provide a forum for those whose message is either unwanted by, or unsuited for, the more mainstream media? In the 1990s, for example, local franchising authorities in Cincinnati, Ohio, Pocatello, Idaho, Jackson, Mississippi, and Kansas City closed or attempted to close the access operations of their communities' cable providers rather than permit the cablecasting of a nationally distributed program by the Ku Klux Klan.

But what of others who hold potentially unpopular opinions? What about animal rights activists? What about pro-choice organizations? What about pro-life groups? What about Chinese dissidents living on the west coast? What about radicals on both the political right and left? You decide. Where would you draw the line? Or would you draw a line at all?

Now, read on and learn about one very successful public access producer whose message is offensive to many and is invisible on commercial broadcast and cable television.

Paper Tiger Television was founded in 1981 in New York City and at that time issued its *Manifesto,* which read in part:

> The power of mass culture rests on the trust of the public. This legitimacy is a paper tiger. Investigation into the corporate structures of the media and critical analysis of their content is one way to demystify the information industry. Developing a critical consciousness about the communications industry is a necessary first step toward democratic control of information resources. (<http://www.papertiger.org>)

The hundreds of episodes produced by Paper Tiger's "volunteer collective of media producers, educators, and activists" have and do appear on hundreds of public access channels across the United States. The collective's goal is to "provide a model and network for other progressive public access programmers." Its guiding philosophy is drawn from the political far left. "The group's name recalls Mao's guerrilla stance against superpower hegemony, and the manifesto's assertion of the importance of the reproduction of ideology is compatible with the ideas of . . . economic Marxism and anarchy" (Boddy, 1994, pp. 357–358). So, what kinds of shows does Paper Tiger produce and distribute? Here are a few examples:

Operation Storm the Media. In the media coverage of the Persian Gulf War even the pretense of separation between the press and the state was abandoned. This show explores the rela-

The second form of concentration in cable is **vertical integration,** wherein a company holds a financial interest in more than one aspect of the industry—production, distribution (the satellite service), and exhibition (the local franchises). As we've seen, systems tend to carry cable networks owned by the MSOs that own them. Critics of cable concentration find this inherently unfair and voice a second concern—that by its mere existence vertical

tionship between corporate sponsorship and media censorship.

Luis R. Beltran Tunes in to Bolivian Miners' Radio. South American scholar Luis Beltran describes the creation and activities of the radical Bolivian miners' radio station poised in the Andes Mountains. For 20 years this radio station has been the voice of miners, peasants, and working people.

Mutiny on the Corporate Sponsorship. This video looks at how mainstream media censors voices, not always through blatant censorship but through the rule of the status quo, which is dictated by the sponsors and mirrored by the corporate print and broadcasting elite.

How History Was Wounded: An Exclusive Report on Taiwanese Media. This is an exclusive report from Taiwan investigating how Taiwanese news media covered the Tienanmen Square massacre. It compares coverage of the mainland clampdown with Taiwanese coverage of their own government's suppression of political movements.

A Cry for Freedom and Democracy. Made in Chiapas, this video follows human rights activists, journalists, and family members as they try to gain access to the region blockaded by the military. Residents of Chiapas who witnessed the Mexican Army's indiscriminate brutality following the New Years Day Zapatista uprising give their firsthand accounts.

Put yourself in the position of general manager of a public access channel. Do you air Paper Tiger on your channel? Why or why not? Now, imagine that you are a local franchising authority staff person. Do you welcome the series to the system under your authority? Why or why not? Now imagine that you are the owner of a large MSO. Do you fight to regain the access channel that's running Paper Tiger Television to put it to more profitable use, or do you support the access channel's use of its time for such programming? Why or why not?

The Paper Tiger logo.

Now, consider this. Paper Tiger's far left ideology is offensive to many Americans. So, too, is the ideology of the KKK. How would you justifying excluding Klan content (if you would) while accepting programming provided by Paper Tiger (if you did)?

Source: All descriptions are from <http://artcon.rutgers.edu/papertiger/catalogue/SHglobal.html>.

integration limits programming competition (Figure 6.3). An MSO is more likely to initiate a new program service because it has guaranteed channel space on at least its owned systems. New, potentially innovative program services have no such guarantee. The counterargument, however, is that guaranteed channel space encourages risk taking. MSOs point to innovative offerings like BET, Lifetime, and the Discovery Channel as examples of channels that never

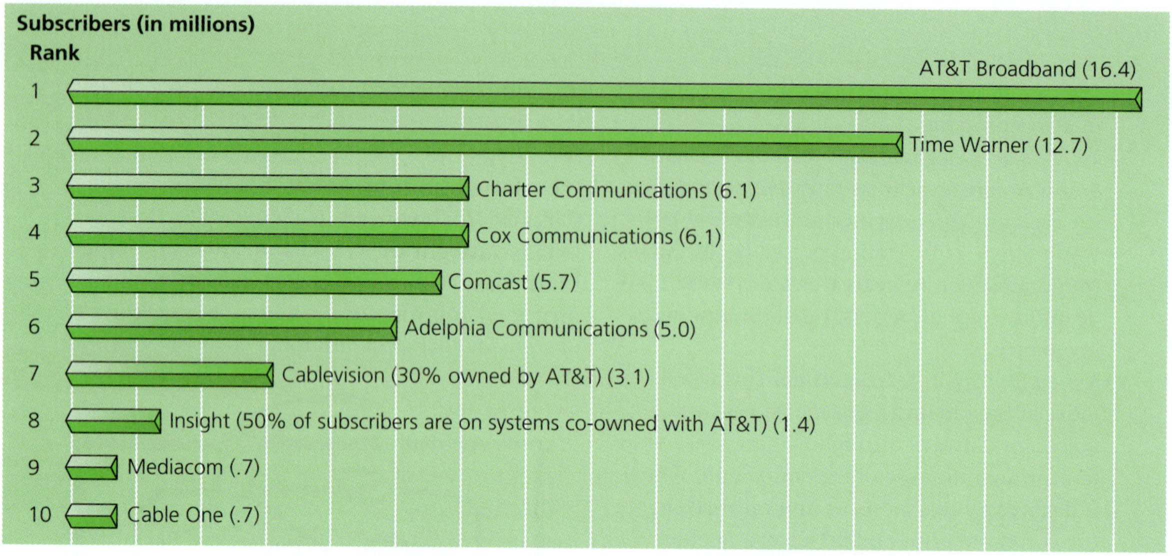

Figure 6.2 Top 10 MSOs Ranked by Number of Subscribers, 2000
Source: Adapted from Higgins, 2000. The big three. *Broadcasting & Cable,* 1 May, pp. 24–52.

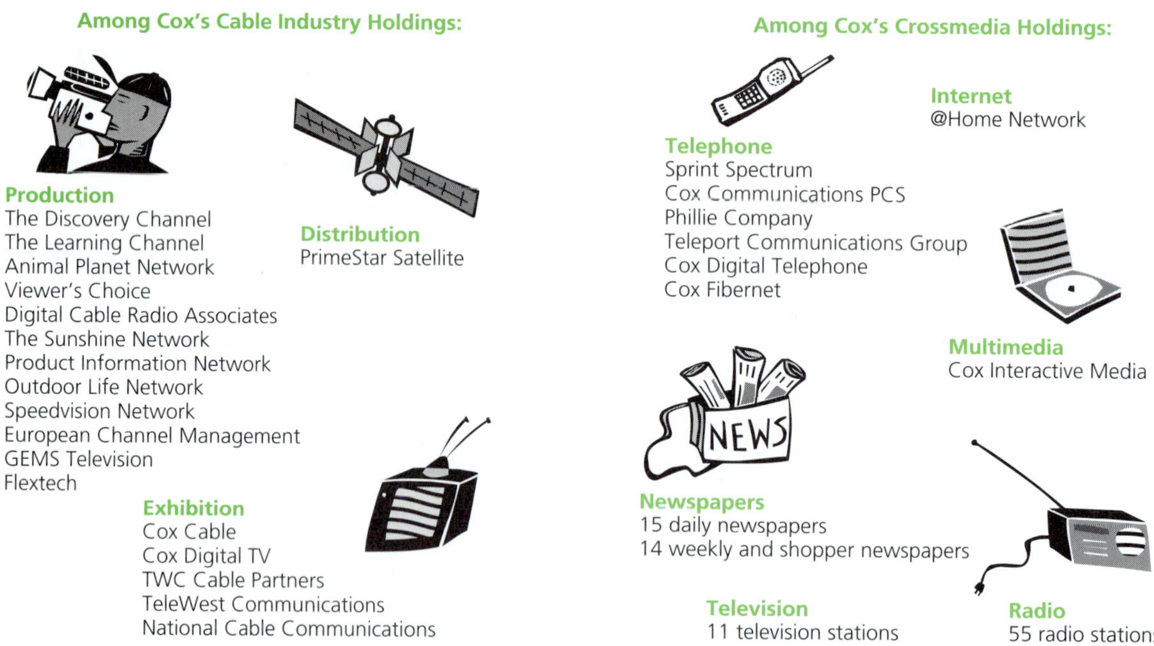

Figure 6.3 An Example of Vertical Integration and Crossownership
Cox Communications, the nation's sixth largest MSO, offers an example of vertical integration and crossownership in cable. *Source:* Adapted from <http://www.cjr.org/owners/cox.asp>, February 1, 2000.

would have been developed without MSO investment. Still, the FCC is sufficiently wary of this concentration of power that it requires operators to dedicate no more than 40% of its first 75 channels to program services owned by their owners.

The third way concentration has come to cable is in the form of **conglomeration,** the ownership of large MSOs by even larger companies having both media and nonmedia holdings. General Electric, for example, owns not only NBC and 13 television stations but also owns outright or in part 27 cable television channels in the United States and abroad (Chapter 1). Figure 6.4 shows the many and varied holdings of giant multinational corporation Seagram's.

Critics of conglomeration fear that the number of voices and variety of expression in the media (in this case, in cable television) will be diminished as ownership is concentrated in fewer and fewer hands. Another fear is that the conglomerate's media holdings will become nothing more than profit centers, no different from the conglomerate's juice or diaper supply businesses. Defenders of conglomeration argue that media companies will only survive in the reality of today's world of converged telecommunications if they are, in fact, parts of larger, integrated entities. One venue where this debate is being played out dramatically today is in the convergence of cable television with telephony and other services provided by the telcos.

The Telcos

The telcos had been banned from participation in cable television until the Communications Act of 1996 eliminated **crossownership rules,** allowing telco/cable crossownership. Since then, there has been a rush by phone companies to either buy outright or strike partnerships with cable operations. Familiar telephone names like Bell Atlantic, GTE, US West, and BellSouth entered the cable franchise business, although far fewer telcos became involved in cable television than anticipated by the FCC.

But the telcos are interested in cable systems for reasons other than their ability to deliver video channels. Ever since 1984, when AT&T was forced to separate from the regional Bell operating companies (**BOC**s, often called Baby Bells) to settle an antitrust suit brought by the federal government, it and other long-distance carriers like MCI and Sprint have been barred from the local phone business. The BOCs were likewise barred from offering long distance.

In an effort to spur competition in the telecommunications industry, the 1996 Telecommunications Act opened all telecommunications services to all comers. But because the BOCs owned the phone lines, companies wanting to provide local service had to find another way to enter people's homes. The cable companies, already wired into 67 million homes, offered the solution. Linking with a cable operation solved an additional problem for the long-distance companies. When they do use the local phone companies' lines to

Seagram's Cable Industry Holdings

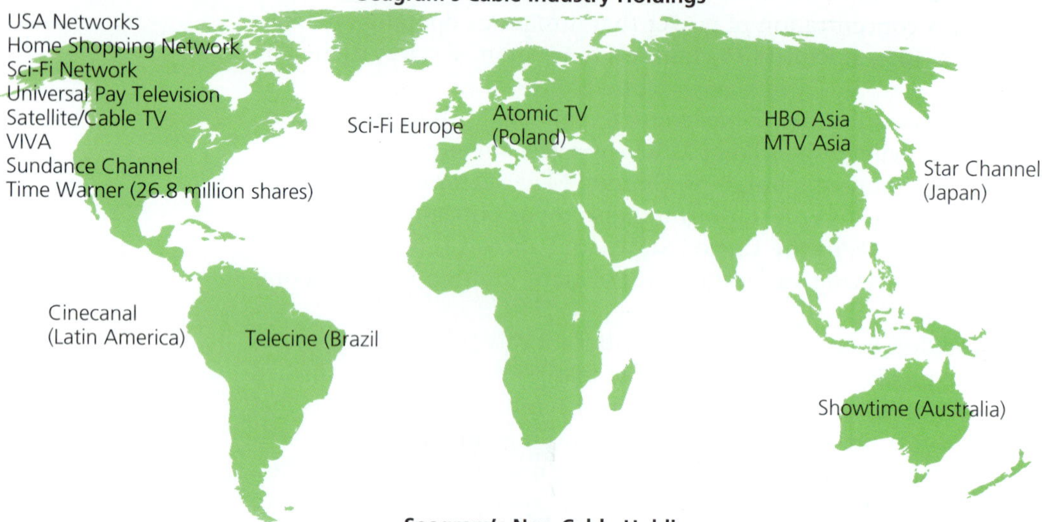

Seagram's Non-Cable Holdings

Universal.com

Universal Studios Online

Spencer Gifts

All Nations Catalogue

Beverages

The Tropicana Beverage Group
Dole Juices
Tropicana
Absolute Vodka
Chivas Regal Scotch
Crown Royal Scotch
Four Roses Scotch
Glenlivet Scotch

Television and Video

Universal Television Group
Brillstein-Grey Entertainment
Silver King Broadcasting
Universal Studios Home Video
35 different film and television production companies around the globe

Mutimedia Entertainment

Sega Works
Universal Studios New Media Group
Universal Interactive Studios
Universal Digital Arts
Interplay

Film Studios

Universal Studios
Universal Pictures
United International Pictures
October Films
Cinema International BV
Universal Studios Hollywood
Universal Studios Florida
Universal Studios of Japan

Recording Companies

Universal Music Group
MCA Records
MCA Records Nashville
Decca Records
GRP Recording Company
Geffen/DGC Records
Universal Records
Rising Tide
Interscope Records
Hip-O Records
PolyGram
A&M Records
Def Jam Records
Deutsche Grammophon
Island Records
London Records
Mercury Records
Motown Records
Phillips Classics
Polydor Records
Verve Records

Theme Parks

Universal's Islands of Adventure

Cinema Corporations

Cineplex Odeon
Cinema International Corporation
United Cinemas International

Concerts

Ticketmaster
Universal Concerts

Music Publishing and Distribution

Universal Music and Video Distribution
Universal Music International
MCA Music Publishing
Interscope Music Publishing
PolyGram Music Publishing
Island Music
Leonard Berstein Music Publishing Co.

Figure 6.4 An Example of Conglomeration
Seagram's, best known for alcoholic beverages, also has important cable industry and other corporate holdings. *Source:* Adapted from <http://www.cjr.org/owners/seagram.asp>, February 1, 2000.

enter people's homes, long-distance providers have to pay a connection fee to the local phone company for every call delivered over those lines. For AT&T alone, these fees amount to $10 billion a year. Cable allows long-distance companies to avoid local phone networks entirely. In the first half of February 1999 AT&T joined forces with the four largest MSOs in the United States. It merged with TCI in a $50 billion deal (renaming it AT&T Broadband), paid $69 billion for MediaOne, and entered into partnerships with Time Warner, which at the time was investing $1.5 billion a year in equipment and cable upgrades, and Comcast, promising to bring its high-speed digital cable to their subscribers. In addition, AT&T spent $2 billion to buy half (it already owned the other half) of medium-sized MSO Lenfest Communications. Together these deals made AT&T the world's largest MSO, with access to more than 60% of all U.S. cable households. Other MSOs provide local and long-distance phone service, most notably Cox Communications, Cablevision Systems, and Jones Communications.

Despite this activity, phone-over-cable will spread slowly, with even AT&T predicting that it will not provide full service to cable homes until 2004. There are two reasons. The first is technological: "the need to upgrade the patchwork of local cable networks to systems that can send and receive data; the need for large quantities of reliable, inexpensive devices to convert a phone's signals into digital cable signals; and the need for equipment to link the phone-over-cable-systems to the existing public phone networks" (Healey, 1999a, p. 1C). The basic technology to meet these needs already exists, so the challenge is getting manufacturers to agree on compatibility standards. The second reason that phone-over-cable may be slow in coming is consumer resistance. Many people, already dissatisfied with the level of service provided by their cable companies, are wary of relying on them for phone service as well. "I want my phone to work in the rain," one consumer told the *San Jose Mercury News,* "I can't think of the number of times a thunderstorm has put (my cable) out of service" (as quoted in Healey, 1998, p. 2C).

But there is another—even more important—reason the telcos are interested in hooking up with cable; that is convergence. If telephone service can be delivered by the same cable that brings television into the home, so too can the Internet. And what's more, if the cable line is high-speed fiber optic wire capable of handling digitally compressed data, that Internet service can be even faster than the service provided over traditional phone lines, sometimes up to 50 times faster than dial-up telephone modems. Cable, in other words, can become a one-stop communications provider: television, audio, high-speed Internet access, long-distance and local phone service, multiple telephone lines, and fax. The industry calls this **bundling.**

How valuable is a bundle-receiving subscriber to a cable/telco combination? In 1998 Marcus Cable was bought by Microsoft cofounder Paul Allen for a price that amounted to $2,313 per subscriber. Allen was ridiculed for paying so much. A year later, when AT&T bought MediaOne, each subscribing home was purchased for the equivalent of $4,700. Cox Communication's acquisition of Media General Cable a month after the AT&T deal cost it $5,400

Critical Issues *The Promise of Cable*

As the industry was morphing from CATV to cable television in the 1970s, "an ever expanding chorus of expert opinion (echoed) a new, hopeful view" for the medium (Streeter, 1997, p. 223). The United States would become the "wired nation," its towns wired cities; cable would return television to the people—it would become the people's medium. Traditional television was the ill; cable television the cure.

> Television was no longer seen as an infant institution, and its problems were no longer interpreted as temporary foibles, amenable to correction within the existing overall structure. People in positions of authority and power were beginning to seek solutions to television's failings not in adjustments to the existing system, but in alternatives to the system itself. (Streeter, 1997, p. 232)

But by the early 1980s, talk had turned to the failed promise of cable. Cable had not become an alternative to dull, unchallenging, three-network-dominated television; rather it had become simply *more* television: more movies, more sports, more commercials, more situation comedies.

But bundled cable service, providing all sorts of converged technologies, has rekindled what sociologist Thomas Streeter (1997) calls "utopian speculation" about cable's very near future. He samples contemporary comment:

- "Futurist" George Gilder predicts that, with the help of interactive television, "the human spirit—emancipated and thus allowed to reach its rarest talents and aspirations—will continue to amaze the world with heroic surprises" (p. 238).
- Mitchell Kapor, cofounder of the Internet advocacy group Electronic Frontier Foundation, predicts that the National Information Infrastructure (the convergence of Internet and cable) will promote "grassroots democracy, diversity of users and manufacturers, true communications among the people, and all the dazzling goodies of home shopping, movies on demand, teleconferencing, and cheap, instant databases" (p. 239).
- The Administration of President Bill Clinton predicts that the convergence of Internet and cable technologies will allow the arts and humanities to "play a vital role in creating a new sense of citizenship and community," it will "bring new opportunities and resources to our nation's disadvantaged youth, allowing them to share their ideas, thoughts and creative energies, and to make new links with other young people throughout the nation," it will "give all Americans, of all races, ages and locations, their cultural birthright: access to the highest quality thought and art of this and prior generations" (p. 238).

Maybe.

Despite all the criticism of television in the days of the Big Three (ABC, NBC, CBS), at least Americans shared a common culture. Yes, it may have been a shallow culture of *I Love Lucy* and *My Favorite Martian*, but it was a widely shared culture. And it was also a culture boasting journalists of the caliber of Walter Cronkite and screenwriters the caliber of Rod Serling (*The Twilight Zone*). For better *and* for worse, precable television was the stock that helped flavor the American melting pot. So, will we really be better off when Americans are fragmented among 500 demographic, taste, and interest channels? Will we be better off when Americans are linked anonymously across fiber optic wires in virtual rather than actual communities, holding virtual rather than real conversations? This is not an argument against the new multichannel, bundled telecommunications universe, simply a reiteration of the classic warning, "Be careful what you wish for. You just may get it."

What do you think? Have you considered what the future will really look like for all wired Americans, for the technology haves and the technology have-nots (Chapter 1), for those who look like you, for those who don't?

Potential revenue per subscriber	Service	Penetration	Average revenue per subscriber
34.36	Basic cable	100%	34.36
8.65	Pay cable	70%	6.06
10.40	Advanced analog	10%	1.04
36.92	PC online	10%	3.69
6.37	TV set online	3%	.19
11.91	Digital cable	20%	2.38
2.50	Pay-per-view	75%	1.88
4.02	Local ads	100%	4.02
55.00	Telephone	20%	11.00
$170.13			$64.62

Figure 6.5 The Value of a Subscriber in a Bundled Cable World
High-speed cable service brings multiple communication services to subscribers. It is the promise of millions of homes buying multiple services that makes cable franchises so attractive to telcos and others wanting to deliver bundled services. Here is one estimate of how much a typical cable home will spend each month in 2002. Revenue per subscriber represents what a household would spend each month if it took every service. Average revenue per subscriber represents what a company would collect, on a monthly average, from each household at the predicted level of penetration (the proportion of all subscribers choosing a specific service). *Source:* Higgins, 1999.

per subscribing home. And what makes cable subscribers worth that much money? As Figure 6.5 demonstrates, industry predictions are that each home connected to a high-speed cable system will buy $64.62 worth of service per month in 2002, as opposed to a 1999 industry-wide average of only $40 a month (Higgins, 1999).

But to cable's competitors in the telecommunications industry—for example, DBS, local phone companies, and Internet service providers—this bundling puts too much power in the hands of the cable operators and denies cable's competitors access to valuable customers. (Cable companies, as opposed to phone companies, are not yet required to make their lines available to competitors.) Cable's competitors have asked, therefore, that the FCC require operators to **unbundle**—that is, to mandate **open access** (cablecasters call it "forced" access)—making cable lines available to others for a fair price. The FCC has promised to rule in the near future, but in the meantime numerous franchising authorities and municipalities have passed their own rules requiring systems to be open. In November 1999 a federal court ruled that such requirements were legal.

The potential of the "new cable television" is the subject of the box titled The Promise of Cable, and careers in cable are the focus of the box titled Working in Cable Television.

Careers *Working in Cable Television*

The organization of a cable operation is in many ways similar to that of a television station. For example, both have specific departments that take responsibility for different aspects of the enterprise, and there is greater flow across positions in smaller operations than at larger ones. The differences that do occur, however, naturally stem from at least three distinct characteristics of cable and broadcast:

- Broadcasters sell viewers to advertisers; cablecasters sell viewers to advertisers, but must also sell themselves to viewers (subscribers).
- Broadcasters, even network affiliates, produce much more local content than do cable franchises (even most access channels are produced, not by the franchisee but by the access programmer).
- Television viewers install their own equipment (they buy a set); cable must be installed by the operator.

The primary divisions at a cable operation are customer and employee relations, advertising sales, operations, marketing and sales, and business.

Customer and employee relations handles subscribers' questions and complaints, plans and implements external or community public relations efforts, and, as in a typical broadcast station, effects the organization's employee functions such as hiring, training, and benefits. Together with the operation's general manager, the director of customer and employee relations will usually have prime responsibility for preparing documentation in support of its refranchising before the appropriate municipal or regional authority.

Advertising sales operates much like the sales department of a typical broadcast station. The sales manager oversees a staff of sales reps or account executives who sell local time on advertiser-supported national channels and on locally originated programming, if any. Many cable franchises dedicate one or more channels to video classified advertising. The sales department, logically, has responsibility for selling this service.

Operations is the department, under the direction of an operations manager, that keeps the system up and running, from the head through the hub to installation in subscribers' homes.

Marketing and sales does not sell advertising but rather sells the system to subscribers. As such, under the direction of the marketing and sales director, this department conducts (or contracts for) market research and direct sales campaigns, either in malls, door to door, by telephone, or by mail. There is no programming department per se at cable operations. Therefore, "responsibility for decisions about what services to carry and how to schedule them varies greatly from system to system" (Eastman, 1993, p. 247). Decisions about tiering, pricing, and the selection of services are usually group decisions, but they invariably involve the marketing and sales director and the general manager.

Business involves such day-to-day functions as accounting and computer and office management. Traffic, in the sense of scheduling and billing for commercials, often falls under the auspices of the business manager.

Satellite

As we've seen, it was cable's convergence with satellite technology that turned John Walson's cottage industry into a powerful medium. Once HBO began satellite distribution, soon followed by Ted Turner's Superstation WTBS, cable went from a means of helping people receive clearer pictures to an important source of programming (and therefore a major telecommunications industry) in its own right.

But satellites are being used for far more than cable television and DBS (Chapter 3). In fact, satellites sit at the center of almost all efforts to converge video, telephony, and Internet technologies. For example, a telco using satellite channels to digitally deliver telephone service can easily and inexpensively piggyback video and Internet service on those same channels. This is one impetus behind AT&T's aggressive expansion into cable television as well as AOL's 1999 investment of $1.5 billion in satellite television provider DirecTV. Satellites, then, can be seen as the technological catalyst for much of the convergence and concentration now occurring not only in cable television but among all the once distinct telecommunications industries.

Chapter Review

The visions of three Pennsylvanians, John Walson, Milton Shapp, and Robert Tarlton, eventually became the telecommunications giant we now call cable television. Initially conceived of as a way to deliver clear signals to people in remote areas, cable quickly became much more than that. Now local and distant signals, as well as a variety of pay channels, come to people's homes on systems composed of a headend, super trunk cable, hub, trunk cable, feeder cables, and drop cables. The operations that maintain these systems are called franchises, and their nature and operation have been defined by often conflicting government regulation.

The FCC entered cable regulation as a result of its Carter Mountain decision and has regulated and reregulated to at first slow, then control, then free cable's growth. Nonetheless, early rules governing local carriage, public access channels, and the power and operation of local franchising authorities remain today.

Cable programming exists in several forms. Basic cable typically fills systems' lowest tiers and pay cable its upper tiers. System operators choose to carry different cable networks for different reasons, among them being audience demand, local avails, no or low cost, rebates and payments, and shared ownership between the system and the network.

These and newer services are made possible by a variety of sophisticated technologies. Fiber optics have greatly increased the efficiency and bandwidth of the cables that make up cable television. Digital cable television, especially when combined with digital compression, makes possible multiplexing, carrying two or more different signals over the same channel. Multiplexing, in turn, permits interactive cable and video-on-demand.

Concentration, in the form of MSOs, vertical integration, and conglomeration, is widespread and controversial in the cable industry and increasingly involves the telcos, eager to enter people's homes over the same wires that deliver television signals. The same wires and satellites that bring television to subscribers can be used for a host of bundled services, from local and long-distance telephone to fax to high-speed Internet access.

Questions for Review

1. What were the contributions of John Walson, Milton Shapp, and Robert Tarlton to the development of cable television?
2. Differentiate between twin lead, coaxial, and fiber optic cable.
3. What significance is there in the passage of the medium's name from CATV to cable television?
4. What are the elements involved in the reception and distribution of cable television signals?
5. What are pass-by rate, density, penetration, and churn?
6. What are the primary capital costs and operating expenses involved in running a cable operation?
7. What is the significance of the FCC's ruling in the Carter Mountain matter?
8. Explain the difference between basic cable, expanded basic cable, pay cable, and pay-per-view.
9. What are digital cable television, multiplexing, digital compression, and interactive cable, and how are they interrelated?
10. What are the three forms of concentration in cable television? What concerns are raised by each?
11. What are some of the factors leading to the telcos' interest in cable?
12. What is bundling? Must-carry? Open access? Why are they important to cable and noncable telecommunications companies alike?

Questions for Critical Thinking

1. What do you think of must-carry rules? Why should a cable operator be forced to carry local stations' signals even if it doesn't want to? What gives the government the right to tell cable operators how to run their businesses?

2. Are you a cable subscriber? Why or why not? At what level? Why that level? Have you ever added to the industry's churn? Why or why not?

3. Does concentration in cable disturb you? Why or why not? What do you think of the critics' fears? Are they realistic or overblown?

4. Have you ever watched public access cable? Have you seen programming that you thought was useful, or at least that was good television? Have you seen access content that you thought was silly? Offensive? Describe your experience.

5. If you were a cable operator, how willing would you be to unbundle your facilities to accommodate competitors? Why do you think it is proper (or improper) for the federal government to require unbundling should it do so?

Engaging Technology

1. *Multichannel News* maintains one of the best Web sites for people interested in cable. In addition to much news about cable and other providers of multiple channels of video and audio content, the site (<http://www.multichannel.com/assoc.htm>) offers hot links to dozens of other valuable sites. Choices can range from the entire text of the Telecommunications Act of 1996 to sites for state and regional cable associations. Check this site out. Connect to some of the different links and determine for yourself what different cable organizations think are their most pressing issues.

2. The primary trade association for cable is the National Cable Television Association. Its Web site (<http://www.ncta.com/>) provides a wealth of data about the business and audience ends of cable,

as well as up-to-the-minute presentations of new and emerging cable technology. Visit this site and determine where those in the industry think cable is headed.

3. We've seen in this chapter that one potential outcome of the 500-channel environment is that people will be divided even more into small, special interest groups. One organization that hopes to minimize the impact of this trend, should it occur, is the National Association of Minorities in Cable. Visit <http://www.namic.com/> to see what this group is about. If you're interested, contact them, either to join or to gather more detail about their operation.

4. Paper Tiger Television and Deep Dish TV are the two most organized and active public access channel content providers in the United States. Visit their Web sites at <http://www.papertiger.org> and <http://www.igc.apc.org/deepdish/> to see what they are up to, and use the e-mail link provided to talk with members of these groups. Report on your exchange for your classmates.

Chapter Seven

The Internet

In March 1999, as Serbian military and police forces were purging Yugoslavia's Kosovo region of ethnic Albanians and as NATO bombs pounded the Serbs in an effort to force them to halt their ethnic cleansing, a 16-year-old girl using the name Adona was smuggling reports out of Kosovo to an anxious world. "Just from my balcony, I can see people running with suitcases and I can hear some gunshots. . . . A village a few hundred meters from my home is all surrounded. As long as I have electricity, I will continue writing to you. Right now, I am trying to keep myself as calm as possible. My younger brother, who is 9, is sleeping now. I wish I will not have to stop his dreams" (as cited in Burress, 1999, p. A14). Adona, herself an ethnic Albanian, didn't need electricity for light by which to write; she needed electricity to keep her personal computer linked to the Internet. "She's like Anne Frank with a laptop," said Finnegan Hamill, a 16-year-old high school student in Berkeley, California. "It's kind of like her diary, except in a real-time environment" (p. A14).

Hamill, a reporter for Berkeley-based Youth Radio, used Adona's e-mails as the basis of numerous reports over a several month period. His stories appeared on NPR and CNN and were distributed worldwide by those news organizations in the form of downloadable audio files on their World Wide Web sites. Listeners learned that

Adona was a typical teenager. "I love listening to Rolling Stones, Sade, Jewel, and R.E.M. . . . You don't know how I'm longing to go to a party, or on a trip, or anywhere" (as cited in Conniff, 1999, p. 108). But unlike most teenagers, she feared being "raped, with no parts of body like the massacred ones" (p. 108).

Why did the saga of this young woman so move the public? Hamill, Adona's electronic correspondent, answered, "It's the perfect aspect . . . who's being affected. It's not what many people think, of hard-liner Serbs or hard-liner Kosovars. . . . She's really an innocent bystander" (as cited in Burress, 1999, p. A14).

Adona went offline and silent at the end of March. But 5 months later Kujtesa Bejtullahu (Adona's real name) arrived in California, her U.S. immigration and education sponsored by a local church. The two friends met at the San Francisco airport on August 25—8 months after she and Hamill first connected online.

Preview

Adona's story highlights much of the wonder surrounding the Internet. Here was a relationship between two young people who had never met except online. Here was information from a war zone, zipping instantaneously around the world, free of military and news organization gatekeepers. Here was a member of a threatened minority given full voice and an international audience. Here was the convergence of radio and digital technology permitting computer users, no matter where they lived, to listen to Hamill's reports. Here was an opportunity for people to hear a personal perspective on that terrible conflict, one offered by someone "on the ground," as correspondents were fond of saying about those unfortunate people caught up in the mayhem. In short, here was the Internet at its best.

In this chapter we study the technology that made Adona's reports and their worldwide distribution possible—the Internet. We begin with the development of the computer, leading to the military applications that ultimately produced the Internet. Then we take a detailed look at the Internet as it exists today. We examine its formats and its capabilities, especially the popular World Wide Web. The number and nature of today's Internet users are discussed, as are the growing efforts of online advertisers and sellers to reach them through e-commerce and e-tailing.

The Internet's future is examined in terms of the evolution of portals and the move toward industry concentration. Emerging technologies, driven by and encouraging convergence, are detailed. We investigate push technology, wireless access to the Net, the capabilities of embedded systems, the audio compression technology MP3, and the marriage of Internet and television.

Technological Roots of the Internet

To better respond to a feared invasion from the Communists in 1950, President Dwight Eisenhower ordered a system of high-speed roads to be built. He based his vision on the German highway system, the Autobahn, which had allowed the Nazis to move their armies so effectively during World War II. With the new superhighways in place, no matter where an enemy might attack in the United States, the military could respond effectively and rapidly.

In 1962 the Air Force wanted another "highway," one that would effectively and rapidly move information instead of armies. The military sought to maintain the ability to transfer information, even under enemy attack. The resulting system was the beginning of what we now call the Internet, or the Net, a global network of interconnected computers that communicate freely and share and exchange information. Before we delve into the details of this network of computers, we should first consider the origin of computers themselves.

Development of the Computer

The title "Father of the Computer" resides with Englishman Charles Babbage. Lack of money and the unavailability of the necessary technology stymied his plan to build an "Analytical Engine," which was to be a steam-driven computer. But aided by the insights of mathematician Lady Ada Byron Lovelace, in the mid-1800s Babbage did produce designs for a "computer" that could conduct algebraic computations using stored memory and punch cards for input and output. His work provided inspiration for those who would follow.

Over the next 100 years, a number of mechanical and electromechanical computers were attempted, some with success. Colossus, developed by the British to break German secret codes during World War II, was the first electronic **digital computer;** that is, it reduced information to a **binary code** made up of the digits 1 and 0 for storage and manipulation. The first "full-service" electronic computer, ENIAC (Electronic Numerical Integrator and Calculator), based on the work of John V. Atanasoff at Iowa State University, was introduced by scientists John Mauchly and John Presper Eckert of the Moore School of Electrical Engineering at the University of Pennsylvania in 1946.

ENIAC hardly resembled the computers we know today. Eighteen feet tall, 80 feet long, and weighing 60,000 pounds, it was composed of 17,500 vacuum tubes and 500 miles of electrical wire. It filled an auditorium and ate up 150,000 watts of electricity. Mauchly and Eckert eventually left the university to form their own computer company, later, in 1950, selling the company to the Remington Rand Corporation. At Remington they developed UNIVAC (Universal Automatic Computer), which became the first successful commercial computer. It was bought and used by the Census Bureau in 1951.

The commercial computer explosion was ignited by IBM. Using its already well-entrenched organizational system of trained business machine sales and service professionals, IBM helped organizations find their way in the early

ENIAC.

days of the computer revolution. One of its innovations was to sell rather than rent its **mainframe computers** to customers. Mainframe computers are large processing units that are connected to many individual terminals. As a result of IBM's success, by 1960 the computer industry could be described as "IBM and the Seven Dwarfs"—Sperry, Control Data, Honeywell, RCA, NCR, General Electric, and Burroughs (Rosenberg, 1992, p. 60).

Military Applications

In 1957 the Soviet Union launched Sputnik, Earth's first human-constructed satellite. The once undisputed supremacy of the United States in science and technology had been overtaken, and its scientists and military officials were in shock. The Advanced Research Projects Agency (ARPA) was immediately established to sponsor and coordinate sophisticated defense-related research. In 1962, as part of a larger drive to promote the use of computers in national defense, ARPA commissioned Polish immigrant Paul Baran of the Rand Corporation to produce a plan that would allow it to maintain command over its missiles and planes in the event a nuclear attack knocked out conventional means of communication. The military thought a decentralized communication network was necessary. In that way, no matter where the bombing occurred, other locations would be available to launch a counterattack. Among Baran's plans was one for a "packet switched network." He wrote:

> Packet switching is the breaking down of data into datagrams or packets that are labeled to indicate the origin and the destination of the information and the forwarding of these packets from one computer to another computer until the information arrives at its final destination computer. This (is) crucial to the realization of a computer network. If packets are lost at any given point, the message can be resent by the originator. (as quoted in Kristula, 1997, p. 1)

The genius of the system Baran envisioned was twofold: (1) common communication rules (called **protocols**) and common computer languages would allow any type of computer, running with any operating system, to communicate with another; and (2) the system required destination or delivery instructions to be embedded in all information sent over it, allowing for instantaneous "detours" or "rerouting" if a given computer on the network became unavailable.

Using Honeywell computers at Stanford University, UCLA, the University of California–Santa Barbara, and the University of Utah, the switching network, called ARPANET, went on line in 1969 and became fully operational and reliable within one year. Other developments soon followed. In 1972 an engineer named Ray Tomlinson created the first e-mail program. In 1974 Stanford University's Vinton Cerf and the military's Robert Kahn published the protocol that allowed any computer on what they called "the Internet" to communicate with any other on that network. In 1979 a graduate student at the University of North Carolina, Steve Bellovin, created USENET and, independently of Bellovin, IBM created BITNET. These two networking software systems allowed virtually anybody with access to a Unix or IBM computer to connect to others on the growing network. By July 1988 there were 33,000 such links, and by July 1998 there were 36,739,000 (Ticker, 1999).

The Personal Computer

A crucial part of the story of the Internet is the development and diffusion of personal computers. IBM was fantastically successful at exciting businesses, schools and universities, and other organizations about computers. But IBM's and other companies' mainframe and **minicomputers** employed **terminals**—stations at which users worked—connected to those larger centralized machines. As a result, the Internet at first was the province of the people who worked in those settings.

When the semiconductor replaced the vacuum tube as the essential information processor in computers (see the box titled From Tube to Microprocessor), its tiny size, absence of heat, and low cost made possible the design and production of small, affordable **personal** or **microcomputers** (PCs). This, of course, eventually opened the Net to anyone, anywhere, who had a PC.

The leaders of the personal computer revolution were Bill Gates and the duo of Steve Jobs and Stephen Wozniak. As a college freshman in 1975, Gates saw a magazine story about a small, low-powered computer, the MITS Altair 8800, that could be built from a kit and used to play a simple game. Sensing that the future of computing was in these personal computers and that the power of computers would reside not in their size but in the software that ran them, Gates dropped out of Harvard University and, with his friend Paul Allen, founded Microsoft Corporation. They licensed their **operating system**—the software that tells the computer how to work—to MITS. With this advance people no longer had to know sophisticated operating languages such as FORTRAN and COBOL to use computers. At nearly the same time, in 1977,

Tech Talk *From Tube to Microprocessor*

The first electronic computers used vacuum tubes. As a result, they were huge, hot, and expensive to run. Today's personal computers employ microprocessors. As a result, they are small, generate no heat, and are relatively inexpensive to own and operate. Here is an explanation of this evolution.

Computers operate using a network of switches that send and stop electricity. The first switches were vacuum tubes.

In 1948 William Shockley developed the **semiconductor**—material that sometimes conducted electricity, and sometimes didn't. Semiconductors quickly became the basis of a new kind of miniature switch called the **transistor.**

In 1958 scientists at Fairchild Electronics developed a system of etching tiny transistors onto a single silicon-based wafer, the **integrated circuit.** Advances over the next 3 decades improved the efficiency of these circuits and miniaturized them even further, to the point that several thousand, sometimes millions, of circuits could reside on a single chip. These chips are called **microprocessors.**

Jobs and Wozniak, also college dropouts, perfected their Apple II, a low-cost, easy-to-use microcomputer designed specifically for personal rather than business use. It was immediately and hugely successful, especially in its development of multimedia capabilities such as advanced sound and image applications. IBM, stung by its failure to enter the personal computer business, contracted with Microsoft to use the Microsoft operating system in its IBM PC, first introduced in 1981. All of the pieces were now in place for the home computer and Internet revolutions.

The original Apple II.

The new iMac.

The Internet Today

Here are two visions of the Internet, both from leaders of the Internet industry. The first is from Bill Gates (1995), founder of Microsoft:

> The revolution in communications is just beginning. It will take place over several decades, and will be driven by new "applications"—new tools, often meeting currently unforeseen needs. . . . [O]ne of the most remarkable aspects of this new communications technology is that it will eliminate distance. It won't matter if someone you're contacting is in the next room or on another continent, because this highly mediated network will be unconstrained by miles and kilometers.
>
> A different metaphor [from the information superhighway] that I think comes closer to describing a lot of the activities that will take place is that of the ultimate market. Markets from trading floors to malls are fundamental to human society, and I believe this new one will eventually be the world's central department store. It will be where we social animals will sell, trade, invest, haggle, pick stuff up, argue, meet new people, and hang out. Think of the hustle and bustle of the New York Stock Exchange or a farmers' market or of a bookstore full of people looking for fascinating stories and information. All manner of human activity takes place, from billion-dollar deals to flirtations. The highway will enable capabilities that seem magical when they are described, but represent technology at work to make our lives easier and better. (p. 6)

John Chambers, president of Cisco Systems, an Internet networking company, offered his view of what he called the "Internet Revolution":

> The Internet will change how people live, work, play, and learn. The Industrial Revolution brought together people with machines in factories, and the Internet revolution will bring together people with knowledge and information in virtual companies. And it will have every bit as much impact on society as the Industrial Revolution. It will promote globalization at an incredible pace. But instead of happening over 100 years, like the Industrial Revolution, it will happen over seven years. (as quoted in Friedman, 1998, p. 6B)

As we read these predictions of where the Internet is going, it is important to remember that we will be instrumental in determining the future of this powerful medium. We are like the radio audience of the early 1920s and the television audience of the 1950s—but with a crucial difference. The public's influence in shaping the development and future of those two media was, at best, indirect. Legislators and regulators, through suggestions and formal rules, negotiated with the radio and television industries to shape their development, diffusion, and operation—all in the name of the people. The people, in their role of audience, could reject or accept the product of that negotiation. That is, they could either tune in or not and like or dislike what they saw and heard. This is what audiences do. But the Internet does not have audiences; it has users. And users use. As such, we will determine the Internet's development and operation—*if we choose to do so.* As former NBC and PBS president Lawrence K. Grossman (1999) wrote, "Gutenberg made us all readers. Radio and television made us all first-hand observers. Xerox made us all publishers. The Internet makes us all journalists, broadcasters, columnists, commentators, and critics" (p. 17). Therefore, we need to learn to use it. Or, if we are already linked and online, we need to learn to use it more effectively.

The Operation of the Internet

It is only a slight overstatement to say that computers are rarely used for computing anymore because the Internet has given the computer so much more versatility.

The Internet is most appropriately thought of as a "network of networks" that is growing at an incredibly fast rate. These networks consist of **LANs** (**L**ocal **A**rea **N**etworks) connecting two or more computers, usually within the same building and **WANs** (**W**ide **A**rea **N**etworks), which connect several LANs in different locations. When people access the Internet from a computer in a university library, they are most likely on a LAN. But when several universities (or businesses or other organizations) link their computer systems, their users are part of a WAN.

As the popularity of the Internet has grown, so has the number of **Internet service providers** (sometimes called **ISPs**), companies that offer Internet connections at monthly rates depending on the kind and amount of access needed. There are 6,500 ISPs operating in the United States, including some of the better known, such as America Online, Prodigy, and Earthlink (Specter, 1999). Through providers, users can avail themselves of numerous services.

The Real World *Using E-mail Effectively*

The Web may be the Internet's most popular application, but its most used feature is e-mail. As a result, it is through e-mail that people most frequently shed their audience role and assume that of user, both as sender and receiver. Even the most frequent e-mail user can improve his or her online skills. Here are some hints.

As Sender

Like the Internet that carries it, e-mail is free and open. But communication, online or otherwise, has ramifications. Users often err in allowing e-mail's ease of use and *apparent* impersonal nature to affect their normal ways of communicating. Although delivered digitally through an unimaginable number of technological links, e-mail is *not* impersonal. When you sign it, it's *yours*. It reflects you, who you are, and what you think of the recipients. To avoid troublesome miscommunication, use care in creating e-mail messages:

1. Before you begin writing, *consider the destination.* Issues of formality, amount of supporting explanation, evidence or documentation for arguments, and other content matters should be determined, in advance, by an understanding of who will read your message—for example, friend or foe, or superior, peer, or subordinate.
2. The Net is open and free. So, before you write, *anticipate unintended readership.* Bosses can read people's e-mail at work. The American Management Association reports that one third of all major U.S. companies regularly access their employees' e-mail (McNichol, 2000). Even in nonwork settings, recipients can transfer your words to others, without your permission or knowledge, with one stroke of a key. These others might be people you wish had never seen what you have to say.
3. *Employ courtesy.* If you would not insult a person to his or her face, don't do it online. The distance between you and your recipients is nonexistent in the cyberworld, because messages are delivered instantaneously. However, there is no instant, direct feedback that enables you to adjust when you sense miscommunication. Craft your messages as if they were face-to-face conversations.
4. *Use care when producing sensitive content.* If you must criticize, confront, evaluate, or otherwise commit important and potentially problematic content to the Net, take a breather. For example, type your message and print it out and read it in hard copy before you send it. This can be a form of "counting to 10." It also lets you read, reread, and edit your words on paper, which many people find more natural than doing so on screen.
5. *Focus on the task.* E-mail's simplicity of use creates the danger of easily made or thoughtless mistakes. When you are creating e-mail messages, make that your only activity. A mispunched key, for example, can send a message to the wrong place. Remember, a problematic face-to-face conversation can be instantly corrected. E-mail lives in people's in-boxes for a long time and can also be forwarded to others.
6. *The rules of grammar still apply.* Many users think of e-mail as a way to send off quick notes, saving more important messages for face-to-face communication or letters. But your e-mail repre-

E-mail (Electronic Mail) The Internet's most frequently used function is **e-mail.** There are more than 400 million e-mail addresses in the world, with 60% attached to machines in the United States (Specter, 1999). In 1998 Americans received 4 trillion e-mail messages, compared to 107 billion pieces of first-class **snail mail,** e-mail users' unflattering name for the kind of mail delivered by the postal service ("Numbers," 1999, p. 23). With an Internet e-mail account, users can communicate with anyone else online, any place in the

sents you. Even a note composed of only a few lines, if it contains some misspelling or a comma or two dropped or out of place, can be taken as a sign of a careless person, one unconcerned about how he or she appears, one unwilling to extend the respect of a well-crafted message to the recipient.

7. *When interpersonal problems arise, get offline.* Many users hide behind their e-mail. Especially when miscommunication leads to confrontation, hashing things out online—where nuance, irony, humor, and other tools of good interpersonal communication are sometimes wasted—may not be in your best interest. At work, walk down the hall and visit the other person. With friends, or where distance intervenes, pick up the phone. Older technologies like the shoe and the telephone do retain some advantages.

As Receiver

Users can never be certain of the level of technological, grammatical, or communication skill possessed by those who communicate with them online. E-mail receiving also requires the development of certain abilities.

1. *Avoid knee-jerk reactions.* We've already seen that despite its benefits electronic communication lacks many of the factors that make more traditional forms of communication flexible and less open to error. Accept the fact that not all e-mail senders are as skilled as you are.

2. *Don't be a slave to e-mail.* Instant, virtually no-cost communication from friends and colleagues around the corner and around the world makes e-mail fascinating, perhaps too much so. Check your e-mail only once or twice a day. Respond in a timely fashion to the most important messages, saving the others for more thoughtful response.

3. *Use the technology.* Several e-mail managing systems can sort or even delete incoming messages based on specified keywords or locations. Claris Emailer, Microsoft's Inbox, Cyberdog, and Eudora are the better known products. These enable you to establish priorities for what gets read and when, saving time and patience.

4. *Use care when subscribing* to lists or joining discussion groups and chat lines. As valuable as they are, they mark you as someone interested in a given issue or topic. You may—in fact you most likely will—become the recipient of many more messages than you had anticipated once you identify yourself in a particular way online.

5. *Separate personal and work e-mail.* The ease of the Net often masks its intrusive power. Your employer most likely will not welcome the use of your work time to send personal messages. Your home life will be richer and less stressful if you apply the same standard of separation there.

6. *Don't be a spammer.* It's simply too easy to send unnecessary copies of your message to everyone in your dorm, at your place of work, or on a preconstructed distribution list. Use discretion. Who *needs* to get your e-mail? They should be the primary recipients. If you insist on sending to others who *might* find the information interesting, indicate "FYI," "spam," or other notation on the subject line to alert them to its low priority.

world, with no long-distance telephone fees (just applicable local phone connection charges). Most e-mail programs allow people to do all of these things:

- List mail received and sent
- Read or delete an item from the list of documents received
- Print or save a document as a file
- Store frequently used names and addresses

- Automatically attach signatures at the end of letters
- Send replies, with portions of the original message in the reply
- Forward mail by simply readdressing it
- Attach other files to mail
- Send a document to any number of people at once

Each person online has a unique e-mail address that works just like a telephone number. There is even an online "Yellow Pages" directory to help users find other people by e-mail. The box Using E-mail Effectively provides some hints on how to use e-mail.

LISTSERVs E-mail can also be used to join mailing lists, bulletin boards, or discussion groups that cover a huge variety of subjects. These bulletin boards are often called LISTSERVs, named after the free software program used to run most of them. Users simply subscribe to a group, and then all mail posted (sent) to that group is automatically forwarded to them by the host computer. The lists are typically produced by a single person or central authority such as a university, foundation, or public interest group. A listing of discussion groups can be obtained by e-mail from <listerv@ubvm.cc.buffalo.edu>.

USENET The Internet is the primary carrier of **USENET,** a system for distributing messages (or articles) around the world, primarily for discussion by members of a given *newsgroup,* identified by a name indicating the subject of interest. For example, news.announce.newusers is a newsgroup dedicated to distributing articles for newcomers to USENET. Articles are posted to a newsgroup and then made available to others who belong to or otherwise access it.

USENET is easy to access. Where users would normally type <http://> on a browser's address line to reach a Web site, they simply type <news:> (no back slashes) and the address of the newsgroup of interest. But what newsgroups are available? A list can be found on USENET, naturally, at <news: news.answers>. Another is available through FTP (file transfer protocol) at <ftp://rtfm.mit.edu/pub/usenet-by-group/news.answers/mail/mailing-lists>. But this raises the question, What is FTP and how is it used?

File Transfer Protocol (FTP) Information other than e-mail and articles can also be transferred to and from the Internet. **FTP** is an Internet application that allows users to send complete files between other computers and their own. **Downloading** describes the activity in which users transfer files *from* an outside computer *to* their own. **Uploading** describes the task whereby users transfer files *from* their own computers *to* others. Some information archives—for example, secret government files and specific corporate databases—are secured and can only be accessed by appropriate personnel. But most information on the Internet is available for public access.

Accessing FTP is similar to accessing USENET and the Web. Again using a browser's address line, users type in <ftp://> and the address of the desired file. You can try out FTP by searching for the list of USENET newsgroups pro-

Londoner Tim Berners-Lee developed hypertext transfer (HTTP) in the early 1990s while working at Cern, the international particle physics laboratory near Geneva, Switzerland. He now serves as the Director of the World Wide Web Consortium at the Massachusetts Institute of Technology.

vided here. Browsers, however, can only download FTP files. To upload files, users must employ FTP software programs such as Fetch and Leech, available on the Internet.

Telnet Remote login (rlogin), or **Telnet,** is a telecommunications software program (available for free on the Internet) that enables people to become users on a remote computer. That is, users can log in to other computers on the Net as if they were connected to them directly. Once the remote computer is accessed, it can be used for its intended purposes. With permission, users can alter content or otherwise manipulate settings on the remote computer.

World Wide Web Another way of accessing information files is via the World Wide Web (usually referred to as "the Web"). The Web is not a physical place, nor a set of files, nor even a network of computers. The heart of the Web lies in the protocols that define its use. The World Wide Web (WWW) uses hypertext transfer protocols (HTTP) to transport files from one place to another. Hypertext transfer was developed in the early 1990s by England's Tim Berners-Lee, who was working at Cern, the international particle physics laboratory near Geneva, Switzerland. Berners-Lee gave HTTP, and therefore the Web, to the world for free.

What makes the World Wide Web unique is the striking appearance of the information when it gets to your computer. In addition to text the Web presents color, images, sounds, and video. This, combined with its ease of use, makes the Web a most popular aspect of the Internet for a large majority of users. The ease with which users can access the Web is a function of a number of components: hosts, URLs, browsers, search engines, and home pages.

Hosts Other than e-mail transactions, most users' Internet activity consists of accessing files on remote computers. To reach these files, users must first gain access to the Internet through "wired-to-the-Net" **hosts,** that is, computers connected to the Internet. These hosts are most commonly called **servers.**

Whether they offer access to other servers or not, all computers *directly* connected to the Internet are identified by numbers that represent location and routing information. To make these identification numbers more convenient and user friendly, they are often given one or more names. For example the number for one of San Jose State University's computers is 130.654.62.2, but its name is athens.sjsu.edu.

When using the World Wide Web, it helps to understand how names are constructed. The rightmost part of the name is its **zone.** There are two kinds of zone information. One is *geographical,* such as *mx* for Mexico, *us* for the United States, and *uk* for the United Kingdom. The other is *organizational* and signifies the nature of the activity conducted by that site's creators. You may be familiar with these abbreviations:

Zone	***Type of Organization***
com	Commercial organization
edu	Educational institution
gov	Government body or department
int	International organization (mostly NATO)
mil	Military site
net	Networking organization
org	Anything that doesn't fit elsewhere, such as professional organizations, especially nonprofit organizations

Once users gain access to a host computer on the Internet, they then have to find the exact location of the file they are looking for on the host. Each file or directory on the host computer connected to the Internet is designated by a **Uniform Resource Locator (URL).** URLs are Internet addresses, which are managed by the nonprofit Internet Corporation for Assigned Names and Numbers (ICANN). They indicate three things:

- The program for accessing a file (or the protocol that will be used)
- The path to that file within the file directory of that computer
- The name of the file or directory in question

Here are some typical URLs:

Use	*URL Identifier*	*Example*
E-mail	mailto:	mailto:sbaran@email.sjsu.edu
FTP	ftp://	ftp://scholar.lib.vt.edu/
Telnet	telnet://	telnet://vtls.vt.edu
Usenet News	news:	news:comp.infosystems.www
WWW (HTTP)	http://	http://www.yahoo.com

As any user of the Web soon discovers, however, sites are recognized by their **domain names** rather than by their URLs. The URL for Internet video software maker Burstware, for example, is <http://www.burst.com>. But the Web site is more widely known by its domain name, burst.com. Domain names are designed to give some indication of the nature of a site's content or owner. For many years domain names were registered and licensed by a single company, Virginia-based Network Solutions, Inc. But in mid-1999 ICANN authorized several other firms, among them AOL and France Telecom, to register domain names in an attempt to spur competition and reduce the cost of registration.

Browsers Software programs loaded onto the user's computer and used to download and view Web files are called **browsers.** Browsers access these separate files (text files, image files, and sound files) and put them all together for viewing. Netscape and Internet Explorer are two of the most popular Web browsers.

Search Engines Finding information on the Web is becoming easier thanks to the growing number of companies creating Web- or Net-search software. These programs are sometimes called **search engines, spiders,** or **Web crawlers.** They all provide on-screen menus that make their navigation as simple as pointing and clicking. Two of the most popular search engines are Alta Vista (<http://www.altavista.com>) and Excite (<http://www.excite.com>).

Home Pages Once users reach the intended Web site, they are greeted by a **home page**—the entryway into the site itself. It contains the information the site's creators want visitors to know, but it also contains **hyperlinks** to other related material in that site as well as to material in other sites on other computers linked to the Net anywhere in the world.

The Internet and Its Users

It is almost impossible to tell exactly how many users there are on the Internet. People who own computers are not necessarily linked to the Internet,

Critical Issues *A Declaration of Independence for Cyberspace*

The fight over the commercialization of the Internet pits those who want to keep this new medium free, open, and chaotic against those who wish to harness its power for economic and commercial gain. One of the best known commentators on this conflict is John Perry Barlow, a founder of the Electronic Frontier Foundation, an advocacy group for free expression on the Internet. At the time of the passage of the Communications Decency Act (see Chapter 12), Barlow published his *Declaration of the Independence of Cyberspace.* In it he forcefully argues for a rejection of "old" notions of class, race, association, and power. Read his call to arms and determine for yourself the merit and accuracy of his manifesto. What kind of world do you think an independent cyberspace might produce?

Declaration of the Independence of Cyberspace

Governments of the Industrial World, you weary giants of flesh and steel, I come from Cyberspace, the new home of Mind. On behalf of the future, I ask you of the past to leave us alone. You are not welcome among us. You have no sovereignty where we gather.

We have no elected government, nor are we likely to have one, so I address you with no greater authority than that with which liberty itself always speaks. I declare the global social space we are building to be naturally independent of the tyrannies you seek to impose on us. You have no moral right to rule us nor do you possess any methods of enforcement we have true reason to fear.

Governments derive their just powers from the consent of the governed. You have neither solicited nor received ours. We did not invite you. You do not know us, nor do you know our world. Cyberspace does not lie within your borders. Do not think that you can build it, as though it were a public construction project. You cannot. It is an act of nature and it grows itself through our collective actions.

You have not engaged in our great and gathering conversation, nor did you create the wealth of our marketplaces. You do not know our culture, our ethics, or the unwritten codes that already provide our society more order than could be obtained by any of your impositions.

You claim there are problems among us that you need to solve. You use this claim as an excuse to invade our precincts. Many of these problems don't exist. Where there are real conflicts, where there are wrongs, we will identify them and address them by our means. We are forming our own Social Contract. This governance will arise according to the conditions of our world, not yours. Our world is different.

Cyberspace consists of transactions, relationships, and thought itself, arrayed like a standing wave in the web of our communications. Ours is a world that is both everywhere and nowhere, but it is not where bodies live.

We are creating a world that all may enter without privilege or prejudice accorded by race, economic power, military force, or station of birth.

We are creating a world where anyone, anywhere may express his or her beliefs, no matter

and people need not own computers to use the Net. They might have access to computers at school, in the library, or at work. Still, recent data indicate that over one half of all U.S. homes have a computer *and* an Internet account (Healey, 2000). Internet research company IDC set the number at 196 million people worldwide, and the *Computer Industry Almanac* counted 349 million (Plotnikoff, 2000).

how singular, without fear of being coerced into silence or conformity.

Your legal concepts of property, expression, identity, movement, and context do not apply to us. They are all based on matter, and there is no matter here.

Our identities have no bodies, so, unlike you, we cannot obtain order by physical coercion. We believe that from ethics, enlightened self-interest, and the commonweal, our governance will emerge. Our identities may be distributed across many of your jurisdictions. The only law that all our constituent cultures would generally recognize is the Golden Rule. We hope we will be able to build our particular solutions on that basis. But we cannot accept the solutions you are attempting to impose.

In the United States, you have today created a law, the Telecommunications Reform Act, which repudiates your own Constitution and insults the dreams of Jefferson, Washington, Mill, Madison, DeToqueville, and Brandeis. These dreams must now be born anew in us.

You are terrified of your own children, since they are natives in a world where you will always be immigrants. Because you fear them, you entrust your bureaucracies with the parental responsibilities you are too cowardly to confront yourselves. In our world, all the sentiments and expressions of humanity, from the debasing to the angelic, are parts of a seamless whole, the global conversation of bits. We cannot separate the air that chokes from the air upon which wings beat.

In China, Germany, France, Russia, Singapore, Italy and the United States, you are trying to ward off the virus of liberty by erecting guard posts at the frontiers of Cyberspace. These may keep out the contagion for a small time, but they will not work in a world that will soon be blanketed in bit-bearing media.

Your increasingly obsolete information industries would perpetuate themselves by proposing laws, in America and elsewhere, that claim to own speech itself throughout the world. These laws would declare ideas to be another industrial product, no more noble than pig iron. In our world, whatever the human mind may create can be reproduced and distributed infinitely at no cost. The global conveyance of thought no longer requires your factories to accomplish.

These increasingly hostile and colonial measures place us in the same position as those previous lovers of freedom and self-determination who had to reject the authorities of distant, uninformed powers. We must declare our virtual selves immune to your sovereignty, even as we continue to consent to your rule over our bodies. We will spread ourselves across the Planet so that no one can arrest our thoughts.

We will create a civilization of the Mind in Cyberspace. May it be more humane and fair than the world your governments have made before.

Source: <http://www.eff.org/~barlow/Declaration-Final.html>.

Commercialization of the Internet

Whichever statistic you believe, a lot of people are using the Internet. That ever-growing number of users—and their apparent willingness to go online to find commercial information and to buy products—is at the heart of a furious debate over the future of the Internet.

The Internet was developed, nurtured, and popularized by **hackers,** people interested in technology, information, and communication through computers. In *Road Warriors: Dreams and Nightmares Along the Information Highway* (1995), Daniel Burstein and David Kline argue that the Internet and business are poor partners because the values each represents clash. Where the Internet is concerned, the hackers, the geeks, and the innovators have ironically become the "traditionalists." They fear that business will turn the Internet into an electronic shopping mall and that this commercialization of what was once the most free of communication technologies will lead to growing privatization and control. Commerce, they claim, cannot function amid chaos and disorder. It is this very anarchy, they argue, that has made the Internet so exciting—nobody needs permission to get on or off the Internet; nobody can tell a user what to say.

Douglas Rushkoff (2000), a communication technology writer, expressed a similar fear:

> Those of us who were enjoying the Internet back in the 1980s and early '90s remember a time when this technology was about communication. Businesses weren't even allowed on the Net. . . . It was a level playing field, where the size of your computer and the contents of your wallet meant nothing. . . . The Internet changed the way people thought about media and power. . . . The introduction of business to the Internet changed all that. Corporate behemoths ignored the indigenous Net population as they colonized our space. (p. A23)

Another perspective on the traditionalists' position can be found in the box titled A Declaration of Independence for Cyberspace.

Online traditionalists point to the history of television. In that medium's early stages there were predictions and promises of a new medium of expression, education, and entertainment. Politics and political discourse, for example, would be transformed as people became aware of and involved in public affairs. The reality, the online traditionalists contend, is that television, with its commercial support, profit orientation, and lowest common denominator mentality, has cheapened politics and political discourse. Given this view, it is no surprise that a 1995 survey detailed in *Advertising Age* reported that two thirds of the U.S. adult respondents said "No" when asked whether advertising should be allowed on the Internet (Fawcett, 1996).

Defenders of e-commerce argue, however, that the Internet will always be accessible and open. There is no spectrum scarcity to limit access as there is in broadcasting (Chapter 2), so the television analogy is inappropriate. In addition, because very small amounts of money are required for individuals to access and use the Internet, especially in contrast to the budget needed to start and run a broadcast or cable operation, or a newspaper or magazine, the commercial orientation of those media will never fully overtake the Net. It is precisely this commercial potential of the Internet, they contend, that will keep the cost of access low and its value high. Regardless of the position we take, there is little doubt that the online world is increasingly characterized by commercialization.

E-commerce and E-tailing

E-commerce (the sale of goods or services over the Net) has been part of the Internet almost from the start of its popular use. E-mail has long been seen as a way to deliver commercial services, for example, booking travel and hotel reservations, online banking, and ordering from online catalogues. But three developments encouraged fuller commercial exploitation of the Net.

The first was the increase in the number of people who were online. Naturally, this increase made Internet advertising and sales more profitable for businesses. Of equal importance to would-be **e-tailers** (online retailers) is that, at least in the United States, those who use the Internet "control more than 80% of America's liquid capital" (Gaitskell, 1998, p. 37). In other words, the people who are online have money. A second development encouraging e-commerce was the introduction and rapid acceptance of the World Wide Web. Exciting graphics, sound, and even motion could be used to promote and display products—a huge advance over the useful but static information available for e-mail-based commerce. The Web's ease of use and its instant links to related products and ordering instructions and information allowed consumers to immediately respond to online sales pitches. The third development was the availability in the mid-1990s of trustworthy **encryption,** or encoding and decoding, technologies that made the online use of credit and bank card numbers, addresses, social security numbers, and other sensitive information safer for both seller and buyer (see Chapter 12 for more on protecting online personal data).

Online commerce has not taken off as rapidly as critics feared or proponents hoped. E-commerce still represents only 1.2% of the total U.S. retail sales of nearly $3 trillion (Tessler, 1999b). Online advertising (Chapter 9) is hampered by relatively unsophisticated animation and sound (compared to television). Many Internet users see advertising as cluttering the screen and delaying their access to the material they really want to see. Newspaper and magazine readers can quickly turn the page when confronted with unwanted advertising or, using a table of contents, they can go directly to the material they want to read. Unwanted online advertising surrounds much material, and because its images and text are downloaded along with the desired content, it slows access. Advances in computer technology that improve the quality of sound and images and speed their delivery will improve this situation for online advertisers and their customers. The box titled Modems looks at one of those technologies.

E-tailing has been slowed by a number of other factors. The first is the nature of online customers. Numerous demographic analyses of the Internet indicate that "serious" users tend to be male and highly educated. These folks tend not to be shoppers, online or otherwise. Because of their computer sophistication, they are also the users most likely to be disenchanted with the relative lack of sophistication of online commerce. A second problem is lingering fear about security. In 1996 several companies, including Visa and Mastercard, agreed to a Net-wide encryption standard called Secure Electronic

Tech Talk *Modems*

Not all modems are created equal, as Internet users wanting to access Web sites offering sophisticated audio and video well know. **Dial-up modems** connect users' computers to an ISP via traditional telephone lines. **Cable access modems** connect users' computers to the Internet over cable television lines via the cable company's designated ISP, for example AT&T's @Home. **Digital subscriber line** (**DSL**) **modems** connect users' computers to the Internet via dedicated telephone lines passing through the ISP offering the DSL service. Each has advantages and disadvantages, but the latter two forms offer the most promise for high-speed downloading of audio and video.

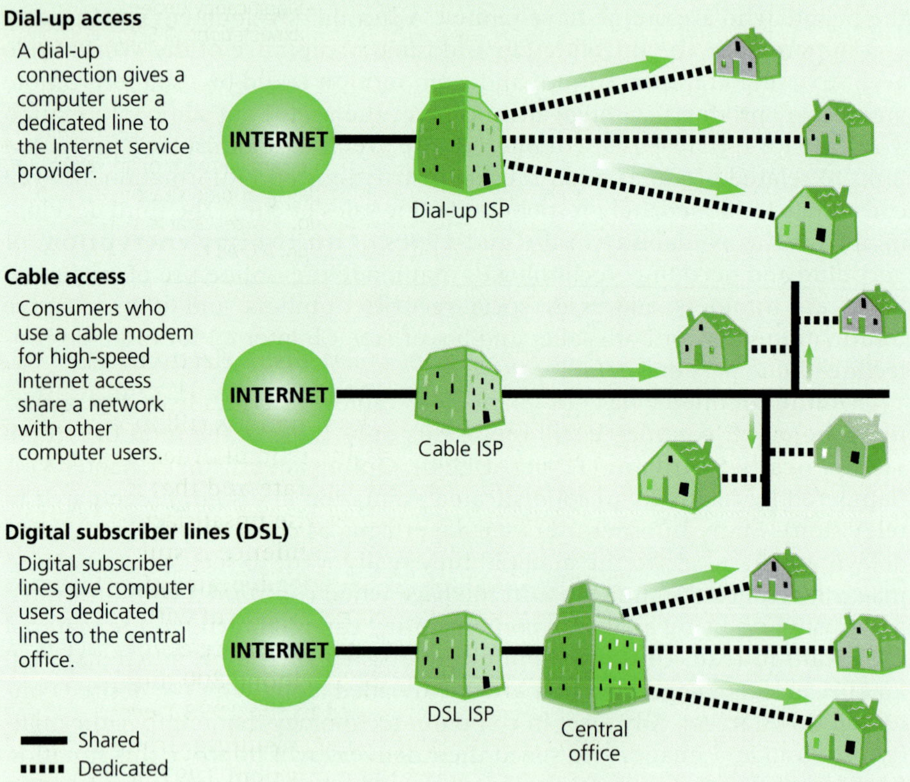

Transaction Protocol. Online sellers hope this industry-assured security will allay consumers' concerns, but the fear lingers.

A third difficulty is buyers' need to see, feel, and even smell many of the products they buy. Music, software, and airline tickets pose little problem for online buyers. But other products, such as clothes and cars, often require a real- rather than a virtual-world test. Online sellers believe many price-sensitive

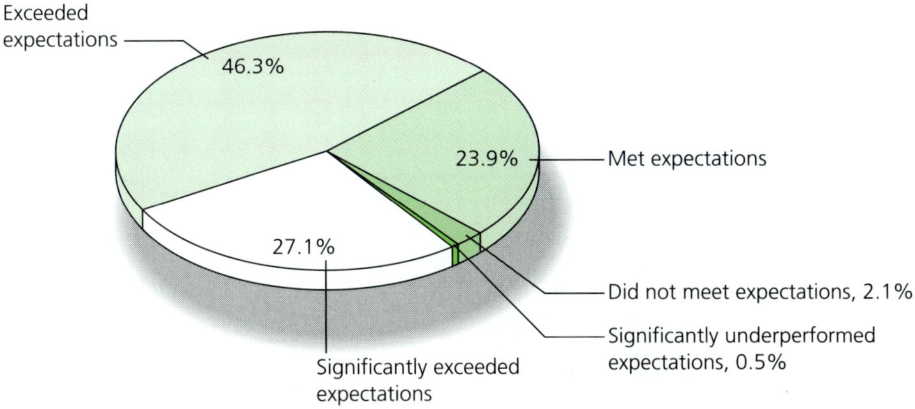

Figure 7.1 Growing Customer Satisfaction
Although slow to develop, the online retail business showed strong growth in 1998 and 1999. Part of the reason is that consumers are becoming more familiar with online buying and, as a result, more satisfied with the experience. The holiday shopping season of 1999 demonstrated just how comfortable buyers were with e-tailing. *Source:* Lazarus, D. (1999, December 28). Christmas cheer for e-tailers. *San Francisco Chronicle,* pp. A1, A11.

shoppers will accept this disadvantage as the benefits of electronic shopping become better known.

Finally, there is the issue of trust. Online commerce is still in its relative infancy. People have little fear that their catalogue mail orders will not be shipped. They believe the accounting will be accurate and that if the goods are not delivered, or if they are damaged or otherwise unsatisfactory, the catalogue dealer will make it right. This level of confidence is still lacking for many people in their online transactions. Trust, say advocates of online selling, will come with time, and they are hastening its arrival with techniques such as live, one-on-one chat sessions with customer service agents and even increased utilization of that old technology, the telephone 800 number.

Many observers believe that that time arrived in the 1998 December holiday buying season. E-commerce advisor Boston Consulting Group reported that holiday shopping grew by 230% over the equivalent 1997 period. Individual companies posted remarkable growth. For example, SkyMal's December 1998 Internet sales ($2.1 million) rose 600% over those from December 1997 ($300,000), and Sharper Image saw its December 1998 online sales exceed those of both its in-store and mail order businesses (Gaw, 1998). In all, 15 million individuals made online purchases from the Web in 1998 (Kong, 1999). The 1999 holiday buying season reinforced this trend (Figure 7.1), with online sales from Thanksgiving to New Years exceeding $4 billion (as compared to $1.5 billion in 1998). Total online retail sales for 1999 were $40 billion, compared to $14.9 billion in 1998 (Tessler, 1999a, 1999b). So important is the promise of e-commerce to the nation's economy that in February

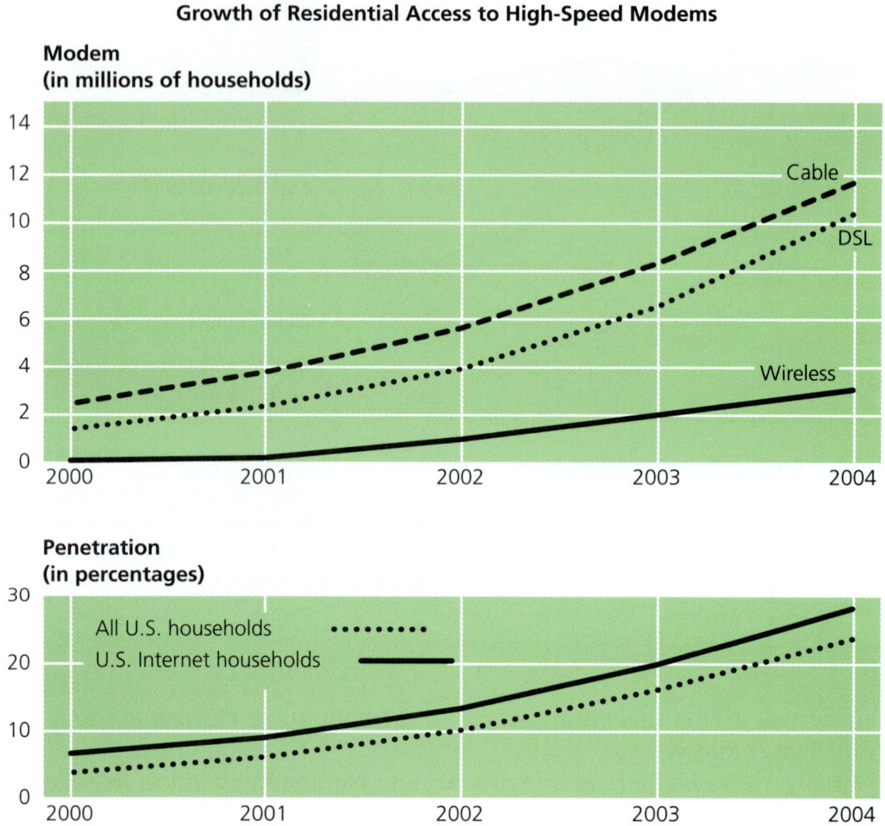

Figure 7.2 High-Speed Modems
Fuller diffusion, acceptance, and profitability of Internet and Web video rely on more people having access to high-speed cable, DSL, and wireless modems. Predictions vary as to how quickly these technologies will find their way into U.S. homes, but some forecasts such as this one offered by *Broadcasting & Cable* place them in fewer than a fourth of all U.S. households by 2004. *Source:* The Strategis Group, 2000.

1999 the U.S. Department of Commerce announced that it would begin tracking annual consumer sales on the Internet in its annual nationwide retail survey (Figure 7.2).

The public's acceptance of e-commerce is founded on a fact of contemporary life—no one has any time; we're all rushed—and a shift in how e-tailers help online shoppers deal with this unfortunate reality. In the beginning e-tailers were content simply to cut out the agent or person in the middle, thereby reducing costs for buyers. The Internet's retail leader Amazon.com, for example, built its success on this model. But now, with increases in the number of sophisticated users, the presence of more experienced e-tailers, and the development of more sophisticated hardware and software, e-commerce is "about bundling services that save time or act as 'intelligent agents' for con-

The Internet's success as a medium of commerce is best seen in the proliferation of online classified ad sites, such as the heavily trafficked job search page *monster.com.* Online classified advertising is detailed in Chapter 9.

sumers" (Ellis, 1999, p. 7B) while at the same time keeping prices low (and value, therefore, high). Car buyers, for example, can visit hundreds of virtual showrooms, comparison shop, digitally preview their potential purchase in numerous colors and with a limitless array of options, and order not only the new car but its financing and insurance—all from a comfortable chair after the kids go to bed.

Online software known as **bots** (short for robots) make this process as simple as clicking a mouse. The bot will search the Web for the requested item, delivering to the user not only the Web location where it can be purchased but information on price, availability, and, in some instances, quality. There are more than 200 shopping bots online, and among the best known are mySimon

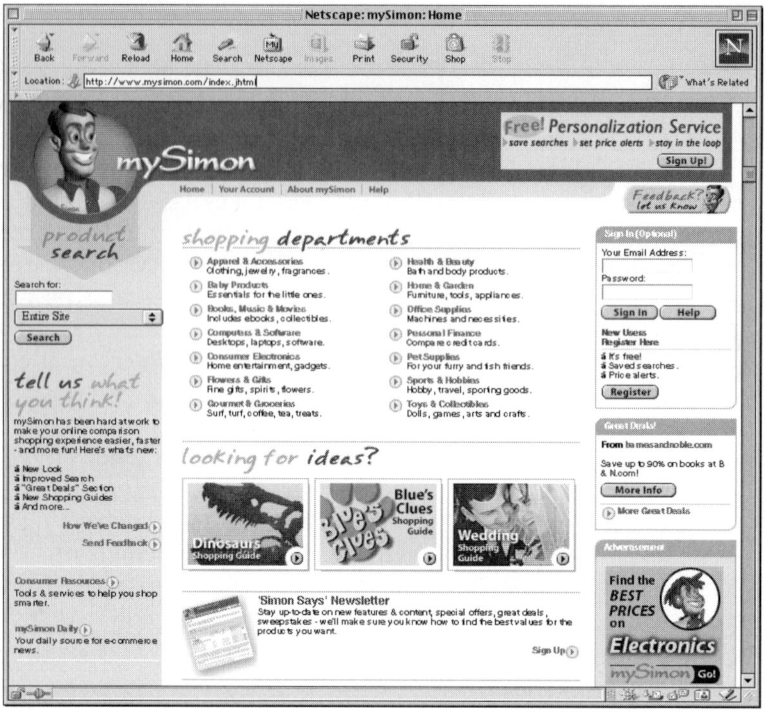

mySimon is the Internet's best known and most used shop bot.

(<http://www.mysimon.com>) and Mata Hari (<http://www.thewebtools.com>). But bots have come under recent criticism. Many do not objectively deliver and rank retail sites, as most bots give higher visibility to retail sites that pay them a fee. In addition, when users enroll as "members" of a specific bot, they are often deluged with **spam** (unwanted Internet messages). You can find information on available bots, with some indication on their quality of service, at <http://www.bots.internet.com>.

Portals

We've already seen that search engines such as Yahoo, Lycos, and Netscape allow users to search the World Wide Web by providing on-screen menus of URLs related to the users' designated topics of interest. But two factors are fueling a reconsideration of how these gateways to the Web can be used to increase their profitability. The first is the rapid growth of e-commerce. The second is the slow growth of online advertising. The majority of online ad space offered to potential advertisers goes unsold (see Chapter 9).

One recent trend in the Internet industry is to rethink search engines as portals: one-stop brand-name entryways to the Web that offer users news, chat, search functions, original content, links to specific sites, advertising, and e-commerce. Where a traditional search engine gathers an array of sites from the Web, leaving the individual user to choose from among the identified sites, a portal limits users' choice to a select number of sites. The portal's operator can then sell the presumably millions of users who want this selectivity to advertisers, much as television networks sell viewers to advertisers. In addition the portal operator can charge e-tailers for screen space on its home page as well as charging for a percentage of all sales that result from a user's click on the e-tailer's icon that immediately takes the user to that e-tailer's site.

Advocates of the portal movement tout these branded starting points to the Net as a service to users. Infoseek's head Harry Motro, for example, explains, "The Net is an anonymous medium. People aren't sure who and what they are dealing with. Trust and confidence is a huge element of what's missing. What we are building (by joining with Disney to create the Go Network portal) is brand" (as quoted in Janah, 1999a, p. 1E).

As we saw briefly in Chapter 1, those wary of the ascendance of portals see this branding as precisely what is wrong with the trend. Yahoo! co-founder Jerry Yang argues, "Independence and the ability to aggregate without necessarily being partial to one brand . . . has been a big philosophy of ours. Our notion of a topic, say news, is an aggregation of the best out there, period, whereas an effort like GO is maybe more particular to just their version of news" (as quoted in Kong, 1998, p. 1C). Stated more directly by HTTP originator Tim Berners-Lee, the Web "is not supposed to be a glorified television channel" (as quoted in Thomas, 1999, p. 19).

Portals' critics point to the example of the three traditional television networks—for decades television's major "brands." When cable offered viewers a greater number of options, they abandoned these "one-stop program providers" for the variety of cable and other technologies. "Why take the Internet backward?" ask portals' detractors. It is precisely the Net's lack of central structure that gives it its power and vitality. They see as ominous the fact that it is precisely those old-line broadcast networks that are leading the drive toward portals, in essence smothering the Internet's user-centered nature with broadcasting's audience-centered operation: ABC (Disney) owns GO, NBC owns Snap!, and CBS is the majority stockholder in iWon.com, a portal that offers daily cash prizes to lure Web surfers.

What portals may represent, however, is the evolution of the Internet into a medium not unlike cable television, where some viewers rely primarily on the brand names like the big networks and cable channels with their familiar and trusted fare, and others prefer to venture out to cable's more challenging options. In other words, portals may well attract users who might not otherwise travel the Net's myriad mysterious routes, but more adventurous or sophisticated users will still have full freedom to surf where they wish.

Nonetheless, there is a movement among online traditionalists to use technology to ensure that users who wish to do so can remain free of portals' power to funnel people to fewer and fewer sites. Lycos and Netscape (<http: www.dmoz.org>) both make **open directory** available to their subscribers. Twenty thousand volunteer Web surfers (called editors) in 229 countries search for and add up to 1,400 sites a week to open directory's "classified

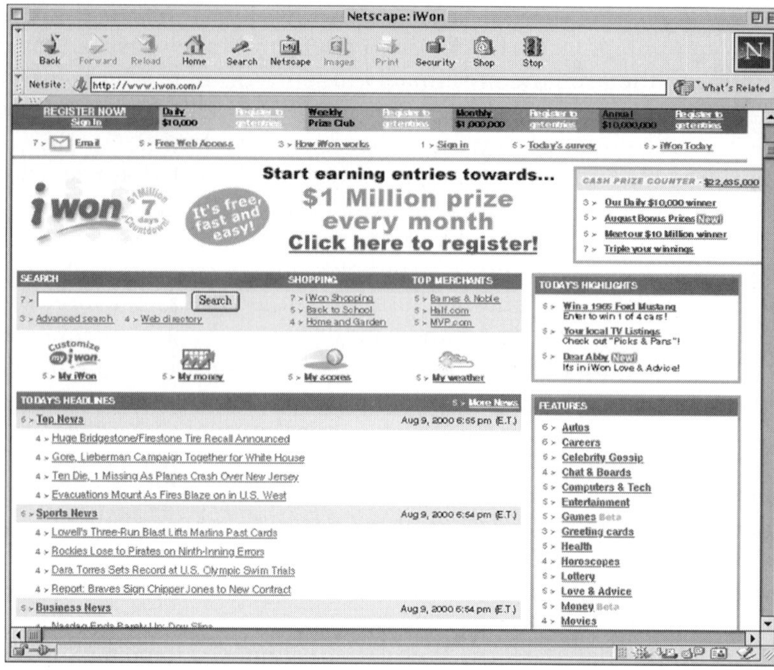

The three old-line television networks each have a portal of their own. CBS's iWon offers cash and other prizes to draw users.

guide" to the World Wide Web. The resulting hundreds of thousands of categorized sites are available and open to all who choose open directory rather than portal provided links (O'Brien, 2000).

Concentration

The commercialization of the Internet is also an important force behind the growing concentration of the industry. The acquisition of search engines by larger media corporations and their transformation into portals—for example, ISP AOL's 1998 $9.8 billion purchase of search engine Netscape—is one manifestation of this trend. A second way that the Net's commercial potential is encouraging mergers and buyouts is **horizontal integration.** That is, companies buy other similar companies in an attempt to control a larger portion of a specific market. In the case of the Internet, online companies are buying other online companies specifically for their retail money-making potential. The most obvious example is the Internet's retail colossus Amazon.com. Begun as an Internet book seller, it now counts among its $600 million in annual sales rare books, music, video, groceries, toiletries, pet supplies, and prescription drug retailing. It also owns a person-to-person auction service estimated to be an $18.7 billion business by 2002 (Buel, 1999). Amazon.com's initial foray into these businesses has typically been through the acquisition of existing sites. Other examples abound. In one 18-month period covering 1998 and

Careers *Working in the Internet and the World Wide Web*

People don't have careers "in the Internet" or "in the Web" like they have careers "in broadcasting" or "in cable," and the reason for this is obvious. Whereas broadcast and cable operations are discrete places with parking lots, buildings, and personnel offices, the Internet and the Web are entities that exist everywhere and nowhere at the same time. They are networks of interconnected computers with no central office or administration. But there is almost no career in modern telecommunications that does not involve going online and interacting with these technologies and others who use them. This makes learning to use the Web and the Net as crucial, possibly more crucial, to your success as learning to use any other telecommunications production and distribution technology. If convergence means anything, it means that the technologies that have traditionally formed the core of telecommunications are increasingly digitally and Internet-based.

Because you can read this book, you already have a head start—English is the language of the Internet. Yes, there are sites in other languages, but the vast majority of Internet activity takes place in English. This is not an insignificant fact as telecommunications becomes increasingly globalized.

The Internet and the Web afford people wanting telecommunications careers the benefit of independence; that is, careers can be built outside traditional organizational structures. Aspiring journalists needn't work their way up through a newspaper or network news operation, and aspiring television and film producers needn't subject their vision to the demands of network and studio executives. Matt Drudge, controversial reporter and former ABC news analyst—never closer to a telecommunications job than working in the CBS gift shop—built an impressive career from his one-person Web site, the *Drudge Report.* Trey Parker and Matt Stone, producers of *South Park* and *Baseketball,* began their careers by building an irreverent Web site for their short animated film (*A South Park Christmas*), eventually turning it into a hit on Comedy Central. Jeff Bezos, founder of Amazon.com, amassed a $9 billion personal fortune because he came to "the deceptively simple conclusion: that the most logical thing to sell over the Internet was books" (Bayers, 1999, p. 173).

Not everyone will have this kind of success. In fact, many people don't want such public careers. But this does not take away from the fact that for the first time since the very beginnings of radio individuals with energy and enthusiasm, technical skill, and a bit of entrepreneurship can make their way in an emerging telecommunications industry and can shape both its present operation and its future.

1999, Excite bought several Internet companies, including rival search engines Magellan Internet Guide and WebCrawler, as well as GoMedia, Kirkwood Interactive, Netbot, Throw, MatchLogic, and Classifieds2000.

Concentration is unwelcomed by Internet traditionalists. Indicative is this commentary from Andrew Shapiro (1998) of the Brennan Center for Justice at the New York University law school.

> Are we already resigned to accepting that much of this new (medium), still less than five years old, will be as concentrated and conglomerated as the rest of the media and communications industry? . . . What's especially depressing about that prospect is that the real promise of the Internet seems to be its ability to allow smaller companies, nonprofits, communities, and even individuals to seize power from hulking behemoths. (p. 7)

A third force making concentration in the Internet industry inevitable, according to its proponents, is convergence. We've already seen that concentration in the telecom and cable industries (Chapter 6) is being driven by companies' desires to converge and deliver digital data—in all its forms—over the Internet to people and businesses. But convergence is also behind much of the horizontal concentration in the Internet industry itself. For example, in early 1999 search engine Yahoo! bought Broadcast.com (a company with losses of $16.4 million the previous year) because it wanted to expand into the online audio and video business. Why would the operators of the Internet's most-visited Web site buy a money-losing site for $6.1 billion? Convergence is the answer.

As we saw in several of the previous chapters, much of the Internet's audio and video potential is hampered by technology—slow modems and lack of bandwidth. Once these hurdles are surmounted with greater diffusion of high-speed **modems** (devices that translate digital computer information into an analog form so it can be transmitted through telephone lines) and fiber optic cable, Yahoo! and many others in the digital world believe that the Internet will expand even more than it already has. As one industry analyst explained, "Yahoo recognizes that the Net today is a prototype of what it will become in the future. (Once modem and bandwidth limits are overcome) the Net will become the dominant form of electronic media, and full multimedia will become an integral part of the Net" (Janah, 1999b, p. 10A). For information about careers in this growing industry, read the box titled Working in the Internet and the World Wide Web.

Emerging Internet Technologies: The Heart of Convergence

Most of the examples of convergence we have discussed in this text have their roots in digitalization and the Internet. For example, Web radio and cable-delivered telephone and data services are only possible because of digitalization and the Net. In this respect, then, we have already studied several emerging and converging digital and Internet technologies. Here, we focus on several others that promise to change the face not only of the Internet but of all telecommunications.

Wireless Internet

Another emerging Internet technology building on the promise of efficiency is **wireless Internet**—delivery of high-speed Internet data to users through the air. Foreseeing the convergence of voice with audio, text, and video data downloaded to users' cellular phones and laptops as well as desktop computers, advocates of wireless Internet use unoccupied radio and television channels, microwave relays, satellites, cellular phone channels, and even a fleet of very high altitude aircraft to deliver the Internet (Platt, 1999). The advantage of wireless access goes beyond the existence of a fully mobile Internet; specifically, that advantage is speed. Bandwidth limitations that currently impede access to sophisticated Internet audio and video material for many users are largely overcome with wireless, as data travel over the air rather than on crowded wires. To put the speed of wireless Internet into perspective, the fastest form of Internet connection in general use is called **T1,** and it operates at 1.54 megabits per second. But wireless Internet delivers data such as Web pages at 7.5 megabits per second—250 times faster than the fastest telephone modem (Marshall, 1997).

Embedded Systems

An **embedded system** is a computer system that adds intelligence to other technologies. It is technologically simple to embed Internet protocols into almost any technology. Internet protocols are already being embedded into what can be called "information appliances," devices such as wristwatches, handheld information managers like Palm Pilot, and music synthesizers. This enables these devices to be attached to the Internet and, among other things, receive information or instructions from distant users. But the Internet is now being converged with an even wider array of technologies: home and office thermostats (enabling users to warm and cool spaces before they arrive), soda machines that send buyer and inventory information to distributors, gas pumps that offer frequent flier miles (automatically credited to your account) with a fill-up, and microwave ovens that download recipes and cooking instructions from the Web. As University of Pennsylvania telecommunications professor Dave Farber argues, Internet protocols are the *lingua franca* of emerging technologies because the Internet "is the standard. . . . You can connect anything and you know it'll work" (as quoted in Gillmor, 1998, p. 1D).

Push Technology

Push technology—software that permits the automatic delivery or downloading of Internet data to even sleeping computers—was supposed to have revolutionized the Internet in the mid-1990s. Once users "taught" their computers what they were interested in, the idea was that those computers would continuously scan the Internet and automatically deliver data to users' machines. The reason push technology never fulfilled its promise should be obvious. Modern computers are extremely smart and efficient, and the amount of downloaded data quickly overwhelmed computers and even networks, despite their size and capacity.

But it is computers' intelligence and efficiency that are resurrecting push. The idea of push has always resonated well with Internet users, especially businesses. But the efficiency inherent in having computers search for pertinent information (rather than having paid employees constantly requesting and browsing for it) simply had to be better harnessed. Two of the more successful new narrowly focused push technologies now available are Marimba (<http://www.marimba.com>) and BackWeb (<http://www.backweb.com>).

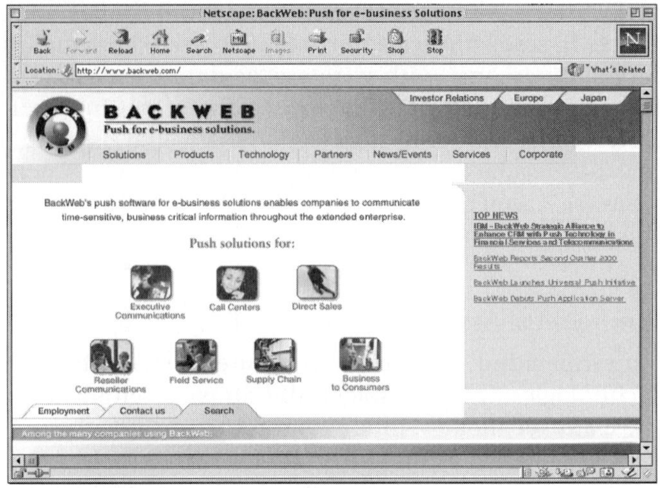

BackWeb is one of the more successful new, narrowly-focused push technologies now available.

Although they are examples of push, their creators prefer to see them as technology that "automates delivery of the important stuff so companies can make sure that the right people see it at the right time" (Gillmor, 1999, p. 10C).

MP3

MP3 (for MPEG-1, Audio Layer 3) is compression software that shrinks audio files to less than a tenth of their original size. Originally developed in 1987 in Germany by University of Erlangen computer scientist Dieter Seitzer in conjunction with the Fraunhofer Institut Integrierte Schaltungen, it began to take off in the early 1990s as more users started hooking up to the Net with increasingly faster modems. This **open source** (freely downloaded) **software** permits users to download recorded music, and it is available at sites such as <http://www.MP3.com> and <http://www.MP3.box.sk>. So popular has MP3 become, that in July 1999 it surpassed "sex" as the most frequently searched-for term on the Web, and in the first 6 months of that year nearly 3 billion MP3 files were downloaded from the Net, or 17 million *every day* (Peraino, 1999). In addition to "brandless" MP3, users can access commercial versions of the software such as RealJukebox and MusicMatch, which offer additional features such as selectable levels of audio quality, the capacity to turn music from home CDs into computer files, and compatibility with other audio streaming technology such as Liquid Audio, a2b, MS Audio, and G2. When RealNetworks made JukeBox available in May 1999, for example, 350,000 users downloaded the software in 2½ days, the fastest download of free software in Internet history (Quittner, 1999). In either its open source or commercial format, MP3 is controversial. The furor surrounding this convergence of digital

Diamond Multimedia's portable MP3 player, the Rio 500.

and recording technologies is captured by the *New York Times* reporter Jon Pareles (1998):

> For utopians, MP3 is a way to liberate music from the clutches of gatekeepers and profiteers, and perhaps to return music to its intangible essence. But the recording business sees MP3 as a Pandora's box of copyright destruction, unleashing anarchy and piracy while robbing musicians of royalties and record labels of capital.
>
> In effect, both sides are correct. There's bound to be serious legal scuffling between people who own copyrights and people who believe in the hacker credo that information wants to be free.
>
> Yet recorded music is considerably more than a corporate revenue source, and the implications of digital distribution go far beyond the particulars of software and gadgets and royalty collection. For listeners, music has never been about its physical form, but about what's in the grooves or magnetic particles or digital bits; it's the information, not the plastic. Digital distribution can turn that sentiment into a reality. And that shift could alter the way music is made, released, sold, stored, and valued. (p. 1, Section 2)

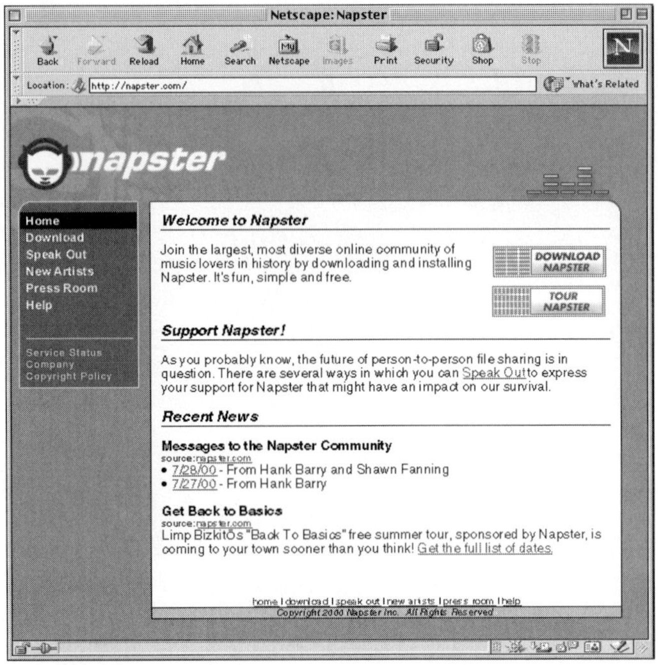

Napster allows users to easily search for and retrieve individual songs and artists from one another's hard drives. The controversy surrounding this popular technology is detailed in Chapter 12.

The crux of the problem for recording companies is that they sell music "in its physical form," whereas MP3 enables music distribution in a nonphysical form.

First conceived of as a means of enabling independent bands and musicians to post their music online where it might attract a following, MP3 became a headache for the recording industry when music from the name artists they controlled began appearing on MP3 sites, making **piracy**—the illegal recording and sale of copyrighted material—of high-quality recordings a relatively simple task. Not only could users listen to their downloaded music from their hard drives, but they could make their own CDs from MP3 files and play those discs wherever and whenever they wished. Matters were made even worse for the recording companies when manufacturers such as Diamond Multimedia introduced portable MP3 players, freeing downloaded music even from users' computers. With the spread of software such as Napster, which allows users to easily search for and retrieve individual songs and artists from one another's hard drives, control over music distribution was further eroded.

Rather than accept MP3, the Recording Industry Association of America (RIAA), representing all of the major U.S. labels, responded with a technological solution. At the end of 1998 the RIAA announced the formation of the Secure Music Digital Initiative (SDMI)—code named the Madison Project—in conjunction with IBM and more than 100 other music, consumer electronic, and information technology companies. In July 1999 the RIAA announced final approval of SDMI software containing copyright screening technology.

But by this time it was too late. MP3 had become the technology of choice among digital audio fans. In January 2000 the RIAA went to court, where it was victorious. Citing MP3.com's database of more than 80,000 albums, a federal judge in New York ruled that MP3 was not used simply to store already purchased music. Rather its primary application was copying and distributing this copyrighted material. Soon after, the RIAA and MP3.com, the Net's primary distributor of MP3 files, came to a settlement in which the online company paid close to $100 million to the record companies in exchange for the right to legally distribute their music.

Apart from—and well before—the recording labels reached their settlement with MP3.com, numerous well-known musical artists embraced the compression technology as a way to bypass the recording companies, reward fans, and spur sales. For example, the Greatful Dead encouraged downloading and swapping its live concert recordings (as long as the music was not sold for profit), Public Enemy released albums online through AtomicPop.com in advance of their appearance in stores, and They Might Be Giants has released MP3-only albums through GoodNoise.com. Openly critical of the record companies and embodying the spirit of those artists eager for digital distribution of their work is Hole's Courtney Love, who scolded the RIAA, "If you want to talk about stealing, let's talk about record-company contracts. Stop the pious crap and figure out how we can use the Net to make the music we want" (as quoted in Surowiecki, 2000, p. 35).

Internet Video

We've already seen that Yahoo!'s purchase of Broadcast.com was predicated on its conviction that video will be a major part of the Internet's future, even though slow modem speeds and insufficient bandwidth are currently limiting users' access to online video. But there are numerous companies vying to enter the online video trade. Their efforts fall into three nondistinct categories, delivering the Internet via home television sets, delivering television channels to homes via the Internet, and delivering video to homes via the Internet.

The Internet on Television Although America Online, @Home, and Yahoo! all announced plans to begin offering the Internet over home television sets, the most aggressive advocate for accessing the Net on home screens is Microsoft's WebTV. WebTV is betting that the convergence of television and the Internet will be sufficiently attractive to viewers/users that they will pay $300 for a set-top receiving box and $20 a month for the service itself.

WebTV turns a television set into a computer screen, permitting access to the Internet as is done on users' computers. As such, it provides access not only to the Web but to several sites of its own, including one for children and another providing local weather, news, and event information. But WebTV also offers several features that enable the Internet to enhance television viewing itself. For example, it provides a program listing feature that takes viewers from a renewable (rather than a scrolling) screen schedule directly to a show with a simple click. Using the same feature, users can program a VCR with a click of the mouse/remote or instruct the technology to alert them that a favorite program is about to start, even automatically changing the channel to that show.

An interesting convergence of the Net and television resides in what WebTV calls its TV Crossover Links. When watching a particular program, a small icon will appear on the screen and, with a click of the remote, viewers are immediately linked to an Internet site providing background or other information relating to that show. The original program remains on the screen in a picture-in-picture format.

At the end of 1999 there were just under 1 million subscribers to WebTV, and another 7,300 subscribed to competitor WorldGate (Haley, 1999). Some observers feel that this service will not become much more popular than that because, even though innovators may converge technologies, they cannot force users to converge Internet use and television viewing habits. The computer and the Internet, they argue, are task-oriented technologies that users are most comfortable using "up close and personal." Television, in contrast, is primarily a broad-based entertainment and information medium that viewers are most comfortable watching from several feet away.

Receiving Television Via the Internet Broadcast.com has been the pioneer in delivering over-the-air television channels to viewers' computers via the Internet. Other companies, most notably AtomFilms and RealNetwork, have also begun this service. Their primary audience is people who want to watch television, for example, for local traffic news or a business news report at work. Viewers will not typically sit still in front of their computers long enough to watch a full-length television program, and both of these companies and scores of others have moved into the video (rather than television program) Internet business.

Video on the Internet Video compression software similar to audio's MP3 makes it possible to download high-quality moving images from the Internet. For example, in 1998 well-funded Digital Entertainment Network (DEN) (<http://www.den.net>) began producing original 6-minute-long programs followed by 15 minutes of related interactive content. It sold advertising on these shows, calculating that it could earn a profit with as few as 50,000 unique, rather than all-at-once, viewers. Despite critics' convictions that personal computer users would not sit still for even short "television" programs, several other producers and distributors of original online content entered the market using the DEN model, notably Pseudo Programs, Broadcast.com, and CNN Interactive. The critics were correct—at least about DEN. The pioneer Internet video site failed in early 2000, with $60 million in losses.

Video over the Net's most successful application to date has been its power to enable users to search for and download video material from a variety of sources. Real Broadcast Network, for example, links users to video clip providers such as CNN.com and ABCNews.com. As a sign of the technology's acceptance, Real Broadcast Network points to December 16, 1998, the day U.S. warplanes bombed Iraq as the House of Representatives voted on the impeachment of President Clinton. The broadcast and cable networks were limited to presenting split-screen coverage of these momentous events, but the online network delivered a wide variety of video clips on both events from a variety of sources to more than 1 million users. This success, as well as that of its summer 1998 Netcast of a live birth (1.4 million viewers/users) and Broadcast.com's 1999 presentation of a 15-minute Victoria's Secret fashion show (1.5 million viewers/users), has convinced some analysts that the true future of converged Internet/television technology is in the presentation of special events. Real Broadcast Network's general manager Mark Hall says, "Event programming is the Number 1 thing. Doing it on the Internet provides a unique benefit that you just wouldn't get on TV" (as quoted in Tedesco, 1999, p. 22). That unique benefit includes not only exclusivity but interactivity.

Chapter Review

The first electronic digital computer was developed during World War II to break secret code. The first full-service electronic computer, introduced in 1946, was ENIAC. Computers quickly made their way into U.S. business and commercial life, largely through the efforts of IBM.

The Internet is the product of the military's desire to maintain U.S. defenses after a nuclear attack. Decentralization was the key to allowing communication to continue no matter where an attack occurred. The solution was a network of computer networks—the Internet. Once personal or microcomputers entered the picture, the Internet became accessible to millions of noninstitutional users.

It is difficult to accurately measure the number of Internet users. Estimates range as high as 200 million. The average Net user is relatively well off and highly educated.

The growing number of people online has inevitably led to efforts to advertise on and sell by Internet. This commercialization of the online world is greeted with disdain by traditional Net users, who fear that their medium will lose its freedom and energy to a corporate takeover. E-commerce and e-tailing, however, are not progressing as fast as proponents had hoped, due to a number of factors including relatively unsophisticated moving images and sound, the intrusiveness of Internet ads, consumer fears about the security of credit card and other sensitive information, and lack of buyer trust in this new way of doing business. Still, there is mounting evidence that e-tailing and e-commerce are gaining acceptance among users. The Net's growth in these areas is fueling development of portals and industry concentration, both upsetting to Internet traditionalists.

The Internet's future will be shaped by a number of factors, primarily convergence and emerging technologies. Push technology, which automatically directs information to users, is being reborn after a rocky start. Wireless

Internet not only frees users from their desktops but enables them to overcome bandwidth problems that often limit access to sophisticated audio and video data. Internet protocols are being embedded in an ever-increasing number of home, office, and communication appliances. The audio compression technology MP3 delights users while it threatens record company profits. The technology necessary for accessing the Internet over home televisions and watching video on home computers is already available. As soon as more users have access to faster modems and greater bandwidth, these applications will help reinvent the Internet.

Questions for Review

1. What is the importance of each of these people to development of the computer: Charles Babbage, John Atanasoff, John Mauchly, and John Presper Eckert?
2. What were the contributions of Paul Baran, Bill Gates, Steve Jobs, and Steven Wozniak to the development and popularization of the Internet?
3. What is the difference between microcomputers and mainframe computers?
4. What are USENET and FTP?
5. What factors have led to the popularity of the World Wide Web?
6. What are the differing positions on the commercialization of the Internet?
7. What factors have delayed the success of e-commerce and e-tailing?
8. What are portals? What advantages do their proponents tout? What drawbacks concern their opponents?
9. What factors encourage concentration in the Internet industry?
10. What are push technology, wireless Internet, and MP3?
11. What are the different forms of Internet/video convergence?
12. What are embedded systems?

Questions for Critical Thinking

1. Are you online? How often do you go online? How well do you fit the basic demographic picture of the typical online user? If you are dissimilar, in what ways?

2. Do you believe commercialization of the Internet is a reasonable price to pay for its continued success and diffusion? Explain.

3. Have you ever shopped online? Describe the experience.

4. Tim Berners-Lee gave HTTP and the Web to the world for free. An Internet traditionalist to the core, he argues that he is simply conforming to his belief that information should always be free. But what about you? Are you sufficiently committed to the Internet community and information freedom that you would have done what Berners-Lee did, knowing full well that you would be denying yourself the kind of wealth and notoriety enjoyed by other innovators such as Bill Gates? Explain your answer.

5. Compare your e-mail habits before you read this chapter with the hints it offers. How effective an e-mail sender and receiver are you?

Engaging Technology

1. We've seen that accurate data on the number of Internet users is hard to come by. Visit these sites offering Internet statistics to keep track for yourself of just how many people are hooking up: <http://www.asiresearch.com> and <http://www.mids.org/mids>.

2. The Electronic Frontier Foundation is the most visible and vocal voice of the online traditionalists' views on the uniqueness of their medium. Visit its site at <http://www.eff.org> to learn about its positions on several Internet issues.

3. Despite its controversial nature, e-commerce is here to stay. For an industry point of view on who buys online and what interests these cyberbuyers, visit the A. C. Nielsen Company's Internet Advertising Marketing Research site at <http://www.nielsen.com.au/AdAudit>.

4. In our discussion of each of the emerging Internet technologies we've identified several industry leaders, for example, MP3.com, and WebTV. Visit the sites of these innovators to assess where their technologies are today and where they will go in the future. As a test of your skill at using the Net, we haven't provided all the necessary URLs. Demonstrate that you can find your way around the Internet by finding as many of the identified companies as you can.

Chapter Eight

Conceptualizing and Counting the Audience

Almost everyone talks about the ratings. We know which is the "number one station in town," that *E.R.* is network television's "top rated show," and that this year's Super Bowl was "the highest rated program in history." A few of us even get to be "in the ratings." We might get a phone call asking about our favorite radio stations or possibly become part of a "ratings family," as did humorist Dave Barry (1990), who wrote,

> So I figured the least I could do, for television, was be a (ratings) household. This involves two major responsibilities: 1. Keeping track of what you watch on TV. 2. Lying about it. At least that's what I did. I imagine most people do. Because let's face it: Just because you watch a certain show on television, that doesn't mean you want to admit it. (p. 23)

But the ratings—the all inclusive, generic name given to various audience measurements—are only one aspect of the audience and its relation to and role in telecommunications. And as you might expect from discussion in previous chapters, convergence is changing not only the role and nature of the audience but even its definition.

Preview

In this chapter we examine different conceptions of the audience and its role in the business of telecommunications. We also discuss how convergence may be changing individuals from audience members into **users,** that is, individuals who use the services of a telecommunications operation rather than simply consume its content.

Whether they are called the audience or users, the success of the various telecommunications industries depends on how many of them there are and on their **demographics** (the characteristics of that audience). We will look at how different audience measurement techniques were developed, discuss the strengths and weaknesses of each, and speculate on their future. This will take us into an examination of sampling and its attendant error.

Our focus is on the two major audience measurement operations: Arbitron for radio and Nielsen for cable and broadcast television. Additional attention will be paid to the difficulties and potential solutions to counting the number of Internet and Web users and the several companies that hope to achieve dominance in the business of counting Internet users.

The Broadcast Versus the Telecom Model

The telecommunications industry depends on people—those who consume the content it distributes and, increasingly, those who use its services. Companies survive and prosper by meeting the perceived needs of these people. But there are important differences in the way audience needs and user needs are met. Likewise, there are important differences in how telecommunications companies relate to audiences and to users.

Traditionally, when we think of the audience, we think of people watching television or listening to the radio. This audience is served through the production and distribution of content, presumably content the audience wants. The audience that is attracted to this content is then sold to advertisers who want to present their products to it. This is the broadcast model of the audience (Chapter 1): audiences are attracted by content and then sold to advertisers. Broadcasters want their audiences to find them and then stay put, because the more people in the audience (or the more people sharing some particular demographic characteristic), the more broadcasters can charge advertisers.

The telcos and Internet service providers operate on a different set of assumptions. In the telecom model audiences do not consume content prepared for them, they *use* the technologies to, in effect, create their own content. You "script" your own phone calls as you speak, and you "produce" your own Internet and Web content through the choices and decisions you make. The phone company earns more profit if its users do not stay put. The Internet and Web prosper precisely because people don't stay put. In the telecom model users are offered *services* rather than *programming*. We will refer to this distinction again in this chapter, especially where convergence blurs the difference between the two models. But first we will examine the broadcast audience.

The Broadcast Audience

The commercial U.S. broadcasting structure is set up to deliver programs to audiences and audiences to advertisers. How we view the position, role, and power of the audience in this enterprise is of vital concern in the judgments we make about how well or poorly broadcasters meet the needs of their audiences. Different conceptions of the audience inevitably lead to different conclusions.

Ask yourself this question: Whose needs must broadcasters meet in order to prosper? Naturally, you say, they must meet the needs of their audiences. If no one listens or watches, stations go off the air and shows are canceled. But the relationship between audience and broadcaster may not be this simple, because we must ask, "Who is the true *consumer* in the broadcast model?" Is the audience the consumer of radio and television products? Is the audience member the individual everyone in the broadcasting business really has to please, or is the true consumer in the broadcast model to be found elsewhere? If we determine that the audience isn't the ultimate consumer of broadcasting's products, just who is?

One common view of the audience argues that it is central to any and all programming decisions. In other words, we get what we want—no more, no less. The audience votes with the turn of the dial or the click of the remote control. The audience is broadcasting's prime consumer and ultimate judge. Broadcasters determine the wishes and needs of listeners and viewers through intensive and extensive measurement of those wishes and needs, that is, the ratings. This understanding of the audience is based on the classic U.S. business model. It proposes that the "production-through-consumption" process is never-ending and circular, with the consumer (in the case of broadcasting, viewers and listeners) holding the single most important position. Here is how this business model operates for a typical consumer product:

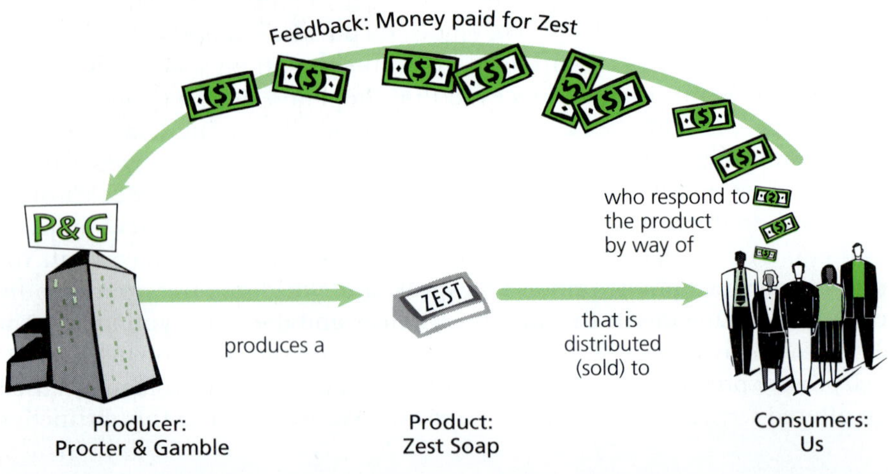

This model works reasonably well with products such as soap, cereal, and automobiles. With these products the purchaser buys the product and consumes it or makes direct, obvious, use of it. There is no question about what the product is or who the consumer is (because the consumption process is literal). The person who buys a bar of soap and washes with it is clearly the consumer of that bar of soap; the person who purchases a box of cereal eats that cereal.

The purchase of a product is registered with the producer in the form of direct income—the purchase of a bar of soap results in a specific and predictable amount of money for that producer. The number of purchases is tallied. This determines the viability of the product. If enough people buy soap (or cereal or cars), the product producer stays in business and, in all likelihood, competitive manufacturers offer similar products based on the success of the original.

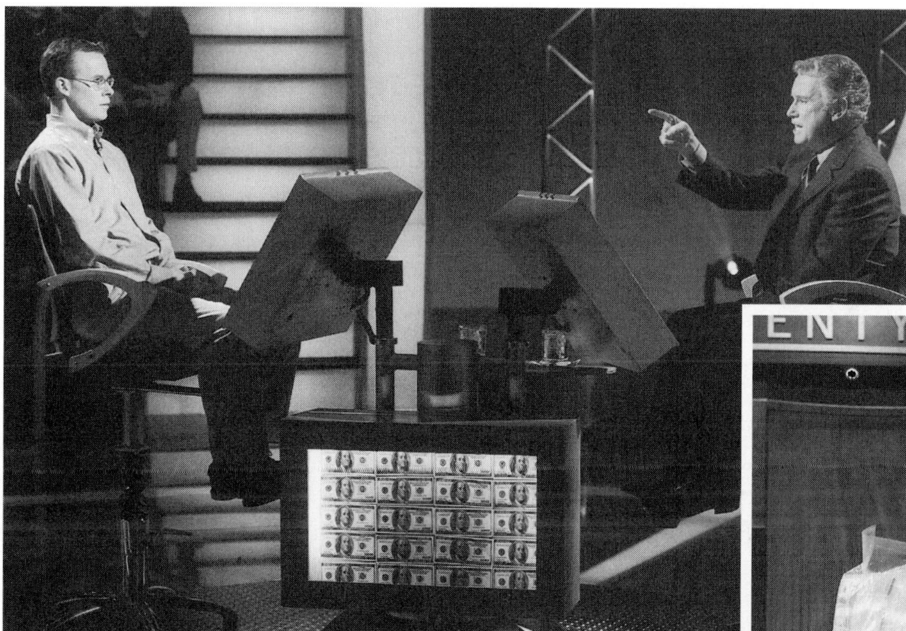

Soon after ABC's *Who Wants to Be a Millionaire?* proved itself a ratings winner, game shows proliferated on all the major television networks. One of NBC's copycats was an updated *Twenty One*, which originally broadcast in 1956 and was soon after at the heart of the quiz show scandal that rocked early network television. Is the cloning of successful programs a sign of broadcasters giving the audience what it wants or an effort to satisfy advertisers intent on buying that audience?

When we apply this business model to broadcasting, the process seems just as straightforward. In radio a format is chosen, that is, a playlist is devised and a disc jockey's style is developed. The popularity of this format is gauged; it succeeds or fails on the basis of its popularity. If a particular format shows up again and again—think of all the Top 40 and soft rock stations you've heard and how alike they sound—this is a direct indication of the wants and needs of the audience. In television a program such as *Who Wants to Be a Millionaire?* is aired, and its popularity with the audience is gauged. It succeeds or fails on the basis of its popularity. If a particular type of program seems to show up again and again—think of all the game shows that flooded the airways after the success of *Who Wants to Be a Millionaire?*—this is a direct indication of the wants and needs of the audience. The U.S. business model applied to network television, then, looks like this:

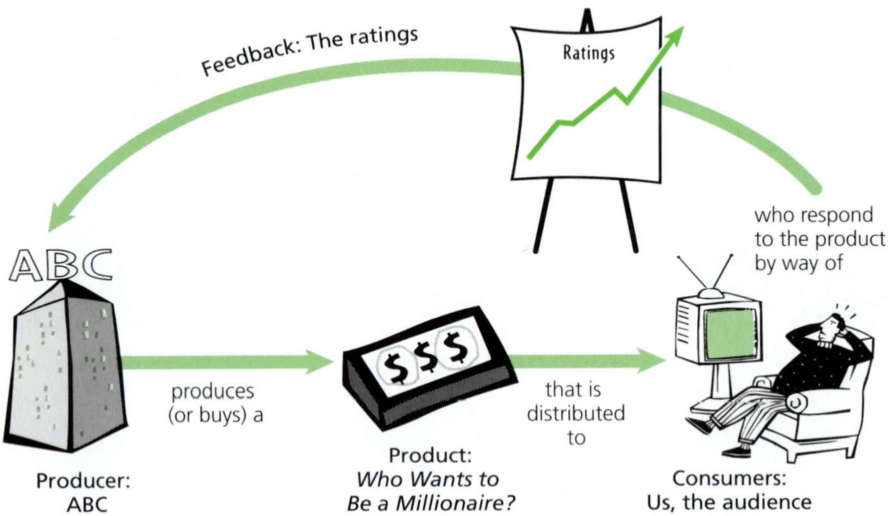

There is, however, another, possibly more accurate view of the role of the audience in broadcasting, one that sees the audience not as the consumer but as the product. Speaking specifically of television, broadcast historian A. Frank Reel (1979) emphatically expressed this view:

> The advertiser on television wants us to buy his deodorant or detergent or automobile; that is, he hopes we will enter into a financial transaction with him in the future. But before we receive his message, a number of purchases have already occurred. The key one involved that same advertiser's paying thousands or hundreds of thousands of dollars in order to get the advertisement to us. Looking at that financial transaction, it becomes obvious that we, the . . . audience, are the commodity that is sold, the product that is bartered. We are examined, analyzed, and dissected—squeezed like the toilet paper, weighed like the steak, tested like the laundry soap. (p. 4)

If, in the classic U.S. business model, the consumer is supreme, this alternative view of the audience—as product not as consumer—calls into question the role and power of the people who make up the audience. This alternative view looks like this; again we'll use ABC and *Who Wants to Be a Millionaire?*

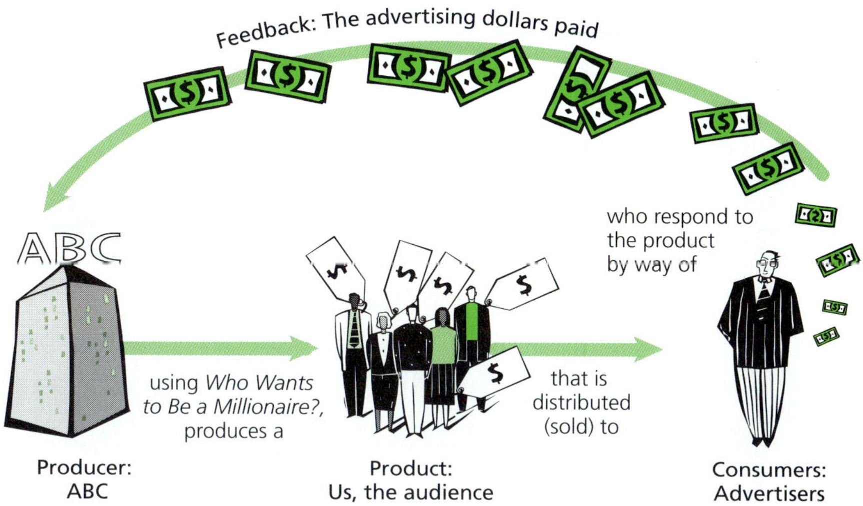

You can decide for yourself what you believe to be the audience's true role in the broadcast system using the box titled Positioning the Audience. Chapter 10 will continue this discussion by examining how broadcast content and programming decisions are shaped by the divergent views of the audience's position in the broadcast model. But in either case—audience as consumer or audience as product—the audience is, as Reel would say, "examined and dissected." That is, it is measured.

The Roots of Audience Measurement

Before the emergence of radio as a mass medium, other media industries were interested in measuring the size of their audiences. The movies had a built-in audience measure, the box office—how many tickets were sold? For an advertising-based medium like magazines, however, a different kind of counting was necessary. In 1914 a booming consumer magazine industry encouraged development of the Audit Bureau of Circulations (ABC) to provide it and its advertisers with accurate and reliable readership figures. When it was determined that advertising would provide the economic basis for radio (Chapter 2), it was not surprising that a similar form of audience measurement was sought.

Individual advertisers attempted to measure the audience themselves, but cost and accuracy were serious problems, especially when advertisers were interested in the number of listeners who heard their spots in a national

The Real World *Positioning the Audience*

On the continuum provided here, place a mark at the spot that you think best represents the audience's true role in the U.S. television system. Defend your response. What happens to the quality of programming as your mark moves either to the right or the left?

Why should anyone in the audience worry about whether he or she is *product* or *consumer*? Aren't there hundreds, if not thousands, of listening and viewing options? Don't they come in all levels of quality, from very bad to very good?

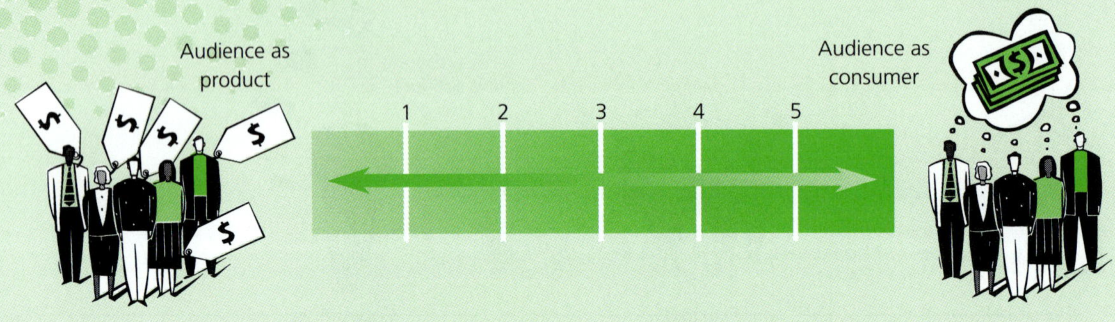

network broadcast. The first systematic rating system, the Crossley Ratings, was begun in 1930 at the behest of the Association of National Advertisers, a group suspicious of broadcasters' self-serving, exaggerated reports of audience size. But there was a problem with the Crossley Ratings. They relied on a survey technique called the **recall method.** Listeners were called randomly on the telephone and asked to remember what they had been listening to, first, on the previous evening, and then later, in the last few hours. Asking people to remember and accurately report their listening produced much imprecision.

Within 10 years Hooper (the Hooperatings) and Pulse were also offering radio ratings. Hooper rejected recall, employing **coincidental telephone measurement.** Rather than being asked to recall what they had been listening to, respondents were asked what they were listening to at the time of the random phone call itself. This was a big advance over the Crossleys, and Hooper is also responsible for developing several audience measures that are still with us today:

- *Sets in use:* the number of radio or television sets turned on
- *Rating:* the percentage of all available sets tuned in to a given program
- *Share:* the percentage of sets in use tuned in to a given program

Tech Talk *Computing Ratings and Shares*

Ratings and shares can be computed using the following formulas:

$$\text{Rating} = \frac{\text{Households tuned in (to a given program)}}{\text{Total number of households with TV (TVHH; all } \textit{possible} \text{ homes)}}$$

$$\text{Share} = \frac{\text{Households tuned in (to a given program)}}{\text{Households using television (HUT; households using television at that time)}}$$

Here's an example. Your talk show is aired in a market that has 1 million television households; 400,000 are tuned in to you. Therefore,

$$\frac{400{,}000}{1{,}000{,}000} = .40\text{, or a rating of 40.}$$

At the time your show airs, however, there are only 800,000 households using television. Therefore, your share of the available audience is

$$\text{Share} = \frac{400{,}000}{800{,}000} = .50\text{, or a rating of 50.}$$

If you can explain why a specific program's share is always higher than its rating, then you understand the difference between the two.

To see how ratings and shares are computed, read the box titled Computing Ratings and Shares.

Pulse, Inc. used a technique called **roster-recall,** in which listeners were interviewed in person and were asked to name the stations they had listened to in the past day and past week. To aid recall, listeners were shown a roster of station frequencies, call letters, and other identifying information such as station slogans.

Each of these services had its flaws. Those that used **random sample** telephone calls inevitably ignored certain segments of the population (for example, people in rural areas, people with unlisted numbers—usually better-off listeners—and those without phones—usually less well-off listeners). In addition, Crossley's coincidental phone calls could not accurately tap mobile use of the medium (such as listening in the car). Another problem with all of these methods is lying. Lying is especially problematic with face-to-face interviews, because the presence of an interviewer can intimidate listeners into answering simply for the sake of answering or into answering in an unnatural manner, for example, to appear more sophisticated ("Of course I listen to *Evening at the Opera*").

A. C. Nielsen

The A. C. Nielsen Company began in 1923 as a product testing company but soon branched out into market research. In 1936 Nielsen started reporting radio ratings after its founder, Arthur C. Nielsen, attended a demonstration at the Massachusetts Institute of Technology of a mechanical metering device called an **audimeter,** which was designed to track radio listening. The audimeter produced a minute-by-minute record of when a radio was on and where the dial was set. According to the company's publicity material, "Nielsen bought the meter practically on the spot, and embarked on a long and expensive research and development effort to improve the audimeter and to use it on a scientifically selected sample of radio listeners" (A. C. Nielsen Company, 1996, p. 1).

In 1942 Nielsen began producing its Nielsen Radio Index for broadcasters and advertisers. The audimeter was a significant improvement over both telephone and face-to-face ratings techniques (although it was not perfect, as we'll see later). Eventually, in 1950, Hooper sold out to Nielsen, leaving it one of the two dominant radio ratings services.

Arbitron

The other surviving radio rating service, begun in 1949, was the American Research Bureau, later renamed Arbitron. The company initially used a mechanical counter like the audimeter, but its primary method of collecting information on radio listenership was and is the **diary sample,** in which listeners are asked to maintain a written diary of their listening for a specified period of time.

Contemporary Radio Ratings

In 1964 Nielsen left radio altogether to focus on television. This left Arbitron as radio's primary audience measurement service. Arbitron collects data on local market radio listening using its Arbitron radio listening diary and publishes a variety of measures in its Arbitron Radio Market Reports. Despite the limitations of diaries, which we'll discuss later in this chapter, Arbitron considers its methodology "the most reliable means of collecting information for *both* radio stations, who need to make programming decisions, *and* advertisers and agencies, who need to figure out where to 'buy'" (Arbitron, 1999, p. 1).

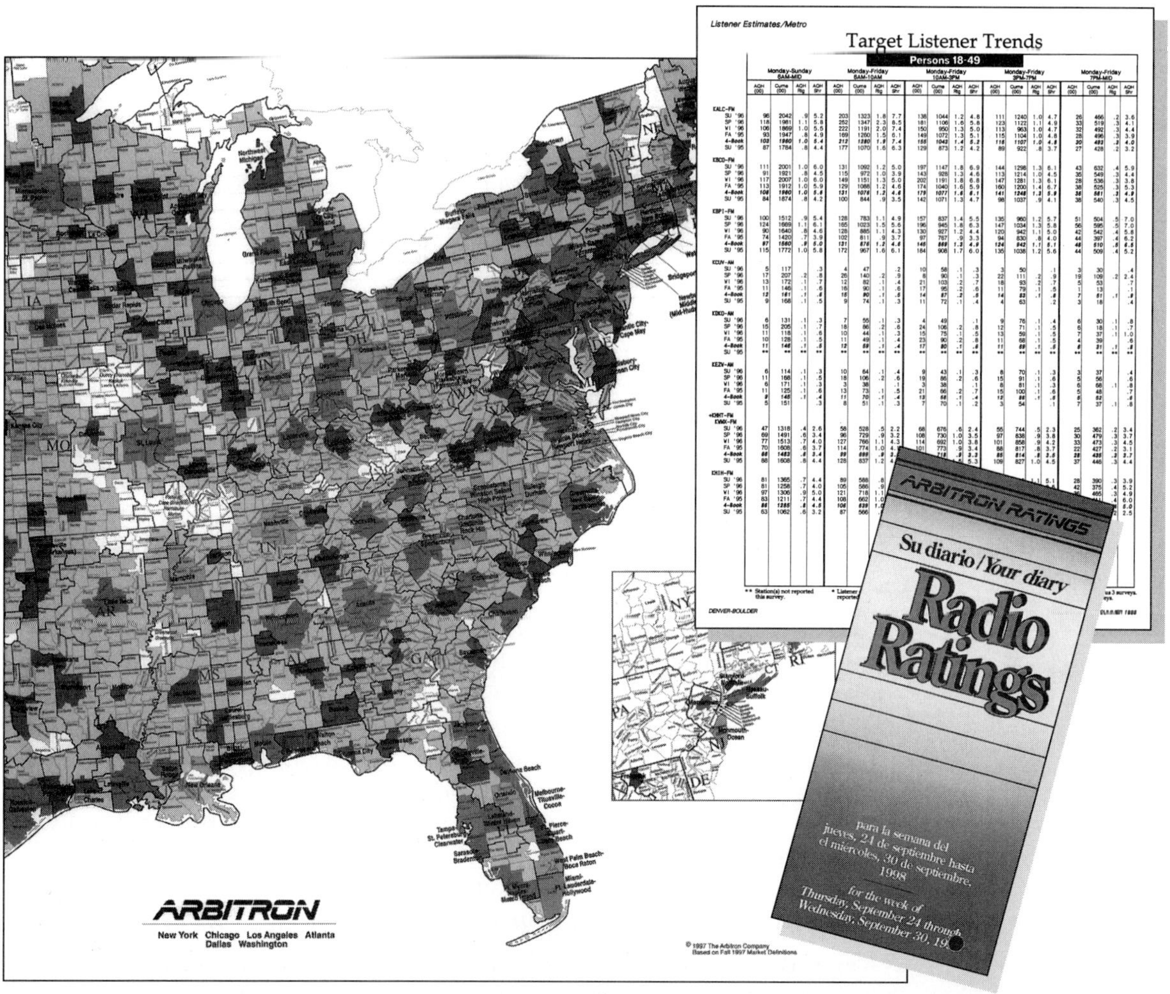

Radio ratings company Arbitron sends its participating listeners a ratings diary. From the data these provide, it produces a variety of statistics, as seen in this sample page from the Denver-Boulder ratings book. Arbitron produces similar books for each of its metros, several of which can be seen on this segment of the company's national metro map.

The company telephones more than 5 million radio listeners a year asking if they would like to be diary keepers, and a full 75% agree. Arbitron refers to this as the Arbitron Consent Rate. Of these, 2.6 million households representing every local market in the country receive diaries. The actual number of diary keepers for each market depends on the size of that market but usually numbers between 600 and 700. Each diary keeper receives these items:

- A 7-day radio listening diary to be maintained by every member of the household more than 12 years old; it can accommodate nine people and is prestamped and preaddressed for easy return
- Completion instructions
- A thank-you letter for participating
- A cash premium, typically quite small

The information requested of diary keepers includes notation of what they listen to every 15 minutes along with this additional information:

- Start and stop times of listening
- Station identifiers such as call letters and frequencies
- AM/FM choice
- Location of listening
- Whether or not there was any listening at all on a given day

Toward the back of the diary there is space to record demographic information such as respondents' age, gender, employment status, and address. Diary keepers in specific markets receive a special Qualitative Diary designed to collect more detailed information on listeners' buying habits and intentions and their demographic, socioeconomic, and lifestyle characteristics. Finally, each diary includes space for listener comments about specific stations, programming, and announcers.

The data from these diaries produce several measures important to broadcasters and advertisers, and they are presented for several diverse areas:

Metros. A city (or cities) whose population is composed of that central city and the county in which it is located. Metros generally correspond to the metropolitan statistical areas (**MSAs**) as defined by the federal government's Office of Management and Budget, sometimes altered or expanded for technical or marketing reasons.

TSAs. Total service areas, which are geographical areas that encompass the MSA and include additional areas outside it meeting certain listening criteria.

DMAs. These designated market areas are sampling areas as defined by the television ratings company A. C. Nielsen.

In addition to sets in use, ratings, and shares, Arbitron reports a number of statistics that can refine the overall ratings of a particular radio station.

Average quarter-hour persons (AQH Persons): The average number of people listening to a particular station for at least five minutes during a 15-minute period.

Average quarter-hour rating (AQH Rating): The AQH Persons expressed as a percentage of the population being measured ([AQH Persons/population] × 100 = AQH Rating).

Cume persons: The total number of different people (the cumulative audience) who listen to a station for at least 5 minutes in any one **daypart** (time of broadcast day, e.g., drive time or evening).

Cume rating: The cume persons audience expressed as a percentage of all people estimated to be in the specified area ([cume persons/population] × 100 = cume rating).

Gross impressions (GIs): The total of the AQH Persons for an advertiser's commercial in a specified period of time; in other words, how many people heard the ad how many times.

Gross ratings points (GRPs): The total of all the ratings points achieved for an advertiser's commercial in a specified period of time.

Cost per rating point: The cost to an advertiser for reaching an AQH Persons audience equivalent to 1% of the population in a specified area.

Cost per thousand (CPM): The cost to an advertiser of garnering 1,000 gross impressions.

Exclusive cume: The number of different people who listen to a given station during a given daypart.

Net reach: The number of different people reached by an advertiser's schedule of commercials.

Frequency: The average number of times a listener hears an advertiser's schedule of commercials.

Time spent listening (TSL): An estimate of the number of quarter hours an average person spends listening during a given period of time.

These measures are sophisticated, but the use of diaries does incur some problems. Lying is one; forgetting is another. Uneven diary return rates among different types of audiences is a third. Nonetheless, as these ratings are the only objective measure of audiences available to both stations and advertisers, both parties agree to use them as more or less true measures of audience size.

Arbitron has also recently begun providing listenership data for online radio stations (see Chapter 2), announcing in late 1998 that it had joined with RadioWave.com to use the latter's sophisticated software to track and report listening for more than 300 Internet radio stations.

Contemporary Television Ratings

Audience measurement for television is conducted primarily by one market research firm, the A. C. Nielsen Company, which provides national audience information to 6 broadcast networks, more than 40 cable networks, 100 program syndicators, and scores of companies and their advertising agencies. It also provides local ratings for each of the 211 U.S. television markets to 1,500 commercial stations, 140 cable companies, nearly 50 syndicators, and more than 2,000 local ad agencies. In addition, Nielsen offers separate ratings for

Introduced in 1987, Nielson's Peoplemeter requires each member of a television home to press its buttons in order to record his or her viewing.

Spanish-language television. Arbitron, reacting to the changing face of the medium, withdrew from the television ratings business in 1994 to focus its resources on radio ratings.

Nielsen offers a variety of services; the best known of these is the Nielsen Television Index (**NTI**). The NTI is what most of us know simply as "the ratings." These are Nielsen's national program numbers. The company offers a variety of services beyond the ratings. In fact, it reports three types of ratings:

Overnights: Nearly instant ratings gathered from homes in several major urban centers.

Pocketpieces: Ratings based on a national sample that are computed and reported every two weeks.

MNA reports: Multinetwork area computations based on the 70 largest markets.

To produce these ratings, Nielsen selects 5,000 households thought to be representative of the entire U.S. viewing audience. From its days as a radio ratings company, Nielsen initially adapted the audimeter, which recorded when the television set was turned on, the channel to which it was tuned, and the time of day. This was a most elementary accounting—it said little about who was watching. In September 1987 Nielsen introduced the **peoplemeter,** a device requiring each member of a television home to press its buttons to record his or her individual viewing. (Parents or guardians are responsible for recording children's choices.) The information recorded is sent to Nielsen by telephone lines, and the company can then determine the program being watched, who was watching it, and the amount of time each viewer spent with it (Figure 8.1).

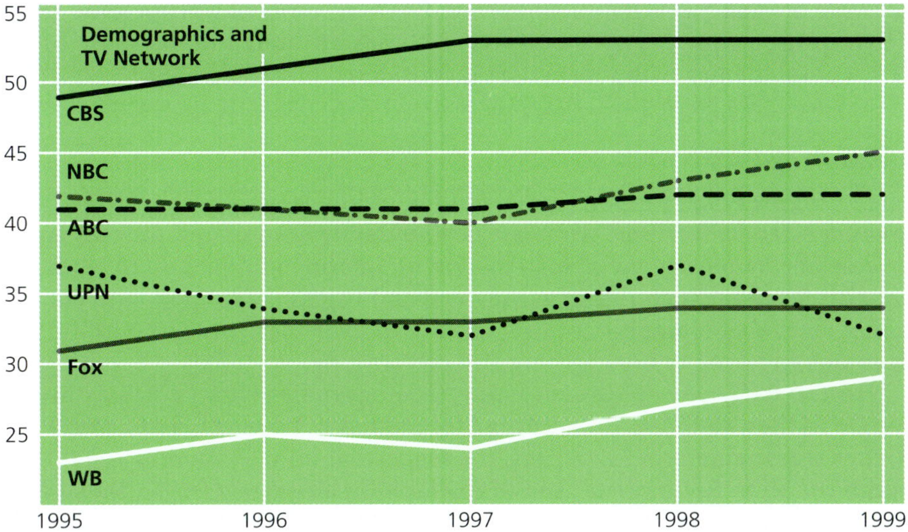

Figure 8.1 The Demographic Divide
Good demographics are vital to the financial success of a television network, and all networks labor to improve theirs. Still, the traditional Big Three networks (ABC, NBC, and CBS) continue to attract older and less demographically desirable viewers to a degree disproportionate to their number in the population as a whole. Here are the median prime time ages (one half of all viewers are above this mark, one half below) for the six national television networks from 1995 to 1999. Note the big jumps between CBS and its traditional cousins NBC and ABC. Then look at the divide between these three and upstarts Fox, UPN, and the WB.
Source: Schneider, 1999.

To draw a more complete picture of the viewing situation and to measure local television viewing, Nielsen conducts diary surveys of viewing patterns 4 times a year. These **sweeps periods** are in February, May, July, and November. During sweeps, diaries are distributed to thousands of sample households in all 211 markets across the country. Viewers are asked to write down what they're watching and who is watching. The diary data are then combined with the peoplemeter data to help stations set their advertising rates for the next 3 months. Market-specific sweeps data are reported in the Nielsen Station Index (**NSI**). Sweeps, however, may soon become a thing of the past, as you'll see in Chapter 10.

Of greater importance to broadcasters, cablecasters, and advertisers than a program's rating is its share. The share is a direct reflection of a particular show's competitive performance. Share doesn't measure viewers as a percentage of *all* television households (as do the ratings). Instead, the share measures a program's audience as a percentage of the *television sets in use* at the time it airs. It tells us what proportion of the actual audience a program attracts, indicating how well a particular program is doing on its given night, in its time slot, against its competition. For example, *The Tonight Show With*

Jay Leno normally gets a rating of around 4 or 5, terrible by prime time standards. But because it's on when fewer homes are tuned in, its share of 15 (15% of the homes with sets in use) is considered profitably high.

Demographics

The ratings services for both radio and television provide broadcasters, programmers, and advertisers with data beyond ratings and shares; they also provide detailed audience member breakdowns, offering information to potential advertisers about such things as viewers' and listeners' age, sex, and educational level. Demographics of this sort are of ever-increasing importance to everyone involved in broadcasting and cablecasting.

In the early days of broadcasting, demographics didn't carry as much weight as they do today. Broadcasting legend has it, however, that it was Paul Klein's reassessment of *I Love Lucy*, the biggest hit on television in 1954, that began the shift. Klein, then head of research for ad agency Doyle, Dane, and Bernbach, advised *Lucy* sponsor Philip Morris to drop its support of the phenomenally successful show. His analysis showed that *Lucy* appealed primarily to (or was at least watched by) children and older women—at the time these groups were not particularly heavy smokers. Klein felt that cigarette maker Philip Morris could get more value for its advertising dollars by sponsoring a lower rated show, *Medic*, which, though viewed by fewer people, reached an audience that was much more likely to buy cigarettes. This was a revolutionary idea, one that would become the broadcast industry standard toward the end of that decade.

In the 1960s CBS was at the top of the ratings. Nonetheless, despite having several big hits, the network lagged behind NBC in attracting urban viewers. Advertisers began to turn to NBC to reach this demographically more attractive (because it spent more money on consumer products) audience. In response to the lost advertising revenue, CBS dropped many of its top-rated shows. Between 1968 and 1970 it jettisoned *The Beverly Hillbillies* (rated 18th at the time), *Gomer Pyle, USMC* (the second highest rated show in the country), *Green Acres* (19th), *Mayberry RFD* (4th), and *Family Affair* (12th). Other popular shows also removed from its schedule were *Hee Haw, Petticoat Junction, The Red Skelton Hour,* and *The Jackie Gleason Show.* Why did these programs fail? They appealed primarily to a nonurban audience and, what's more, to children and "over-50s."

This practice continues today. In 1998 CBS canceled the critically acclaimed and popular *Dr. Quinn, Medicine Woman.* Although the show regularly won its Saturday night time slot throughout its 6-year run, its audience was demographically undesirable. It was composed largely of viewers more than 50 years old. CBS felt that it could earn more revenue from its commercial spots if they were housed in a more demographically friendly program, one appealing to men 18 to 34 years old. *Dr. Quinn*'s replacement was the kung fu–oriented action show *Martial Law.* Where "bad demographics" killed *Dr. Quinn,*

Although *Dr. Quinn, Medicine Woman* consistently won its Saturday night time slot during its 6-year run, its audience was demographically undesirable, composed largely of viewers over 50 years old. CBS canceled the critically-acclaimed show, replacing it with the more demographically friendly *Martial Law.* What did *Martial Law* have that *Dr. Quinn* did not? Heavy viewership among men 18 to 34 years old.

"good demographics" save programs such as *Beverly Hills 90210* and *Melrose Place*. These shows rarely garner high ratings and shares, but their demographics are enviable (affluent, urban 18- to 34-year-olds). You can learn more about the meaning of *Dr. Quinn*'s cancellation in the box titled One Dollar, One Vote.

So powerful have "good demographics" become that in late 1998 *Variety* was able to report that, although CBS was about to "end NBC's three-season winning streak" as television's overall ratings leader, "given a choice, NBC will take a victory in adults 18–49 over households any day—advertisers pay handsomely to reach those 18–49 viewers and care nothing about household figures. So 18–49 success means revenue, while household success, especially

Critical Issues *One Dollar, One Vote*

We've seen that the ratings have their defenders and their detractors. Criticism of the ratings is often aimed at their methodology, the inferences drawn from them, and the broadcast industry's continued reliance on them despite its own misgivings. But there is another, possibly more important complaint about the ratings, one that speaks to their bias against certain groups and types of people. The charge is that they are undemocratic.

When questioned about the quality of their fare, broadcasters routinely respond, "We give the people what they want" and "The people vote with their dials." As former CBS president Frank Stanton told the United States Congress in one of its investigations of the television industry during the 1960s, "In the business of broadcasting, perhaps more than in any other business, the ultimate decision is the public's. In our business the process is one of pure democracy. . . . For television, the public is the ultimate monitor—the monitor-in-chief" (as quoted in Reel, 1979, p. 13). But many critics reject these claims, challenging their implicit suggestion that those who would criticize what the people have chosen are, themselves, elitist and undemocratic.

Let's examine what happened to *Dr. Quinn, Medicine Woman.* The ratings made it clear that the people did, in fact, want this program, voting for it with their dials in sufficient numbers to render it the winner of its time slot for several years running. But CBS and the advertisers it hoped to attract were unhappy with the demographics of that audience, despite its size. The critics' complaint? In a democracy, voting means one person, one vote. But because the broadcast industry favors some "voters" over others, it turns that tradition on its head, making it one dollar, one vote. That is, the "vote" of a listener or viewer that can be sold for more money than that of another counts for more in the "election."

This means that certain members of the public are rendered second-class citizens. Rural folks, people at the low end of the income scale, and older people—because they are demographically unattractive—are especially disadvantaged in broadcastings' elections. Critics find this ironic. Not only do they see it as unfair (as we saw in Chapter 2, the airwaves belong to *all* the people and broadcasters promise to serve *all* the public to win and keep their licenses) but as financially foolish. Every day, 10,065 Americans turn 50 (Harper's Index, 1999, p. 2C), making them the single fastest growing demographic group in the United States. In addition, because these people tend to be further along

---

if reached through a skew toward older viewers, can mean bragging rights that carry little weight by the industry standard" (Bierbaum, 1998, p. 33). Older people, despite controlling more money than do younger people, are unattractive to television advertisers because advertisers believe the buying habits of mature people are more resistant to commercial-induced change.

Advertising rates, then, are set according to the *size* of a program's audience, but also by the *perceived quality* of that audience based on demographic characteristics. Top rates are charged for programs appealing to people between the ages of 18 and 34 who are prime buyers of kitchen and cosmetic products as well as automobiles and over-the-counter drugs—all products that are heavily advertised on television. After this group, rates drop. The next most valuable commodity a show can deliver is an audience between the ages of 18 and 49, typically comprising young, often upwardly mobile

in their careers, they also control a majority of the nation's wealth. Why, ask critics, would broadcasters work so hard to reject this plentiful and well-heeled audience?

Although these viewers and listeners do have more money than the more demographically attractive 18- to 34-year-olds, the broadcast industry explanation is that the buying habits of older people are more entrenched, making them less likely to be moved by advertising. As Robert Masur, president of the Screen Actors Guild, explained on the cable television program *Assignment E!* (1999), advertisers "have identified people over 35 as too difficult to shift in terms of brand loyalty; therefore, their advertising dollars are better spent on people under the age of 35." Les Moonves, president and CEO of CBS concurred, "There is a myth on Madison Avenue that the 18-year-old is someone, that if you reach them when they're 18 with Crest toothpaste, they will continue using Crest toothpaste until they die. So at the age of 50, (people) aren't important anymore. . . . Older viewers have been abandoned, by and large, by everyone" (*Assignment E!*, 1999). There is a parallel and even less flattering interpretation of broadcasters' disinterest in the over-34 age group. Because much broadcast advertising is at best information-free, and at worst manipulative, older and presumably smarter, more experienced audience members do not respond to it.

But you decide. Does one dollar, one vote trouble you as an audience member? Do you think you'd feel differently if you were a radio or television sales manager? Can you find a relationship between your answer here and your placement of the audience on the scale in the box shown earlier in this chapter titled Positioning the Audience?

There is another question too. Does any of this really matter in today's technologically converged, 1,000-channel telecommunications universe? Decades ago, when there were only three television networks, it could be argued that unserved audiences were clearly being treated undemocratically—there was nowhere else for viewers to go. But with so much content available today, is there any audience that is truly unserved? If broadcasters were forced to meet audience needs on a one person, one vote basis (that is, simply in terms of audience size), how long would they survive in a mass communication environment in which every other media outlet seeking advertising revenue is selling its audiences on a demographic basis?

---

families. The lowest rates are charged for time on programs appealing to young children and to those more than 50 years of age. We'll revisit this issue in detail in the next chapter.

Sampling

All audience measurement companies base their numbers not on data collected from the entire viewing or listening audience (an economic and practical impossibility) but on **samples** drawn from those audiences. These samples are considered representative of the **populations** (or audiences) from which they are drawn because those selected as part of the sample are chosen in a (theoretically) random fashion; that is, every member of the population has an equal likelihood of being selected.

Tech Talk *Computing Sampling Error*

The ratings are often criticized for the generalizations to large audiences (populations) they make from seemingly small samples. Surprisingly, however, effective sample size is not related to the size of the population. Instead, it is a function of the amount of uncertainty researchers are willing to accept in their results. Scientists have mathematically determined that a predictable, built-in level of error accompanies all samples of a given size. This is variously called **estimated sampling error, error term,** or most commonly, **margin of error.**

For example, radio show A may lead show B by 6% to 4% in the ratings, but in reality B may lead A. This means that when ratings data are reported it would be wise to consider the sample size and the margin of error.

A sample of 600 respondents for a local radio survey, for example, comes with a margin of error of 4 points. In our example above, if the ratings service had used a sample of this size, show A's rating might be as high as 10% or as low as 2% (6 + 4 = 10; 6 – 4 = 2). Show B's rating could be anywhere between 8% and 0% (4 + 4 = 8; 4 – 4 = 0).

The convention, then, is to choose a sample size that is both economically and efficiently measurable while still giving audience researchers a margin of error they can tolerate.

Here are some sample sizes and their accompanying margins of error. Remember, except in the special circumstance when some small, particular population is under scrutiny, the size of the population is unimportant. Mathematically, samples of these given sizes are always accompanied by the given levels of variation or error if the sample is randomly selected. Therefore, each of these samples would have the same validity.

Sample Size	*Margin of Error*
400	±5%
600	±4%
1,100	±3%
2,400	±2%
9,600	±1%

The ratings, as opposed to other survey data frequently reported in the popular media, are almost never accompanied by their attendant error term when cited. Yet this margin of error is important when evaluating two shows that are separated by only a few ratings (percentage) points. In the case of radio, in some markets as many as a dozen stations with ratings below 5 may be separated by 2 ratings points, making judgments about their relative audience size difficult. Consider Arbitron's use of 600 diaries for a local market. This sample size brings with it an error term of ±4%. In a statistical sense, then, the ratings for these stations are virtually identical.

Sampling brings with it a number of questions, among which are the size of a sample and its attendant sampling error. Sample size is discussed next, and sampling error is the focus of the box titled Computing Sampling Error.

Ratings Problems and Controversies

The most common complaint from viewers and listeners about the ratings is that it seems illogical that a few thousand households can represent the listening and viewing tastes and habits of the entire country or even an entire market. Congress investigated this "problem" in the 1960s, only to be assured that the ratings "mean what they say." They may or may not, but sample size is not the biggest problem with the ratings. Let's take the national television ratings as an example.

Some people complain that 5,000 households is simply too small to reflect the viewing patterns of millions of Americans. But this sample is large enough, if it is selected carefully enough, to ensure *statistical* accuracy within a few percentage points, which may be satisfactory for the big-audience broadcasters and cablecasters. But what of the smaller channels? Their success and failure can rely on as few as one or two households. Take, for example, "tiny cable network Court TV . . . (which) on a recent December night, drew an average 0.1 rating—equal to three homes—between 8 p.m. and midnight. If one of those household members forgot to push that (peoplemeter) button, the numbers would swing wildly" (Jensen, 1999, p. 88). In other words, 5,000 households in a 500-channel environment may mean very little for many television operations.

Other important questions about the ratings have to do with what it is they are measuring in the first place and how the results are used. Audimeter-collected ratings did not measure whether anyone was actually *watching* at a given time. Its replacement technology, the peoplemeter, also has limitations. Punch-in protocols for nonfamily members are sufficiently complex that many users simply fail to acknowledge the presence of additional viewers, or they substitute a family member's code for the guest code. Imagine this scenario:

> Want to watch TV? First you turn the set on, then you punch a button telling Nielsen Media Research that it's you, a 27-year-old college-educated male making $59,000 a year, sitting in front of the tube. Then you can choose your channel. Walk into the room where your 25-year-old wife is already engrossed in *Friends*? Push the button to tell Nielsen you're there. Run to the bathroom during the commercial? Yep, that button should be pushed again, as it should be when you return. And if you have a baby, once she hits two, she too will have to learn to push her own special Nielsen button. The whole family will do this for two years, all for a token reward. (Jensen, 1999, p. 87)

The diary, as we've seen already, is flawed as well; its value is dependent on the active involvement of viewers. Lack of interest, forgetfulness, and lying can and do occur.

All ratings systems have an additional problem. Participation in the ratings distorts how and what people listen to and watch. Knowing that their viewing and listening choices are being scrutinized, or in the case of the diary, writing down bits of information before and after they listen and watch, changes people's consumption decisions. Moreover, 50% of those who are asked to participate in the Nielsen ratings and 25% of those asked to keep a diary by Arbitron refuse to do so. In addition, there is the problem of low **in-tab rates,** that is, the number of people who agree to participate in the ratings who complete their diaries accurately or who employ their peoplemeters properly on a daily basis.

Both response rate and in-tab rate raise the question of the sample's representativeness; that is, how well does it match the entire viewing population? As we've just seen, a sample's representativeness is dependent on every member having an equal likelihood of being included. As more people refuse to participate and as more of those who agree to participate do not, randomness is reduced, which reduces representativeness. In simpler terms, there is, by definition, something different about viewers and listeners who agree to participate in the ratings and those who don't—those people media critic Todd Gitlin (1983) calls "too proud or independent to allow the monitoring of his/her behavior" (p. 49).

The *use* of the ratings is also controversial. They are a device for setting advertising rates, not measures of the popularity or worth of a given piece of content. As such, ratings say nothing about what programming means to viewers and listeners.

Broadcaster/Ratings Service Conflict

Considering all these questions surrounding the ratings and their importance to the financial well-being of almost everyone involved in radio, television, and cable, it should be no surprise that conflict sometimes arises between those who rate and those who are rated.

For example, when the radio industry first began losing audience to television, it immediately identified the problem. It was not television, it was . . . the ratings! At the September 17, 1953, meeting of the NBC Radio Affiliates in Chicago—after 15 years of using the ratings to make huge profits—NBC President David Sarnoff (1953) offered his analysis of the dramatic drop in radio listenership.

> Our industry from the outset has been plagued by rating systems which do not say what they mean and do not mean what they say. They develop figures which give an appearance of precision, even unto decimal points, until you read the fine print.
>
> Unhappily these figures are seized upon by the advertising community as a substitute for analysis and judgment. They are used as the main standard for advertising values in broadcasting, and millions of dollars are spent or withheld each year on the basis of a drop or rise of a few ratings points! . . . Ratings, today, simply do not reflect the *real* audience. (p. 108)

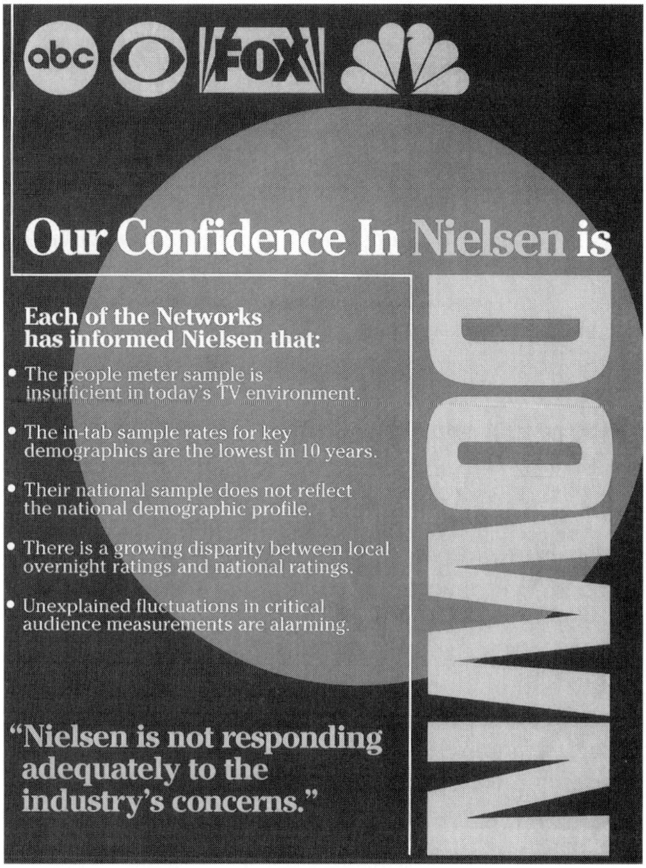

So dissatisfied had the broadcast networks become with the Nielsen ratings that in December 1996 they bought this full-page ad in all the major trade journals to let the world (and Nielsen) know.

The very same situation arose when television began to lose audience to other video technologies. The problem for the traditional networks and stations was not VCR and cable, it was . . . the ratings! The ratings should be "open to interpretation," they said. The sample is "too small," they complained. So in early 1996 Nielsen increased the number of metered homes from its long-traditional 1,500 to 5,000. The networks then protested that the new ratings overrepresented the young and technically savvy. The peoplemeter was more complex than either the audimeter or diary, and the young were more skilled at using it. This criticism may have been true, but these young people were in fact the very viewers who were abandoning the networks for cable, VCR, and satellite. So, the national networks responded by *weighting* or *adjusting* the ratings to represent the numbers of viewers who *should have been counted as theirs* if the ratings were *really* accurate. Naturally, the advertising community reacted negatively. For 7 decades it had been paying for audiences with the price of airtime based on the ratings' estimate of audience size and demographics. Why, advertisers asked, should that system be altered because the smaller audiences now documented by the ratings meant lower costs for them?

The networks have abandoned adjusting the ratings, but there is still tension between broadcasters and Nielsen. NBC and Fox have threatened Nielsen with lawsuits over their falling numbers. In December 1996 ABC, CBS, NBC, and Fox ran a full-page ad in the media industry's most important publications challenging Nielsen's accuracy, and in 1999 a number of local stations actively considered canceling their Nielsen contracts, something that Miami's Fox affiliate WSVN-TV actually did (McClellan, 1999b). In response, Nielsen has begun conducting what it calls **unified analyses.** This technique measures the same samples of viewers a year apart in an effort to demonstrate that the falling numbers for traditional broadcasters are a function of actual declines in viewership rather than methodological flaws inherent in its sampling or diary and peoplemeter methodologies.

Why the Ratings Are Used

Nicholas Schiavone, senior vice president for research for NBC, says, "It's a forty-year-old system that ran out of acceptability ten to twenty years ago" (as quoted in Jensen, 1999, p. 88). CBS's David Poltrack, executive vice president for planning and research, calls the Nielsen's "a system for which we are collectively spending about $300 million, that has lower-than-acceptable cooperation rates, lower-than-needed sample levels, archaic software, a local component that is an embarrassment to us all, no out-of-home measurement, and no qualitative measurement at all" (as quoted in Lorando, 1999, p. 7E). If there is so much apparent dissatisfaction with the ratings on the part of broadcasters, why do they continue to subscribe to Nielsen's estimates of audience size? The answer is simple. Accurate or not, as long as everyone involved agrees to use the ratings, they are the best and only means of setting advertising rates. Television writer Douglas Davis (1993) explains the phenomenon this way:

> This author has never once found a midlevel media professional who believes in the integrity of the Numbers. Some shake their heads in mirth, or disgust. But almost to a man or woman they end up accepting the Nielsen ratings because they provide a "standard," a measure of judgment in an otherwise supposedly chaotic field. . . . "Of course I believe in the Numbers," Shad Northshield, the founding spirit of CBS's richly praised *Sunday Morning* news, once told me. "I have to. I believe them like I believe in Relativity." (p. 199)

ABC researcher Ted Harbert agrees, "We really don't have time, or it wouldn't be practical really to think about if these numbers are right or not" because "everybody is dealing from the same deck" (as quoted in Nova, 1992).

Why do advertisers continue to accept the ratings as the basis for setting the prices they will pay to reach broadcast audiences? Alice Sylvester of ad agency J. Walter Thompson explains: "The Nielsen ratings right now represent an agreement between buyers and sellers to use them as a currency of

Careers *Working in Audience Measurement*

People interested in careers in audience measurement can work at a research company, an advertising agency, or at a cable or broadcast operation. Research companies like Arbitron and Nielsen need people skilled in audience analysis, including those with a familiarity with sampling, questionnaire design, research design, and statistics. But so, too, do advertising agencies that make time buys based on research and cable and broadcast companies that pay for the research and set their rates according to its findings.

At an ad agency the department most frequently involved with audience numbers is the **media department,** which is composed of professionals who make decisions about where and when to place commercials. The media department then buys the appropriate time. The **account executive** for a specific agency client will also have to have a working knowledge of audience measurement, as she or he will ultimately have to defend the decisions of the media department (Chapter 9).

Large cable and broadcast operations typically have a research or research and planning department that not only interacts with the research and ratings companies but also conducts audience analysis for the organization itself. In all these cases an advanced degree in some area of research or marketing is generally required for a management or other leadership position.

negotiation" despite their deficiencies (as quoted in Nova, 1992). They are, in other words, the coin of the realm. Problems notwithstanding, many good people make their careers in audience measurement. Career opportunities are presented in the box titled Working in Audience Measurement.

Counting Internet and Web Users

In our examination of the nature of the audience near the beginning of this chapter, we discussed two models of telecommunications, the broadcast and telecom models. The broadcast model relies on the delivery of people to advertisers. Therefore, even with their flaws, the ratings—measures of how many and what kind of people—remain the basis for setting the price advertisers must pay to reach those viewers and listeners.

The telecom model, however, relies on the delivery of services to people. Counting those people should be simple enough. Those who use the services pay for them. How many users are there? Just count how many people are paying. The phone companies know how many users they have. Subscriber-based DBS and cable networks know how many users they have. They simply count the number of people who pay for their service.

The Internet converges the two models. People pay for and use the service (the telecom model); people who use the service are often sold to advertisers (the broadcast model). This presents a number of unique user/audience measurement problems. As Arbitron's Robert Patchen and Ronald Kolessar (1999) note:

Arbitron has developed and is testing its Personal People Meter, a device that can electronically note all content consumed from all electronic media.

> The current norm of producing separate ratings for each of the electronic media will prove inadequate as the lines separating consumers' mental images of the media continue to blur. Am I listening to radio if the sound is coming from my computer? Am I watching TV if I'm surfing the Web from my easy chair, or downloading a favorite old movie? (p. 42)

Even if all we want to know is how many users there are (the telecom model), the first problem is that it is almost impossible to tell exactly how many people are on the Internet. People who own computers are not necessarily linked to the Internet, and people need not own computers to use the Net. They might have access to computers at school, at the library, or at work. Making matters even worse, more than one person, at home or at work, may use a single computer. This imprecision in counting users can be seen in Internet user estimates that range from 20 million to 100 million and estimates of Web users set at somewhere between 42 million and 57 million, both fairly broad estimates (Chapter 7).

As the Internet in general and the Web in particular become increasingly dependent on advertising dollars for support and profit, more reliable measures—not only of online users but of online viewers and listeners—become essential. But there are problems here. Between 35% and 45% of all people who visit the World Wide Web do so from work, and use from work may vary considerably from use at home. What do those differences mean when relating the ratings from one location (the office) to another (the family room)? Another difficulty for Web raters is that many companies and individuals are unwilling to load use-tracking software, especially from an outside source, on their computers. They fear invasion of their personal and corporate privacy. When Nielsen surveyed potential Web ratings participants, 70% were willing to install tracking software on their home machines but only 10% were willing to do so at work (Buel, 1998). With so many users accessing the Internet from work, however, what does a 90% refusal rate say about the randomness and, therefore, the generalizability of this sample?

A major obstacle to getting good demographics of Web users is that much Web use occurs at work, and most companies balk at the installation of tracking software on company machines.

Another problem area is demographics. Television and radio ratings companies rely on the United States Census to provide necessary information on the demographics of specific populations in specific areas. Although such data do exist for Internet users in general, no such information exists for Web users who access sites from work. Because so many do so, this leaves a big hole in the kind of information advertisers want before they place their commercials.

A third problem area revolves around the Net's globalness. The United States Census says nothing about the millions of users from overseas who may or may not be of interest to advertisers. Another demographic-related difficulty is that high-income users (very attractive to advertisers) working in conservative businesses such as banking, insurance, and strategic planning are the very demographic *least likely* to install tracking software.

Traditional means of measuring use do little to solve these problems. All Web sites can record the number of visitors they have through the use of click streams (Chapter 7). But click streams do not carry the demographic information necessary to attract advertising dollars. Surveys of users, either by phone or e-mail, are also deficient. One of the beauties of the Internet and the Web is that a user can view or listen to a limitless number of sites in any one online session. Reliance on user recall, problematic in traditional broadcast audience measurement, is almost absurd in the world of cybermeasurement.

Despite these difficulties, three primary contenders hope to become for the Web what Nielsen is to television and Arbitron is to radio. Nielsen's NetRatings, in fact, is one. Jupiter Media Matrix and PC Data Online are the others. All rely on a similar methodology. They conduct surveys of the general population to generate information on the size and demographic profile of Web users and the Web audience. They then recruit by phone and e-mail people who constitute a sample called a **panel.** Panel members are offered incentives for participation. Incentives typically include a choice of a modest cash payment, frequent flyer miles, or participation in drawings for laptop computers and other smaller prizes.

The people in the panel agree to install tracking software on their machines that records every place they visit on the Internet. These data are then sent (weekly for NetRatings, monthly for the others) to the ratings firms, which record what sites were visited, for how long, and the ads viewed. The companies regularly add new panel members to replace those who change jobs, move, or otherwise cease to participate, and to better reflect the changing demographics of Internet users in general. NetRatings uses the smallest panel, 3,800 users, and Jupiter Media Matrix the largest, 50,000, of which 10,000 live outside of the United States (Ledbetter, 2000; Buel, 1998).

None of these firms yet dominates the business of measuring Internet and Web users, but their work so far has produced something of a standard for counting Web audiences. **Unique visitors,** or the number of people who visit a specific site in a specific period of time, is the preferred unit of measure. For Jupiter Media Matrix, which publishes monthly reports, for example, the "rating" is called "monthly unique visitors." This count tallies the number of discrete users that visited each of its rated sites in the previous 30 days.

Chapter Review

The two models of telecommunications—broadcast and telecom—embody different assumptions about those who use telecommunications technologies. In general, the broadcast model relies on audiences and the telecom model on users. The broadcast audience can also be viewed as either the consumer or the product of the broadcast industry's offerings, with differing interpretations about its role embodied in each view.

Regardless of which view of the audience one takes, that audience must still be measured. Among the data produced by the ratings companies are ratings, shares, and sets in use. Arbitron began ratings in 1949, soon becoming the dominant provider of radio ratings. Using a diary method, it produces data such as average quarter-hour persons, average quarter-hour ratings, cume persons, cume ratings, gross impressions, gross ratings points, cost per rating point, cost per thousand, exclusive cume, net reach, frequency, and time spent listening. Arbitron reports these data for geographic areas designated as Metros, TSAs, and DMAs.

Television ratings are the province of the A. C. Nielsen Company. The Nielsen Television Index (NTI), reports ratings, shares, and other data in the form of overnights, pocketpieces, and MNA reports. To collect these data, Nielsen uses the peoplemeter. Regardless of the type of ratings being collected and reported, the demographic composition of the audience has become as important as its pure size.

The ratings are not without their critics both inside and outside the broadcast industry. Critics complain of a too small sample size to represent large audiences, the difficulties and biases inherent in the use of diaries and the peoplemeter, their misapplication (they do not measure what people think they measure), the representativeness of those who do agree to participate in the ratings, and their lack of democracy. These problems and the decline in viewership for traditional broadcasting have brought the ratings services and broadcasters into open conflict, but the ratings remain the only objective measure of audience size, however flawed, upon which all parties can agree.

Counting Internet and Web users brings with it a different set of problems. Among them are the difference between home and work usage, the fact that ownership of a computer does not necessarily equate with use of the Internet and the Web (nor does use of the Internet and Web equate with computer ownership), the refusal of people to have their Net use monitored, and the lack of solid demographics for work-based Internet users. Several companies are attempting to overcome these difficulties using a panel method to track use.

Questions for Review

1. What are the broadcast and telecom models of telecommunications? How do they differ?
2. What is the significance of the Crossley, Hooper, and Pulse radio audience measurement companies?
3. What are the strengths and weaknesses of these ratings methodologies: recall, roster-recall, and coincidental telephone measurement?
4. On which methodology does the Arbitron company rely to produce its radio ratings? What are its strengths and weaknesses?
5. What are Arbitron's Metro, TSA, DMA, AQH Persons, and AQH Rating?
6. What are cume persons, cume rating, GI, GRP, cost per ratings point, CPM, exclusive cume, net reach, frequency, and TSL?
7. What does it mean to say a sample is random? What is the benefit of a random sample?
8. What are sampling error, generalizability, and a representative sample?
9. What are the audimeter and the peoplemeter? What are the advantages and disadvantages of each?

10. What is the distinction between a rating and a share? Which is the more important indicator of a program's success? Why?
11. What are Nielson's Metros, MSAs, TSAs, and DMAs?
12. What are overnights, pocketpieces, and MNA reports? Who generates these measures?
13. What are some of the difficulties in measuring the number of Internet and Web users? How do the Web ratings companies attempt to overcome them?
14. What are unique visitors?

Questions for Critical Thinking

1. Do you equate a program's ratings with its popularity? Despite all you've read here, can you defend this "error"?

2. Imagine yourself as an advertiser. Would you appeal differently to broadcast audiences and Web users? How and why?

3. Do you believe all survey data reported in the popular press, whether it is television ratings or presidential preference polls, should be accompanied by sample size and error term information? Why or why not?

4. Do you agree with the idea that the combination of ratings and demographics has produced a broadcast system that is undemocratic? Why or why not?

5. You've read the criticisms of the ratings and both broadcaster and advertiser dissatisfaction with them. But you've also seen that there is an understanding that as long as all parties agree to use them, they're as good a measure as can be had. If you were in broadcasting or advertising, would you accept this way of doing business? Why or why not?

Engaging Technology

1. Ratings giants A. C. Nielsen and Arbitron both have excellent Web sites where you can learn more about their history, operation, methodology, and plans for future audience measurement. Visit these sites and determine for yourself which company seems better prepared to count viewers and listeners as media converge. A. C. Nielsen can be reached at <http://www.nielsenmedia.com>; Arbitron's site resides at <http://www.arbitron.com>.

2. There is a fortune to be made by the person or company that can devise an effective user/audience measurement protocol for the Internet and the World Wide Web. Visit the sites of the companies now working on the problem. See if you can improve on their methods. Try to devise ways around the problems in rating Web usage we've discussed. Maybe there's a job for you with one of these companies. Possibly you can become the "Nielsen of the World Wide Web."

Jupiter Media Matrix can be reached at <http://www.mediamatrix.com/>, NetRatings is at <http://www.netratings.com/>, and PC Data Online is at <http://www.pcdataonline.com/>.

Chapter Nine

Advertising

Sy Frolick was discharged from the Coast Guard in 1946 and found his way to New York, hoping to become an advertising copywriter. His first job, a 30-day trial, was writing radio spots. But the agency he went to work for was moving into television, and among its clients was U.S. Rubber Company, makers of Keds sneakers and U.S. Royal tires. Frolick's agency had just developed a weekly science show for U.S. Rubber, *Serving Through Science,* the company's corporate slogan.

One evening the young and eager Frolick went to see the show in production. Aired on DuMont's WABD and produced in a small rented corner of Wanamaker's Department Store, the program consisted of little more than short films produced by the Encyclopedia Britannica. At commercial time the film was stopped and an announcer stepped before a live camera and read from a script. Frolick thought that was odd, inasmuch as that was how commercials were done on radio and this was television with pictures. So when a representative from U.S. Rubber mentioned to Frolick that the announcer needed to remember to tell viewers that Keds were washable, Frolick had an idea. He set up a washing machine in the make-shift studio, hooking it to a sink in the men's room down the hall. When the show opened, the dirty Keds were dropped in the washer, and then, as the show ended, they were removed, clean and pretty (Matthei, 1997).

Frolick went on to a 40-year career as one of television's preeminent advertising professionals and, more to the point of this chapter, the demonstration commercial was born.

Preview

In this chapter we examine advertising as it exists in the various telecommunications industries. In doing so, we look at how it has developed on those different media, the business of buying and selling it, and how it has been transformed as the economics and regulation of telecommunications have shifted over the decades.

Contemporary radio advertising is explained, paying particular attention to the different types of radio sales and how those sales are aided by tools such as rate cards. The advantages of radio advertising, as well as its disadvantages, are discussed. This same pattern is applied to contemporary television and cable advertising. In doing this we discuss issues such as barter and co-op advertising, program length commercials, and public service announcements.

The new kid on the telecommunications advertising block, online or cyberadvertising, is given a good deal of attention. Various forms of presentation are discussed including online classified ads. Though still young, cyberadvertising has attracted as many critics as proponents, so we detail the plusses and minuses of online advertising.

These different forms of advertising operate not only within the traditions and conventions of the media that carry them but within the overall advertising environment as well. Issues in modern advertising such as increased demographic and psychographic audience segmentation, the regulation of advertising, measuring commercials' effectiveness, and criticism of advertising *per se* are discussed in depth.

The Roots of Broadcast Advertising

The first radio ad was broadcast on WEAF in 1922. After that debut, radio and advertising became important to one another in many ways. First, although some people inside and outside the government were opposed to commercial support for broadcasting, the general public had no great opposition to radio ads. In fact, in the prosperous Roaring Twenties many welcomed them; advertising seemed a natural way to keep radio "free." Second, advertising agencies virtually took over broadcasting. A battery maker's 1923 variety show, *The Eveready Hour,* was the first regularly broadcast sponsored series. Ad agency Blackett-Sample-Hummert even developed a new genre for its client Proctor & Gamble—the radio soap opera. As historian Harry Matthei wrote,

> The admen ruled radio as they never could newspapers and magazines, controlling the editorial content. Companies like Procter & Gamble, Ford, and General

In the early days of television, network programming was paid for and produced by its sponsors. Here, viewers stand outside a department store window, awaiting the 1949 premiere of the *Ed Wynn Show,* which was sponsored by watchband maker Spiedel.

> Foods paid advertising agencies to create and produce—or buy from independent producers—hours of network radio programs every day. The broadcasters functioned only as an electronic pipeline. They generally accepted whatever programming the advertiser chose. (p. 64)

Early Television Advertising

There was no shortage of consumer products after World War II ended. The nation's manufacturing capacity had been greatly expanded to meet the needs of war, and that manufacturing capacity was soon producing consumer products for people who found themselves with more leisure and more money. People were also having more children and, thanks to the GI Bill, were able to think realistically about owning their own homes. They wanted products to enhance their leisure, please their children, and fill their houses.

Advertising was well positioned to put products and people together, not only because advertising agencies had expanded during the war but also because there was television. Radio's formats, stars, and network structure moved wholesale to the new medium. So, too, had its role as the country's primary national advertiser. A thousand advertisers bought television time in 1948, a

Among the earliest demonstration ads, Timex took many a licking but kept on ticking.

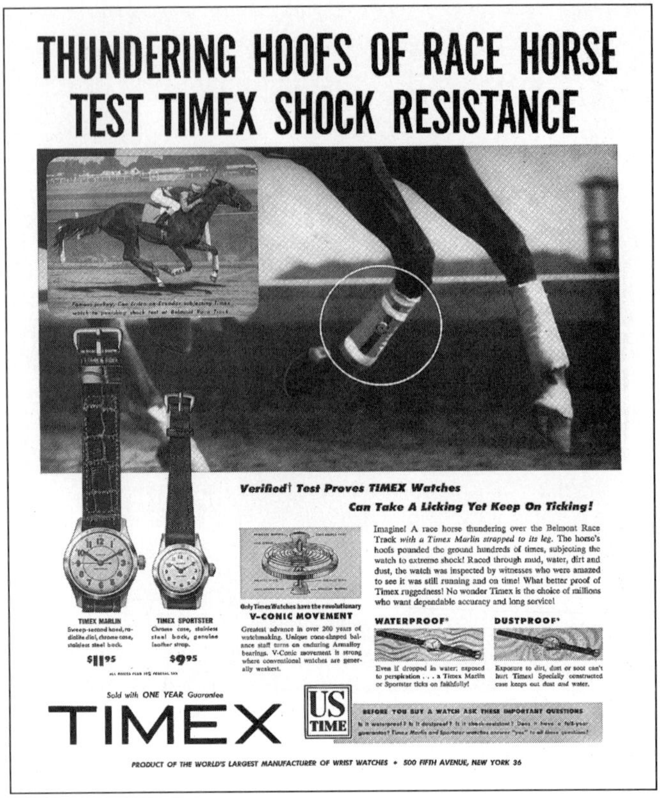

fivefold increase over the previous year. Advertisers bought $12 million in television time in 1949. In 1951 they spent 10 times that much, $128 million.

Television commercials, by virtue of the fact that consumers could see and hear the product in action, were different from advertising in all other media. The ability to demonstrate the product—to wash those Keds, to subject a Timex watch to a torture test such as strapping it to a galloping horse's hoof—led to the development of the **unique selling proposition (USP),** showing the aspect of a product that set it apart from other brands in the same product category. Once an advertiser discovered a product's USP, repeated demonstration commercials could drive home this message. Inasmuch as most brands in a given product category are essentially the same—that is, they are **parity products**—advertisers were often forced to *create* a product's USP. Candy is candy, for example, but M&Ms are unique—they melt in your mouth, not in your hand.

The Evolution of Television Commercials

We saw in Chapter 3 how advertisers and their ad agencies produced early television shows, only to lose that function after the quiz show scandal in

1959. Once advertisers no longer owned program content, it became impossibly expensive to be the sole sponsor of a network broadcast. Networks were demanding a great deal of money to cover their investment in a program. It was now more efficient and more profitable for advertisers to spread commercials across a number of programs, thereby reaching many more viewers. The networks, profiting from this new state of affairs, began selling commercial time in 60-second segments. Because there were now many different "spot" commercials for many different products in the same show, the nature of the commercials themselves changed. A spot had to stand out and be remembered—and greater creativity came to television advertising.

In addition, a number of commercials in an individual program or on a given night of television were selling the same or similar products, and many of these products were essentially the same in quality and cost. The USP became even more important, as did the need to boost **brand awareness,** or the identification with a particular manufacturer of a given product. Brand identification was achieved through slogans and jingles: "You'll wonder where the yellow went, when you brush your teeth with Pepsodent."

Commercials in television's early days were highly product oriented, but this changed in 1971 as a result of government regulation. Cigarette companies had been television's biggest advertisers, but because of mounting challenges from health groups, Congress banned cigarette commercials from the airwaves in 1971. With the loss of this important category of advertisers, the networks quickly discovered that too few advertisers could afford to buy a 60-second spot on a popular prime time program.

Lowering the price of advertising time was a possible solution, but rising production costs for network shows made this difficult. In 1953, for example, a 1-hour episode of *Studio One* cost CBS $30,000 to produce. In 1975 *Gunsmoke* cost that same network $230,000 for 1 hour. The networks' solution was to split the 60-second spots into more affordable 30-second segments. Now, an even greater number of advertisers could afford television, and at the same time, the networks could make more money by selling two 30-second spots at more than the price of one full 60-second spot, making each minute of advertising time more profitable. The networks had solved their problem, but advertisers had a new one—their commercials were now aired among twice as many competing messages. In 1965 every network television commercial was a full 60 seconds, although 23% were **piggybacked;** that is, a single sponsor presented two products in the same minute. By 1975 only 6% were a full minute long.

The number of commercial spots grew and so, too, did the number of commercial minutes. In 1967, for example, there were an estimated 100,000 commercial minutes aired on the networks; in 1974 there were 105,622 commercial minutes—5,600 additional minutes on the three commercial networks alone. As the number of spots grew and their length shrank, commercials became less about the products and more about the people who use them. There was too little time to give any relevant information, and image advertising came to television.

Today the price of commercial time continues to rise, production costs for programming continue to increase (the average prime time drama now costs $1.5 million an episode to produce), the number of commercials on the air continues to grow, the airtime dedicated to advertising continues to expand, and, as a result, the length of commercial spots continues to shrink. Fifteen- and even 10-second spots are now common. This means that more and more spots are crammed into a commercial break, producing **clutter.** How bad is television's clutter? In 1999 the four major broadcast networks each averaged over 16.5 minutes of nonprogram material (commercials, announcements, station identifications, and public service announcements) per prime time hour, a 21.8% increase over 1991. The most cluttered were ABC's *Norm,* with 10 minutes and 19 seconds of ads, and the same network's *Sabrina the Teenage Witch,* which contained 10 minutes and 4 seconds of commercial time per 30-minute episode. The networks' daytime schedules averaged 21 minutes and 2 seconds of advertising per hour. On cable television the problem was just as severe. For example, Fox Family gave its viewers an average of 18.25 minutes of non–program material per hour, and MTV 17.19 minutes (Jessell, 2000).

Combating Clutter

How does a television sponsor get heard and seen above the clutter? The answer for many, if not most, contemporary advertisers is to give even less information about the product and place greater emphasis on style, image, graphics, and look. More specific strategies abound, including creating spots that are visually and aurally disorienting. Examples of the use of visuals to cut through the clutter are heavy use of black and white, computer graphics, and atypical camera angles. Indicative is Gap's 1999/2000 campaign for its khaki pants. The commercials contain no announcer's voice-over and no product information. But in spots entitled "Gap Soul," "Gap Country," and "Gap Swing," beautiful khaki-clad people are shown against a bland background dancing, sometimes shot from above, sometimes from below, and sometimes in computer-assisted stop-and-go action.

Striking use of sound comes in at least two forms: attention-getting sounds and silence. Increasingly, television sponsors are turning to celebrity voices (not the typical imploring television pitchman) to grab viewers. Among them are Donald Sutherland for Volvo, Richard Dreyfus for Honda, Candice Bergen for Sprint, and Jerry Seinfeld for American Express. Nostalgic music, too, is heavily employed, for example, Burger King's long-running campaign of spots featuring '60s and '70s rock and rhythm and blues classics. Silence is used not only to attract attention but to avoid making direct claims about a product that might otherwise not be completely true. Good examples are the 20 seconds of complete quiet in the 30-second 1999 Land Rover spot in which a little boy waits for his friend's mother to find him in her spacious vehicle, and the 15 seconds of soundlessness as a dog plays fetch with herself in an abandoned city to demonstrate that everyone has left town because of British Airways' low fares.

Interep is one of two major national rep companies. Katz is the other.

Another way to be noticed among the crowd is to be offensive. In a 1999 spot for the paging company SmartBeep, a pretty woman passes gas on a blind date. Online computer retailer Buy.com ran spots in that same year of a man sniffing a dog's rear end. Another online computer retailer, Outpost.com, countered with three complementary spots. One has the company's name being tattooed on a little girl's forehead; the second shows a pack of wolves attacking a marching band and tearing its members to shreds; a third has a gerbil being fired from a cannon and splattered against a wall. These contemporary clutter-cutters may say little about the products they tout, but that is not their intention. "If you're going to go conservative," says Buy.com's founder Scott Blum, "you're going to get lost" (as quoted in McCollum, 1999, p. 2C).

Contemporary Radio Advertising

Since the introduction of locally based format radio, the medium has experienced increases in annual billings in every year (see Chapter 2). We saw, too, that radio time is sold on local, national spot, and network bases, with local billings accounting for approximately 80% of all billings, national spots accounting for 16% to 18%, and network sales accounting for the remainder.

Radio stations sell their time to advertisers on three levels. First is **retail sales.** Sometimes called **direct sales,** this involves direct, one-to-one contact between the stations' salespeople (most often called account executives) and advertisers in the stations' signal area. Account executives generally earn 15% commission on retail sales, and these sales account for the large majority of stations' billings. The second level is **local sales,** that is, selling time to advertisers through local advertising agencies. Station account executives normally realize from 6% to 8% commission on local sales. Finally, there are **national spot sales,** when a national advertiser places commercials on a number of stations across the country. National spot sales typically involve a **national sales representative** company, commonly referred to as a station's "rep." Local stations use their reps—companies such as Katz and Interep—to sell their local time to national advertisers, and the rep typically receives a commission of between 5% and 12% on sales. The use of reps by virtually all radio stations has led to the development of **unwired networks,** which are groups of otherwise unrelated stations that are sold to advertisers as if they were in fact actual broadcast networks. In other words a rep "joins" large numbers of its client stations into a "network" for the purpose of placing a national sponsor's spots.

The Internet has created another way that radio stations can sell their **avails,** or unsold commercial spots. Companies such as broadcastspots.com register stations (radio and television) and cable channels as affiliates. Then, at any time of the day or night potential advertisers (local or national) can go to broadcastspots.com on the Web and instantaneously buy avails (as listed by these affiliates). The advertiser pays no fees; these are paid by the affiliate. Once an advertiser buys time, its commercial can be aired by the affiliate within 12 hours. (Of course, the advertiser must get the commercial to the carrying outlet in that time.)

Radio Sales

Regardless of the level of a given sale of time or how it is purchased, the price charged for that time is detailed on a station's **rate card,** which is the station's statement of advertising fees and services. What a station can charge is a function of the size and composition of its audience (see Chapter 8), but the cost is also affected by how much time the sponsor buys and for how long a period. Therefore, rate cards typically include several pieces of information important to potential time buyers:

1. *The fee schedule for airtime for various parts of the broadcast day.* Most rate cards present this information in the form of a grid, with the most expensive time—typically 6 a.m. to 10 a.m., or morning **drive time**—designated AAA, and other dayparts designated as AA (afternoon drive time), A (daytime), B (evenings, 7 p.m. to midnight), and C (midnight to 6 a.m), with fees descending as the grid moves toward C.

2. *Details of the station's policies regarding discounting.* For example, a station might offer an advertiser this or a similar deal: if on air for 52 consecutive weeks, all announcements are provided a 10% discount, and a 5% discount if on air 26 consecutive weeks.

3. *Details of the station's* ***rate protection policy.*** This means that if an advertiser buys time for a given rate for a 6-month period and the station changes its fee schedule during that time, the sponsor is guaranteed the original rate until the end of those six months.

4. *The station's policy regarding terms of payment and commissions.* Common terms include that invoices are due and payable when rendered, cash in advance is required unless credit is approved beforehand, and a 15% commission is paid to recognized ad agencies on net charges.

NEW WAVE BROADCASTING, L.P. – RATES PER STATION – AS OF MAY 11, 2000

CLASSICAL KBOQ/95.5 FM

DAYPART	MON	TUE	WED	THU	FRI	SAT	SUN
5AM-10AM	$35	$35	$45	$45	$45	$20	$20
10AM-3PM	$35	$35	$45	$45	$45	$30	$30
3PM-8PM	$35	$35	$45	$45	$45	$25	$25
8PM-12MID	$25	$25	$20	$20	$20	$20	$20
5AM-12MID	$25	$25	$35	$35	$35	$25	$25

KBTU/THE BEAT OF THE BAY/101.7 FM

DAYPART	MON	TUE	WED	THU	FRI	SAT	SUN
5AM-10AM	$35	$35	$60	$60	$60	$25	$25
10AM-3PM	$35	$35	$60	$60	$60	$40	$40
3PM-8PM	$35	$35	$60	$60	$60	$30	$30
8PM-12MID	$20	$20	$20	$20	$20	$20	$20
5AM-12MID	$25	$25	$35	$35	$35	$25	$25

MODERN MUSIC/KCDU/CD93/93.5 FM

DAYPART	MON	TUE	WED	THU	FRI	SAT	SUN
5AM-10AM	$45	$45	$70	$70	$70	$25	$25
10AM-3PM	$45	$45	$70	$70	$70	$40	$40
3PM-8PM	$45	$45	$70	$70	$70	$30	$30
8PM-12MID	$25	$25	$25	$25	$25	$25	$25
5AM-12MID	$30	$30	$45	$45	$45	$30	$30

NEW ROCK KMBY/104.3 FM

DAYPART	MON	TUE	WED	THU	FRI	SAT	SUN
5AM-10AM	$25	$25	$35	$35	$35	$20	$20
10AM-3PM	$25	$25	$35	$35	$35	$30	$25
3PM-8PM	$25	$25	$35	$35	$35	$20	$20
8PM-12MID	$15	$15	$15	$15	$15	$15	$15
5AM-12MID	$20	$20	$25	$25	$25	$20	$20

KPIG/107 OINK 5 FM

DAYPART	MON	TUE	WED	THU	FRI	SAT	SUN
5AM-10AM	$50	$50	$80	$80	$80	$30	$30
10AM-3PM	$50	$50	$80	$80	$80	$60	$40
3PM-8PM	$50	$50	$80	$80	$80	$40	$30
8PM-12MID	$25	$25	$25	$25	$25	$25	$25
5AM-12MID	$40	$40	$50	$50	$50	$40	$40

Narrowing of any daypart...add $10 to rate

With any two (2) station buy, client receives a 5% discount.

With any three (3) station or more purchase, client receives a 20% discount.

*(Note: Discounts require even distribution over all stations purchased.)

New wave broadcasting offers potential advertisers a rate card showing the cost of time during different dayparts, on different days, and on its different stations in central coastal California.

5. *The nature of the ad copy acceptable to the station, as well as the timing required.* For example, all copy must be consistent with the public interest, and copy must be at the station at least 48 hours prior to the scheduled broadcast date.

6. *Information on the station's feature and spot rates.* **Features** in radio include content such as traffic and business reports, newscasts, sports, and weather. Advertisers favor features for two reasons. Features are attached to the kind of programming to which listeners pay particular attention, and they are usually presented live by a DJ or announcer who is a recognizable (and presumably favored) personality. For this reason, stations typically require regular ongoing sponsorship for features; for example, the rate card may stipulate that all features are subject to 4 weeks' notice for renewal or cancellation.

Radio as an Advertising Medium

In 1999, 12.51% of the nearly $133 billion spent to buy advertising time and space in the major U.S. media was spent on radio, second only to newspapers

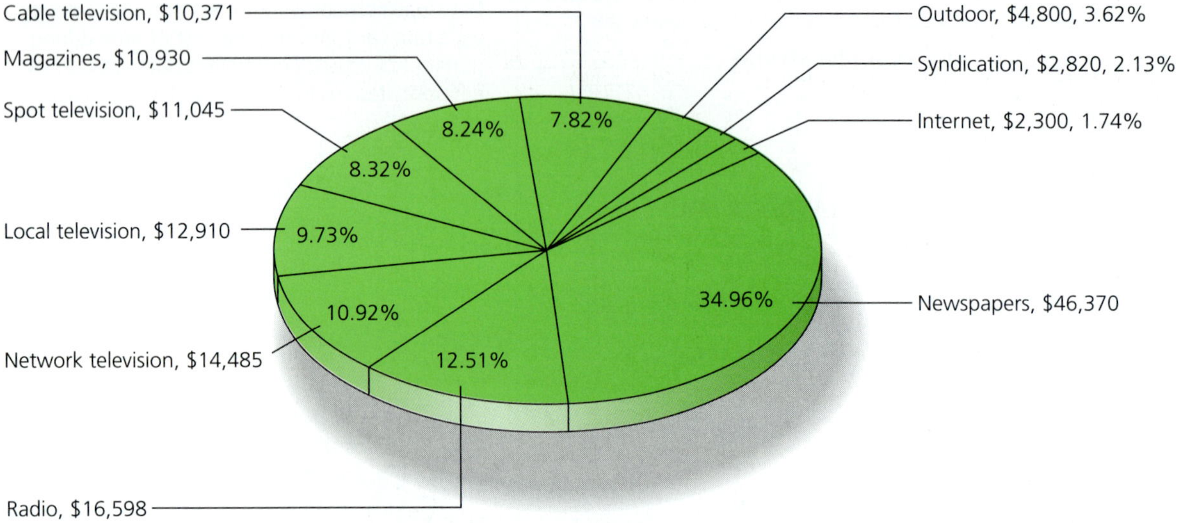

Figure 9.1 Consumer Advertising in Different Media, 1999
Source: "Advertising Outlook," 1999.

("Advertising Outlook," 1999). Figure 9.1 shows a detailed breakdown of advertising dollars spent in 1999.

Why did advertisers spend $16.5 billion on a medium that, after the introduction of television, was declared by many to be dead? Charles Warner and Joseph Buchman (1993) offer several good reasons:

1. *Radio effectively and efficiently targets desirable, discrete demographic groups.* Radio's narrow formats are particularly timely in the contemporary era of increased audience fragmentation (Chapter 1). Advertisers can be confident of reaching upscale classical music fans, older easy listeners, or teen alternative rockers.

2. *Radio is uniquely intrusive.* As Warner and Buchman put it, "Listeners have to make a conscious effort not to hear a commercial" (p. 189). Television viewers go to the refrigerator during commercials, and magazine readers flip past ads. But radio spots tend to blend seamlessly into the content that surrounds them, especially on features such as sports, news, and talk programming.

3. *Radio has the lowest commercial production costs of any mass medium.* Audio is inexpensive to produce, and many radio spots are simply read by an on-air personality or announcer.

4. *Radio typically has the lowest rates and most efficient cost-per-thousands of all media.* This permits sponsors to buy a greater number of spots over a wider period of time, increasing their net reach, frequency, and gross impressions, further lowering their cost per thousand, or CPM (Chapter 8).

Radio is mobile.

5. *Several stations can be combined at a low cost for more effective reach.* Advertisers can buy time on a large number of stations in a given area to increase their reach. When a sponsor runs its spot at exactly the same time on multiple stations, it is called a **roadblock.**

6. *Radio is the most flexible of all media.* A spot can be ordered, produced, and aired in one day. In addition, a single sponsor, a hamburger joint, for example, can have its morning commercial ("Get a breakfast burrito and a cup of coffee for $1.99"), a midday spot ("Bring the whole gang in on the way to the beach"), a dinnertime spot ("After a hard day's work, you deserve a break"), and a late night commercial ("After the show, why not drop in for a burger and a shake?").

7. *Radio is mobile.* No other medium can reach people in as many different locations—the beach, the car, the ball park, the running track, the camping trip.

8. *Radio usually has good suburban reach.* This is important when selling radio time against newspapers, which are the primary competitor for local advertising dollars in most metropolitan areas.

9. *Radio is more emotional than print media.* Disc jockeys can make listeners laugh, add emphasis, play a hummable jingle, and alter their inflection. Print ads can do none of these things.

Radio has three additional strengths as an advertising medium:

10. *Radio offers timeliness.* The low production cost of spots (especially announcer-read commercials) and the potential availability of large amounts of commercial time enable the creation and placement of almost instant commercials. For example, a car dealer can attach its name and business to welcoming home the victorious high school synchronized swimming team within hours of the championship match.

11. *Radio engages the imagination more easily than do print media.* There is virtually nothing that a good audio production person cannot create or conjure up for listeners who "see" radio with their mind's eye. Print ads are limited to what can be presented visually.

12. *Radio is particularly well suited to reaching the teen audience.* This group has a low level of newspaper readership but is a desirable demographic for advertisers.

But radio also has limitations as a commercial medium:

1. *Radio is ephemeral.* A spot is gone as soon as it airs. Listeners normally lack the time (as well as handy paper and pencil) to write down sponsor information. Newspaper and magazine readers can refer to a print ad; radio listeners can't ask the DJ to repeat the information. Radio attempts to overcome this particular deficiency through repetition and frequency, both made possible by radio's relatively low cost.

2. *Radio is fragmented.* Especially in major markets, there are many, many radio stations from which sponsors can choose. As a result, advertisers are often forced to buy time on more than one station. But again, radio's relatively low cost makes this practice common.

3. *Radio listening is open to distraction.* Because only one sense (hearing) is involved in listening, people are frequently doing something else while they listen. As a result, their attention may not always be on the spots (especially when the radio is used as background noise).

Contemporary Television Advertising

Despite continued loss of audience to competing video technologies such as cable and satellite, television remains the "massest" of the mass media. No medium simultaneously delivers as many people to advertisers as does television (Figure 9.2). In fact, despite constant predictions of the demise of over-the-air television, in 1998 television stations and networks had combined billings of $39.4 billion (*Broadcasting & Cable Yearbook, 2000*), and in the sum-

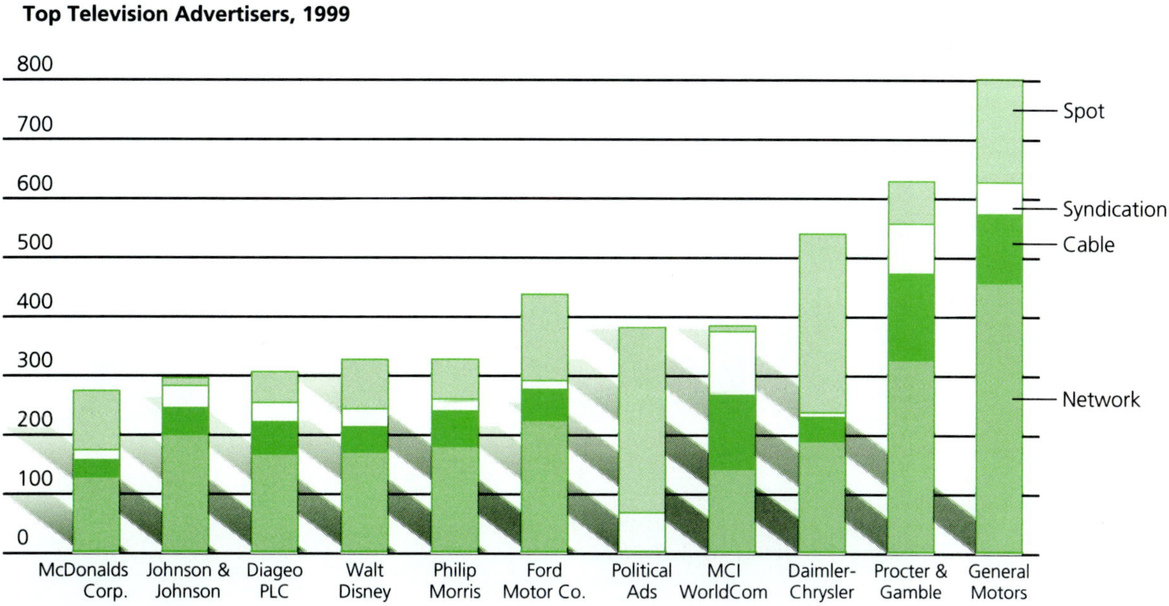

Figure 9.2 **The Biggest Television Advertisers, 1999 (in millions of dollars)**
Source: McClellan, 1999c.

mer of 2000 the national networks experienced a record level of **upfront sales** for their upcoming 2000–2001 prime time schedules, selling off a full 75% of their inventory in one week for $8.3 billion, a $1 billion increase over the previous year (McClellan, 2000).

Television Sales

In many respects, what we've already said about radio advertising holds for television advertising too. Both employ rate cards and reps, and television and radio stations both sell local, national spot, and national network time. Both radio and television depend on ratings to set commercial rates, and both industries dislike the audience-counting methods on which they rely (Chapter 8). But there are important differences as well. For example, network advertising is far more important to the television industry than it is to radio, and radio spots are more likely to employ an on-air, live announcer.

Television time is sold in 10-, 15-, 30-, and 60-second spots, with 30 seconds being most common. When a spot runs between two shows, it is called an **adjacency;** when it airs inside a program, it is a **participation;** when it airs inside a show, but stands alone, without other commercials, it is referred to as an **island.** As we've already seen in this chapter, islands are rare, with most spots residing among the clutter of several others. Such groupings or

Hallmark Hall of Fame's *Winter's End. The Hallmark Hall of Fame* is one of the last sole-sponsored network television programs, although it runs irregularly, sometimes less frequently than once a year.

clusters of commercials, regardless of where they appear in a program schedule, are called **pods.**

Typically, stations and networks guarantee advertisers that their spots will not occupy the same pod with ads for similar products and services. What they cannot guarantee, however, is that there will be no conflict between spots aired by a local affiliate that abuts those carried by the network. What is the best (and therefore the most expensive) location for a commercial in a pod? The generally accepted answer is the last position "in an in-program cluster that appears toward the end of the program, just before the climax or denouement of the plot" (Warner & Buchman, 1993, p. 196).

Unlike radio, television time is not sold in dayparts but on a specific program basis. Often, though, spots are bought on a scatter basis as well. That is, an advertiser's spots are scattered or rotated throughout programming that appeals to its targeted demographic. For example, television advertisers often buy, and television account executives often sell, in terms of a "18–49 female scatter" or a "preschool kid scatter."

Individual program sponsors are rare (Chapter 3). *The Hallmark Hall of Fame* is one enduring sole-sponsored program, but it runs irregularly, sometimes less frequently than once a year. Occasionally, though, a program will have only a few participating sponsors. This is not uncommon in sports, where a beer company, an airline, and a car maker will join to "bring you this game."

Another difference between radio and television—and between network and local television—is the use of **guaranteed CPMs.** That is, television networks, both broadcast and cable, normally guarantee their advertisers a certain return for their money, computed as the cost per thousand (CPM). After a sponsor's schedule of ads has run, its advertising agency and the network conduct a **postbuy analysis** using the Nielsen ratings to determine if the network has delivered the agreed upon number and type of viewers. If it has not, the network will run for free a specified number of additional commercials, frequently called **make-goods,** to compensate for the deficiency. You can see why the networks are among the most ardent critics of the ratings (Chapter 8). Not only do the numbers affect what networks can charge up front for commercial time, but the inadequacies of the ratings can cost the networks money even after the commercial time is sold.

Television as an Advertising Medium

Television is a powerful advertising medium. One example is the success of Hotjobs.com. The fledgling online job board bought 30 seconds on the 1999 Super Bowl for $2 million. In the 30 days of January before the spot ran, the site counted 1.2 million visitors. In the 2 days following the January 31 broadcast of its day-dreaming security guard spot, it counted 6 million visitors (Rosenwein, 1999b). But of course, not all ads appear on the Super Bowl. Still, according to Warner and Buchman (1993), television has many strengths:

1. *Television is the most persuasive medium.* Television combines images, sound, motion, and color, and it permits the demonstration of products.

2. *Television is intrusive.* Commercials live in and around the content people are watching. People must make a conscious effort *not* to see them. The use of celebrities and interesting audio and visual elements sometimes make commercials the best part of television.

3. *Television has the greatest appeal of all media.* Television is everywhere—homes, schools, bars, stores. Its appeal is universal.

4. *Television engages people's emotions.* You may hate the Energizer Bunny, but he moves you. We get a lump in our throat when AT&T shows us family

and friends reaching out to touch one another. Television's images, sound, motion, and color can be effectively utilized and combined in limitless creative ways to touch our hearts and minds.

5. *Television commercials are memorable.* Remember these? "Be all that you can be." "Grab the gusto!" "Just do it." "Got milk?" "You deserve a break today." "Did somebody say McDonald's?" "It's the real thing."

6. *Television is credible.* Industry research indicates that the public considers television commercials the most believable of all forms of advertising.

7. *Television is flexible.* Commercial time can be bought in various lengths, at various times of the day, in different types of content. Only radio advertising is more flexible.

8. *Television advertising conveys prestige and glamour to its sponsors.* The best brands advertise on television. Why else do we often see "As seen on TV" even in print ads? In addition, association with special or quality television content boosts advertisers' images.

There are, however, some disadvantages to television advertising:

1. *Television ads often appear amid clutter.* But, as we've already seen, there are strategies to cut through the noise.

2. *Television is expensive.* Where its CPM is often quite good because of its reach, it is still a very expensive medium in real dollars, both in the cost of the production of the spot and in buying time to air it.

3. *Television forces advertisers to pay for wasted coverage.* Everyone watches television, but not every viewer is of interest to all advertisers. Because of its enormous reach, television is simply more difficult to target than are radio and magazines. The convergence of digital technology and television may help overcome this particular disadvantage. Digital retrieval systems such as TiVo and Replay Networks (Chapter 3) permit viewers to select the specific versions of commercials that most meet their needs.

4. *Television viewing is seasonal.* More people watch when the weather is bad; fewer watch when it is good. So advertisers of summer-oriented products may not be as well served by the medium as some others. Of course, those smaller summer audiences cost less to reach than do January and February's larger audiences.

Contemporary Cable Television Advertising

Advertising on cable television networks operates just as it does for network broadcast television, except the audience for any given cable channel will be significantly smaller. NBC's top-rated *ER* will usually draw a 19 rating. Cable's ratings champ, World Wrestling Federation, has a good day if it draws a 3.

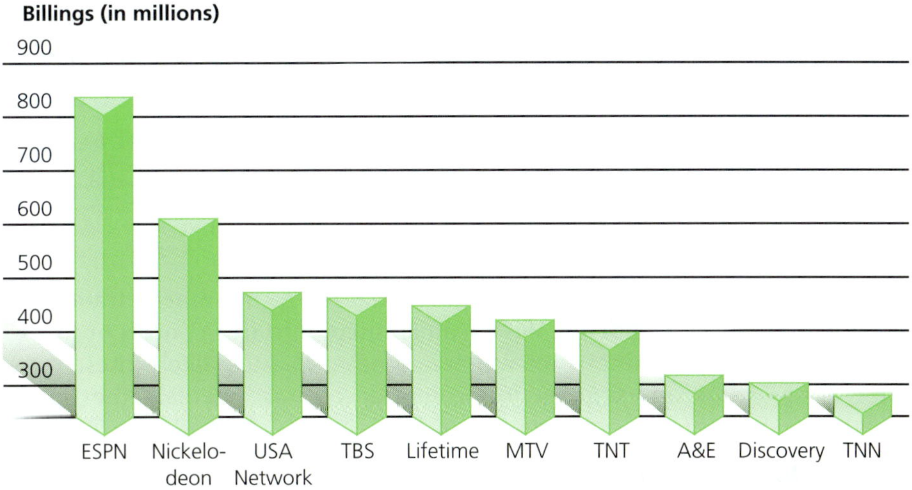

Figure 9.3 Cable's Billings Champions, 1998
Source: McAdams, 1999.

However, cable channels offer more homogenous audiences than does over-the-air television. Local cable operators also sell time, and they do so in a manner not unlike their local broadcast counterparts.

Cable as an Advertising Medium

Cable experienced record billings in 1998, taking in more than $10 billion in national network sales (Figure 9.3). In addition, cable, like radio and broadcast television, also earns advertising dollars from local and national spot sales.

What are cable's strengths as an advertising medium?

1. *Cable is demographically targeted.* There is little "wasted" reach for advertisers who wisely buy cable's demographically narrow content.

2. *Cable's audience is young and upscale* when compared to that of broadcast television. Cable households have higher incomes and levels of education than do noncable homes.

3. *Cable has a high inventory of available commercial time.* Few national cable networks sell all their avails. Local cable operators can offer an advertiser spots on 70 or more channels.

4. *Cable has relatively lower costs.* With such a high inventory, cable advertisers benefit from lower prices for advertising time.

5. *Cable is a good supplementary advertising medium.* Because of its low cost and a ready supply of commercial time, advertisers can use cable advertising to back up or support their ad efforts in other media.

What are cable's deficiencies as an advertising medium?

1. *Low ratings.* Compared to broadcast television, cable stations have a very limited number of viewers.

2. *Lack of penetration.* Despite its tremendous growth (Chapter 6), cable still reaches far fewer homes than does broadcast television.

Barter and Co-op Advertising

Not all radio, television, and cable time is sold for cash. Some is traded for products or services in **barter.** We are all familiar with free concert or movie tickets offered by the stations we watch and listen to. Most likely, these were given to the station in exchange for promotion of those events. Other sponsors that make heavy use of barter are airlines, hotels, and cruise lines. These companies have **perishable inventories;** that is, once the plane leaves the ground, its empty seats are wasted. An empty hotel room brings in no income, and empty cabins on a ship produce no revenue. Why not, then, exchange these things, which otherwise have no value, for airtime that does? Product manufacturers, too, sometimes have an incentive to barter. Especially in the computer and other electronic industries where products rapidly become obsolete or "last year's model," sponsors have goods with little value to them that may be of much value to a station.

But what incentive is there to the broadcaster or cablecaster to barter? Unsold commercial time is as useless as an empty airline seat, hotel room, or cruise ship cabin. Therefore, the advantage of barter to both time buyer and time seller is obvious. There is, however, an important disadvantage to the buyer. Bartered time is invariably shunted aside in favor of sold-for-cash time, and many large market stations regularly sell out their inventory of commercial time, especially in times of economic growth. One way advertisers overcome this handicap is to add some cash to the deal; that is, they pay for airtime with a combination of barter and cash.

Not all airtime that is sold for cash is paid for by the advertisers themselves. Two ways this happens are **advertising allowances** and **cooperative advertising allowances.** Although there is a subtle difference between the two, both are universally called **co-op advertising.** Here is how it works. Advertising allowances are computed as a proportion of a retailer's purchase order from a manufacturer. A typical allowance is 3%. Therefore, if a local retailer orders $200,000 worth of products from the manufacturer, the manufacturer will give that retailer $6,000 toward advertising those products. Computer and high-end electronics companies frequently support their local outlets in this way.

Cooperative advertising allowances are matching funds arrangements between local retailers and manufacturers. Up to a specified limit, the manufacturer will match dollar-for-dollar the local merchant's commercial time buys. Apparel, packaged goods, and outdoor and lawn equipment manufacturers commonly provide co-op support for local merchants.

Public service advertising allows advertisers to use their skills to serve society.

Other Forms of Broadcast and Cable Advertising

Not all the nonprogram material we see and hear in radio and television are spot announcements or commercials. Common but controversial are **program length commercials (PLCs),** often called **infomercials.** These 30- and 60-minute commercials tout the benefits of products and services such as weight loss schemes, psychics, and cosmetics. Broadcasters and cablecasters like them because the PLC's producers pay to air them (sometime for cash, sometimes for a split of the money made from sales originating in the advertising medium's service area), and they are good filler in otherwise hard-to-sell time periods such as late night and Sunday morning.

Another form of broadcast and cable "advertising" is the **public service announcement** (**PSA**), spots promoting nonprofit causes and organizations. Stations make time available to groups such as the United Negro College Fund, the American Heart Association, and others for free, sometimes helping in their production, sometimes simply airing those that are provided.

Broadcasters and cablecasters also run **promotional spots,** or **promos.** These commercial-like announcements promote their own or their networks' upcoming schedules. Production of these spots is frequently done in-house, but just as often the spots are provided by the networks or cable channels.

The Real World *Identifying and Evaluating Transaction Journalism*

Transaction journalism comes in a number of forms: placing links to businesses along side of and even in Web editorial content, "ownership" of given Web content by advertisers, and Web sites selling the access history of specific users to advertisers. All are controversial, and that controversy came to light in the celebrated 1999 flap over Amazon.com's practice of selling prime space on its Web pages to book publishers intent on pushing specific titles without informing users that these were, in fact, commercials and not objective or professional recommendations.

Many e-tailers defended Amazon.com, arguing that this was no different from the common practice of a potato chip manufacturer paying a supermarket for ideal placement of its product or book publishers paying book stores to display specific titles more prominently than others. But others took the opposite view. That is, they argued that transaction journalism raises significant questions of credibility, not only of the improperly (in their view) placed commercials but of the sites that use them—and of the Web itself as a medium of commerce. Self-interest, argue these critics, demands either control of the practice or, at the very least, informing Web users of its existence and use.

The Internet Content Coalition (<http://www.netcontent.org/>), a trade group representing a wide variety of Web content producers, and the American Society of Magazine Editors (ASME; <http://asme.magazine.org/>), the primary trade group for magazines—both hard copy and virtual—have each developed guidelines for the insertion of commercials in otherwise noncommercial content. The Internet Content Coalition proposes these guidelines:

- Any nonbanner advertisement, including portals, **buttons** (graphically presented links), windows, and special advertising sections must be clearly and conspicuously identified with the words "advertising," "advertisement," "advertiser link," or "special advertising section."
- Links to editorial content must not appear in commercials or other advertising (in order to maintain the separation of the two forms).
- Sites that use buttons in editorial content and for commercial purposes must clearly differentiate between the two.

The ASME proposes these guidelines:

- Links that appear inside editorial content must be under the control of the site's editors (rather than the advertisers).
- Online publications cannot sell outright or make a condition of any advertising sale a link from editorial content to any other site.
- Preferential treatment given to an advertiser in any search function such as a search engine or a "related sites" feature cannot be allowed as it is a betrayal of readers' trust.

What do you think of these guidelines? Are they reasonable? Are they even necessary? Why should the rights of Web site creators to do business as they see fit be limited because some people might confuse editorial content with the legitimate practice of advertising? And what about you? Can you always identify commercial and editorial content when you're online?

What about the convergence that is rampant on the Web? For example, many Web music sites are part magazine, providing readers with a wealth of information, and part record store, allowing users to order their favorite discs. Should sites such as this be subject to the same rules as more traditional online publications such as the *New York Times* or *Time* magazine? Shouldn't these more traditional journalistic sites be allowed some leeway in their competition with the more commercially oriented online sites?

Contemporary Online Advertising

The spread and success of online or **cyberadvertising,** the placement of commercials on various online sites, has been slowed for several reasons, including its relatively static look, unproved effectiveness, and an intrusiveness that many users find objectionable (see Chapter 7). Yet in 1999 sponsors spent $4.4 billion on online advertising, up from a mere $312 million in 1996 ("Advertising Outlook," 1999; "Online Ad," 2000). Beyond the obvious fact that more people were going online, making the medium more attractive to advertisers, what changed?

First, the form itself has matured, moving beyond the static **banner,** the online billboard placed across the top of a Web page that accounted for 56% of all Web ad spending in 1999 (Emert, 1999). Users are now likely to confront **interstitial ads,** images that seem to mysteriously appear and then disappear on the screen as users click from one page to the next; **pop-outs,** ads that appear in a smaller window at the border of the Web page being read; and **extramercials,** columns of ad content that slide down or move over the page's content.

Also becoming more common is Web advertising in which the commercial intrudes on the editorial content itself, as in a cyberad for the Intel Corporation in which Homer Simpson runs out of the ad proper and across the page. But another form of cyberadvertising that intrudes on editorial content is much more controversial. It is **transaction journalism,** in which the direct linking of editorial content to sales takes several forms. The *New York Times* online newspaper places links to bookseller barnesandnoble.com next to its book reviews. The *Times* gets paid every time a reader links from its pages to that of Barnes & Noble. Other online newspapers and magazines are even more bold. Some embed links to advertisers in the editorial content they run. Others sell specific stories or pages, for example, sports or business, to advertisers and those advertisers have exclusive rights to all links from those pages. Still others sell users' search histories to advertisers. If sufficiently interested in a given user, the advertiser pays the online content provider for the right to have its banners and pop-outs appear on that user's screen as he or she searches through the site (Effron, 1999b). There is more to learn about this form of online advertising in the box titled Identifying and Evaluating Transaction Journalism.

Other advances in online advertising are **intermercials** and **targeted keyword buys.** Intermercials are attractive, lively commercials that run while people are waiting for Web pages to download. Typically, users must download advertiser-supplied, free audio and video software to quickly and cleanly receive these intermercials. Targeted keyword buys occur when a sponsor buys the right from an Internet service provider to have its ads pop up on the screen every time a user types in the sponsor's name. Punch in Chevy, hit "search," and a Chevy ad, complete with links to dealers, appears.

Online Classified Advertising

Many online advertisers work to make their ads look and sound more like those found on radio and television. Other advertisers are expanding a form of advertising that is far more at home on the Net than it is on radio and television, classified advertising. In fact,

> the Web in many ways is a much better medium for classifieds than newspapers. The interactive functions of the Web enable users to move through massive databases in ways not possible in a pulp-based medium. Why plow through all the ads for Buicks and Oldsmobiles if what you want is just a red Honda no more than five years old with less than 60,000 miles? Plug in the variables, hit "search" and presto! Classified ad sites such as Autobytel.com and Monster.com were springing up on the Web; they offered faster and handier ways of posting and looking for anything you wanted—jobs, cars, apartments, houses, land in Florida, saxophones, and every thingamajig in grandmother's trunk. (Brown, 1999, pp. 54–55)

Sites such as Classifieds.2000.com, match.com, and messagemates.com offer general classifieds (for those thingamajigs in grandmother's trunk). Kbb.com, cars.com, and autoweb.com sell cars. Monster.com, careerpath.com, and careermosaic.com present job listings. Realtor.com, homefair.com, and apartments.com list real estate.

In 1998 online classified advertising represented only 1.3% of the total of all print-based classified advertising and 14% of all online advertising. But industry experts predict that by 2003 online classifieds will comprise 27% of all cyberadvertising (Stone, 1999) and one third of all classified advertising, pulling in more than $6 billion a year (Brown, 1999).

The Web as an Advertising Medium

There is no question that Internet ad billings are steadily rising, but like all the media we've discussed in this chapter, the Web has its advantages and disadvantages as an advertising medium. Here are some advantages of online advertising:

1. *Online users can be connected to an advertiser's own sales site through a link between the ad and that site.* Once there, users can make purchases with a few keystrokes. The frequency with which people move from the ad to the ordering site is called the **click through rate.**

2. *Online ads can be personalized with consumers' names and localized with advertisers' addresses.* Cyberadvertising permits the ultimate in audience fragmentation and direct marketing.

3. *Computer networks make possible ads on demand,* reducing wasted reach.

4. *The online audience is worldwide.* Traditional media are geographically located and regulated. But users are just as likely to access a Web ad residing on a server located in London as they are one in their own town. No other medium offers the Web's reach.

All commercial Web sites strive to be sticky. Qdecor achieves stickiness through a combination of good visual design, high levels of interactivity, and the offer of expert advice.

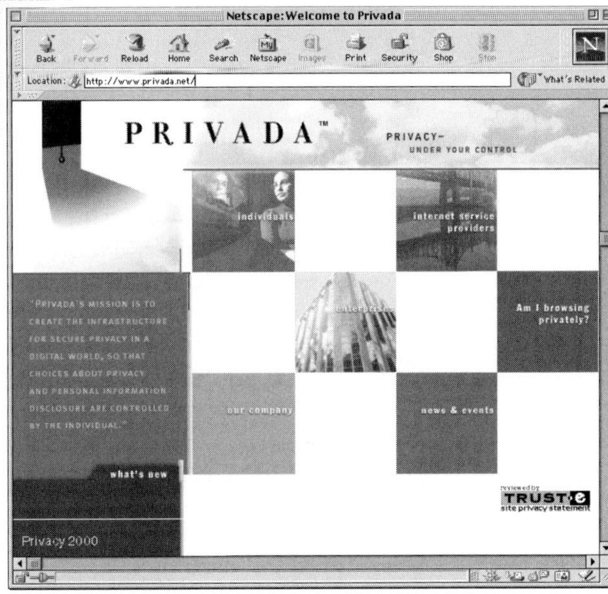

A technological solution to the problem of protecting personal information while engaging in e-commerce is the use of software that permits anonymous browsing. Zero-Knowledge is among the most widely used and respected.

5. *Online advertising has no schedule; its audience is distributed in time.* A big difference between online and more traditional media advertising is that with cyberadvertising there is less interest in such things as CPM, ratings, and shares. The Web is a **personal audience venue,** people use it individually on their own rather than on the programmer's schedule. Therefore, less important than the pure number of people that "tune in" is the relationship built between the advertiser and the consumer. This is why cyberadvertisers strive for **stickiness;** that is, they want their ads to hold and engage users. Still, in 1999, according to NetRatings, 40% of all online ad spending was based on CPM, 6% on performance measures such as click through rate, and the remaining 54% on a combination of the two (Emert, 1999).

6. *In cyberadvertising space and time are inexpensive.* As opposed to virtually all other advertising media, there are no constraints on the total space or time available to advertisers. A Web commercial site can be 1 or 20 pages deep; it can exist for 1 hour or 10 years, all for roughly the same cost.

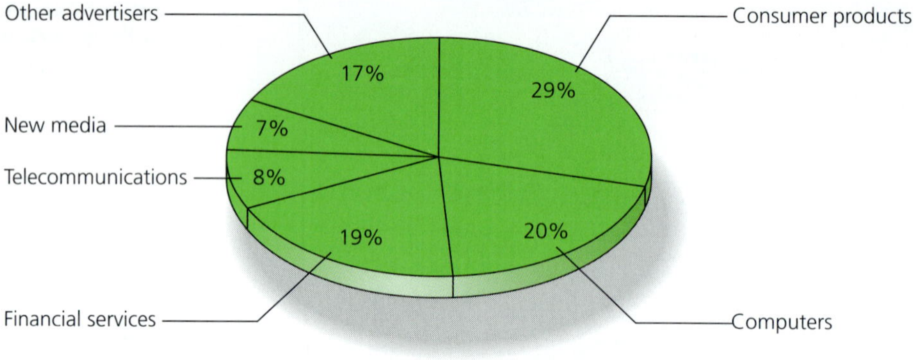

Figure 9.4 Who Advertised Online in 1999?
Source: Emert, 1999.

7. *Online advertising provides advertisers with information about users.* Despite the anonymity online users think they enjoy, every click of a key—the **click stream**—can be, and often is, recorded and stored. This happens whether or not the user actually enters information, for example, a credit card number to make a purchase or a social security number to verify identity. Software records what sites users visit, what they look at in those sites, and how long they stay at a given place in that site as they click from Web page to Web page. In and of itself, this is not a problem, because all the Web site reads about the user is a computer address. But once a user enters personal information, that "anonymous" address now has a name. This is an advantage for advertisers, but as you'll see here and in detail in Chapter 12, many users fear this loss of anonymity and privacy.

Online advertising, as we've already briefly seen, does have its limitations:

1. *Online commercials are relatively static when compared to other visual media.* But as bandwith increases and as more people hook to the Internet with more advanced cable and DSL modems (Chapter 7), the reception of attractive, clear audio and video commercials will become commonplace.

2. *The reach and effectiveness of online advertising is difficult to measure.* We saw in Chapter 8 that several companies are striving to overcome the particular problem of measuring reach. Measuring an online ad's effectiveness might be as simple as counting the number of people who click through to the sponsor's own site, but this is deceptive. There is no way to tell how many people were influenced by the online ad but chose to buy the product or service offline, in the actual world. With click-through rates averaging one tenth of one percent in 1999 (Emert, 1999), cyberadvertisers are convinced that the eventual effectiveness of their commercials may well lie in influencing real-world buying decisions (Figure 9.4).

The Advertising Environment

All media advertising takes place within the larger advertising industry, and the nature of and trends in that industry shape the advertising that appears online and on radio, television, and cable. Four important aspects of the contemporary advertising environment are regulating advertising, increasing audience segmentation (Chapter 1), measuring advertising's effectiveness, and criticizing advertising itself.

Regulating Advertising

The Federal Trade Commission (FTC) is the primary federal agency responsible for regulating advertising, and it filed 150 deceptive advertising actions in 1999 (Savan, 2000). But there is other regulation of advertising as well. The FCC regulates the commercial practices of the broadcasting industry, and individual states can police deceptive advertising through their own regulatory and criminal bureaucracies. In the deregulation movement of 1980 the FTC's oversight of the industry changed. In the past the FTC had initiated investigations of unfair and deceptive advertising. With deregulation they now only investigate and enforce regulations when complaints against deceptive advertising are made to them. As a result, most "regulation" of advertising is done by industry practitioners themselves. This self-regulation is discussed in detail in Chapter 13.

The FTC can impose a number of sanctions when it determines that an advertiser is guilty of deceptive practices. It can issue a **cease-and-desist order,** demanding that the practice be stopped. It can impose fines. It can order the creation and distribution of **corrective advertising**—a new set of commercials produced by the offender that corrects the original misleading effort. Offenders can challenge the commission's decisions in court, and they are innocent until proven guilty. Meanwhile, the potentially unethical advertising remains in the marketplace.

One of the FTC's greatest difficulties is finding the line between false or deceptive advertising and **puffery**—that little lie that makes advertising more entertaining than it might otherwise be. "Whiter than white" and "Stronger than dirt" are just two examples of puffery. On the assumption that the public does not read commercials literally—we know that the Land Rover–owning mother doesn't need 20 seconds to find her child in the vehicle's back seat—the courts and the FTC allow a certain amount of exaggeration.

The FTC and the courts recognize that commercials can be false in a number of ways. The most common advertising deceptions fall into these three areas:

- *Outright lies.* For years Wonder Bread commercials told us that this was the bread that "builds strong bodies 12 ways." When the FTC asked Wonder Bread to name them, it could not. Listerine mouthwash was long advertised as "preventing colds and sore throats or lessening their severity." It does neither.

- *Not telling the whole truth.* The FTC recently punished Winston cigarettes for its advertised claim that its cigarettes "contain no additives." The FTC ruled that although this statement was factual, it did not tell the whole truth—the absence of additives does not make one cigarette safer than another. Winston was required to insert corrective statements in all its advertising.
- *Lying by implication.* That is, using words, design, production devices, sound, or a combination of these to enhance a product's image. Television commercials for children's toys now end with the product shown in actual size against a neutral background. This is done because production techniques such as low camera angles and close-ups can make these toys seem larger or better than they actually are.

To protect themselves from charges of false advertising, advertisers must keep in mind three basic rules, according to FTC Associate Director for Advertising Practices Lee Peeler:

> 1) an ad cannot be deceptive—that is, mislead consumers [by any of the means detailed above] to their detriment. 2) Objective claims must be supported with competent studies. And 3) Advertisers are responsible for the reasonable implications of their ads to consumers. If an advertiser says, "I didn't really mean to convey that," well, that doesn't get them off the hook. All of advertising law is based on what consumers take from the ad. (as quoted in Savan, 2000, p. 62)

Increasing Audience Segmentation

As the number of media outlets for advertising grows, and as audiences for traditional media are increasingly fragmented, advertisers have refined their ability to reach and speak to ever-narrower audience segments, producing and distributing commercials along demographic and psychographic lines.

Demographic segmentation—appealing to audiences composed of varying personal and social characteristics such as race, gender, and economic level—has long been part of advertisers' strategy. Many products are touted in several different commercials, one featuring an African American family, another a Hispanic family, and a third a Caucasian family, for example. But advertisers are making increased use of **psychographic segmentation**—appealing to consumer groups of varying lifestyles, attitudes, values, and behavior patterns.

Psychographics entered advertising in the 1970s and are receiving growing attention as advertisers work to reach disparate consumers in increasingly segmented and cluttered media. A number of respected psychographic segmenting tools are frequently employed in creating, buying, and selling broadcast, cable, and online advertising. ACORN, ClusterPlus, PRIZM, MOSAIC, and VISION employ U.S. Census geographic information such as neighborhoods and zip codes to segment consumers on the assumption that people who live in a given narrowly defined area share similar lifestyles and values. **VALS** is a psychographic segmentation strategy that classifies consumers according to **v**alues **a**nd **l**ifestyle**s**, and this is perhaps the most frequently used tool. Developed by SRI International, a California consulting company, it divides consumers into nine VALS segments. Each segment is characterized by specific values and lifestyles, demographics, and—of greatest importance to advertisers—buying patterns. Advertisers try to create and place ads appealing to specific segments in the specific media and content consumed by the people occupying those segments.

Measuring Advertising's Effectiveness

It might seem reasonable to judge the effectiveness of a commercial campaign by a subsequent increase in sales. But many factors other than advertising influence how well a product fares. For example, changes in the economy, product quality, breadth of distribution, and competitors' pricing and promotion strategies all influence potential buyers. As a result, "for most advertisers most of the time, the answer to the essential question, 'What are you getting for your advertising expenditure?' must be 'I'm not sure'" (Sandage, Fryburger, & Rotzoll, 1989, p. 370). Advertisers find this a less than comforting response, and they often turn to research to provide greater certainty.

A number of techniques are used before a commercial or an ad campaign is released. **Copy testing,** measuring the effectiveness of commercial messages by showing them to consumers, is used for all forms of advertising. It is sometimes conducted with **focus groups,** which are collections of people brought together to see the commercials and discuss them with an advertiser's agency and own personnel. Sometimes copy testing employs **consumer juries.** These are people considered to be representative of the target market who review a number of approaches or variations of a campaign or commercial. **Forced exposure,** used primarily for television advertising, requires advertisers to bring consumers to a theater or other facility (typically with the promise of a gift or other payment) where they see a television program complete with the new commercials. People are asked their brand preferences before the show and are asked again after the show, providing a gauge of the commercials' effectiveness.

Once the campaign or commercial is before the public, a number of different approaches are called for. In **recognition tests** people who have seen a given program or evening of shows are asked, in person or by phone, whether they remember seeing specific commercials. In **recall testing** consumers are asked, again in person or by phone, to identify which broadcast commercials they find easiest to remember. This recall can be unaided, where the researcher offers no hints ("Have you seen any interesting commercials lately?"), or aided, where the researcher identifies a specific class of products ("Have you seen any interesting pizza commercials lately?"). In recall testing the advertiser assumes that an easily recalled commercial is an effective commercial. **Awareness tests** make this same assumption, but they are not aimed at specific commercials. Their goal is to measure the cumulative effect of a campaign in terms of a product's "consumer consciousness." A likely question in an awareness test, which is usually done by telephone, is "What brands of laundry detergent can you name?"

Careers *Working in Advertising*

When people interested in telecommunications careers consider working in advertising, they generally think of producing commercials, that is, employing audio, video, and digital production technology to make commercials. Most radio, television, and cable operations do produce commercials for their advertisers, but these tend to be for local advertisers and are prepared on relatively low budgets. More sophisticated national or regional commercials are typically produced by independent production houses hired by an advertiser's ad agency. Web ads are produced either by an ad agency with a Web department or an independent Web production house hired by the ad agency.

For someone interested in working in advertising, but not necessarily in the production of commercials and Web ads, a position at an advertising agency is an attractive option. The typical ad agency will have opportunities in a variety of areas determined by the size and services of the operation:

Administration is the agency's management and accounting operations.

Account management is typically handled by an account executive who serves as liaison between agency and client, keeping communication flowing between the two and heading the team of specialists assigned by the agency to the client.

The *creative department* is where the advertising is developed from idea to ad. It involves copy writing, graphic design, and sometimes the actual production of the piece, for example, radio, television, and Web spots; but production is more typically contracted out.

The *media department* makes the decisions about where and when to place ads and buys the appropriate time or space.

Market research tests the product's viability in the market, the best venues for its commercial messages, the nature and characteristics of its potential buyers, and sometimes the effectiveness of the ads.

Public relations departments can be found in many large advertising agencies.

Virtually any meaningful career in advertising requires a college or university degree, but the greatest success will come to people who can speak and write well. Advertising, in its many forms, is about communication—communication with clients and communication with consumers on behalf of clients. Solid writing and speaking skills, even for those involved in production, are necessary for the kinds of communication clients expect and demand.

What these research techniques lack is the ability to demonstrate the link that is of most interest to the advertiser—did the commercial move the consumer to buy the product? The value of these tests is in helping advertisers understand how people react to specific commercials and advertising strategies, aiding advertisers in avoiding costly mistakes, and assisting advertisers in planning and organizing the immediate and later campaigns. For more information about careers in advertising, see the box titled Working in Advertising.

Criticizing Advertising

Broadcast and cable commercials enter people's living rooms and are difficult to avoid. This advertising does sometimes offend, and it is often the focus of criticism. But advertising's defenders argue that advertising is important for these reasons:

- Advertising supports our economic system. Without it, new products could not be introduced and developments in others could not be announced. Advertising allows new products and businesses a chance to compete in the marketplace. As the "engine" of our economy, advertising fosters economic growth and creates jobs in many industries.
- People use advertising to gather information before making buying decisions.
- It is advertising's support that makes possible the "free" mass media we use not only for entertainment but for maintenance of our democracy.
- By showing us the bounty of our capitalistic, free-enterprise society, advertising increases the nation's level of productivity (as people work harder to acquire more wealth) and improves the standard of living (as people spend that wealth to acquire more of these products).

The first defense is a given. Ours is a capitalistic society whose economy depends on the exchange of goods and services. Complaints about advertising have less to do with the existence of advertising and more to do with its conduct and content. The second defense assumes that advertising provides *information,* but much broadcast and cable advertising is void of useful information about the product being advertised. The third defense assumes that the only way media can exist is through commercial support. But many nations around the world have built fine media systems without heavy advertiser support (Chapter 11), and the true cost of our "free" media is certainly included in the prices we pay for advertised products. To advertising's critics, the fourth defense—that people work hard only to acquire more things and that people's standard of living is measured by what material things they have—draws an unflattering picture of human nature.

Specific complaints about advertising are that it is often intrusive, deceptive, and in the case of children's advertising, inherently unethical. Advertising is said to foster monopolies and demean or corrupt the culture.

Commercials Are Intrusive Critics fault advertising for its intrusiveness. Few people welcome the number, length, and placement of radio and television commercials. Many Internet users complain about the commercialization of the new medium and fear advertising will alter its free, open, and freewheeling nature.

Advertising Is Deceptive Critics say that much advertising is inherently deceptive in that it implicitly and sometimes explicitly promises to improve people's lives through the consumption or purchase of a sponsor's products. Jamieson and Campbell (1997) described this as the "If . . . then" strategy. They wrote, "A beautiful woman uses a certain brand of lipstick in the ad, and men follow her everywhere. Without making the argument explicit, the ad implies that if you use this product you will be beautiful, and if you are

Is *Pokemon* a kid's television cartoon show or a 30-minute commercial for various toys and games disguised as a kid's television cartoon show?

beautiful (or use this product), you will be more attractive to men" (p. 242). They called the opposite strategy "If not . . . then not." When Hallmark says "When you care enough to send the very best," the implication is that when you do not send Hallmark you simply do not care.

Advertising promises health, long life, sexual attractiveness, companionship, popularity, success, and acceptance. Industry defenders argue that people understand and accept this as allowable exaggeration, not as deception.

Commercials Exploit Children The average child sees more than 20,000 television commercials a year (Stead, 1997). Critics contend that children are simply not intellectually capable of interpreting the intent of these commercials, nor are they able before the age of 7 or 8 to rationally judge the worth of their claims. This makes children's advertising inherently unethical. Television advertising to kids is especially questionable because children consume it in the home—with implicit parental approval—and most often without parental supervision. The question critics of children's commercials ask is this: If parents would never allow living salespeople to enter their homes to sell their children products, why do they allow the most sophisticated salespeople of all to do it for 20 minutes every hour every Saturday morning?

Critics see another example of commercials' exploitation of children in **toy-based children's television programming.** Young children have a hard enough time understanding television's genre conventions, but when they are presented with programs that feature characters that are also toys, that difficulty is exacerbated. How does a 5-year-old, for example, tell where the program ends and the commercial begins in shows like *Voltron, Men in Black, 101 Dalmatians, Animaniacs,* and *Pokemon*? These appear to be little more than 30-minute commercials disguised as kids' shows.

Advertising Is Conducive to Monopolies Advertising's defenders argue that advertising is necessary to create the informed, discriminating consumers necessary to sustain our capitalistic economic system. Yet critics claim that advertising, as it is practiced, can have just the opposite effect and can lead to the creation of monopolies. Ad critic Richard Draper (1986) explained how information-light advertising can lead to monopoly:

> Perfect competition requires highly standardized products, many buyers and sellers, and "perfect information." Every buyer knows what every seller proposes to charge for every product and feature. Price competition, the inevitable result, forces sellers to use their resources efficiently. Consumers benefit doubly—first, because all their products are as cheap as possible and, second, because the sum of society's wealth is greater than it would be in an imperfect market.
>
> The theory of monopoly concedes that advertising deflects consumers from the rational aim of buying in the cheapest market and makes them, instead, want a certain brand, irrespective of cost. (Since the company that produces the brand is the only one than can do so, it has a "limited monopoly.") (p. 15)

In other words, Nike can demand and get $180 for a pair of basketball shoes and Evian can make a profit selling water at $2 a bottle because they have a commercial-created, USP-defined monopoly on Nike and Evian.

Advertising Demeans and Corrupts Culture In our culture we value beauty, kindness, prestige, family, love, and success. As human beings we need food, shelter, and the maintenance of the species, in other words, sex. Commercials succeed by appealing to these values and needs. The basis for this persuasive strategy is the **AIDA approach:** to persuade consumers, commercials must attract *attention,* create *interest,* stimulate *desire,* and promote *action.* According to industry critics, however, problems arise when important aspects of human existence are reduced to the consumption of brand-name consumer products. Freedom is choosing between a Big Gulp and a canned soda at 7-Eleven. Being a good mother is as simple as buying a bottle of Downey Fabric Softener. Prestige is driving a BMW. Success is buying a Lexus. Love is giving your husband a shirt without ring-around-the-collar or your fiancée a diamond worth 2 months' salary.

Critics argue that ours has become a **consumer culture** in which personal worth and identity reside not in ourselves but in the products with which we surround ourselves. The consumer culture is corrupting because it imposes new definitions on traditionally important aspects of people's lives—

Critical Issues *The Right to Advertise Nothing*

The Media Foundation operates out of Vancouver, British Columbia. Its goal is to increase public awareness of overconsumption and overcommercialization, and advertising and commercials are among its primary targets. The group publishes a quarterly magazine called *Adbusters,* maintains an active Web site at <http://www.adbusters.org>, and sponsors events like the annual late-November "Buy Nothing Day" and the mid-April "TV Turnoff Week" (26 state governors officially endorsed this event in 1997). The group also makes video *uncommercials.* These public service announcements challenge well-known actual commercials and are available at no cost to broadcasters and cablecasters who wish to use them to "culturejam," or challenge, the prevailing commercial culture.

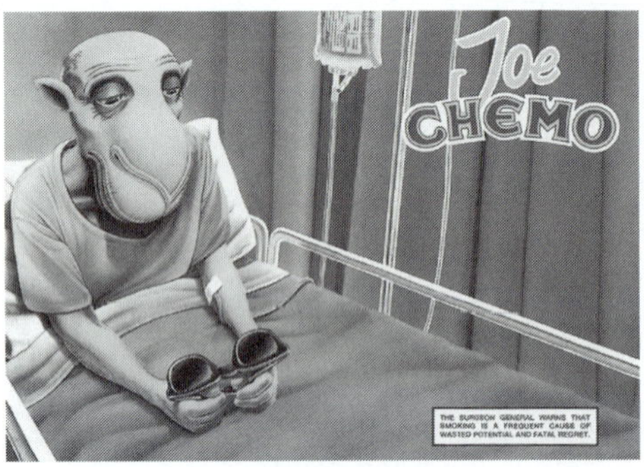

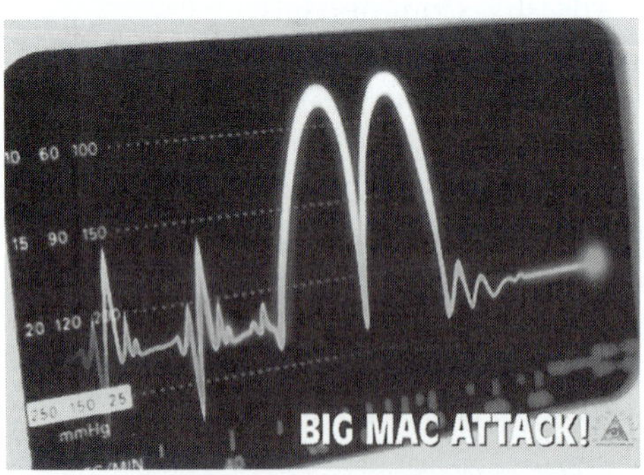

Sometimes this anticommercial advocacy has trouble getting its message out. In 1996, for example, ABC, CBS, and NBC refused to sell airtime for its 30-second uncommercial announcing that year's Buy Nothing Day, even though the Media Foundation was prepared to pay the full commercial rate to air the spot. Said an NBC spokesperson, "We don't want the business. We don't want to accept any advertising that's inimical to our legitimate business interests." In 1997, when the three networks refused to run an uncommercial for TV Turnoff Week, for which the Foundation was again willing to pay top commercial rates, an NBC executive said, "Our business is about turning on TV, not turning off TV. We have absolutely no intention of running these spots."

Sometimes, however, the Media Foundation is quite successful. In the Fall of 1997 it was the subject of an hour-long PBS documentary called *Affluenza.* Its Joe Chemo uncommercials and antismoking posters (parodying Joe Camel) appear in thousands of stores, schools, medical and health offices, and other public places. Many cable access channels run the uncommercials, and they occasionally appear on local noncommercial and commercial stations as well.

Whether or not you believe this effort is effective or even necessary, if you believe in freedom of expression, you have to ask yourself whether it is *proper* for commercial television networks to refuse to carry the Media Foundation's uncommercials. (The networks do have the *legal* right to refuse them, or any advertising for that matter.) Overcommercialization of telecommunications in particular and of our culture in general is a controversial public issue. If broadcasters feel comfortable refusing to air a side of the debate that they feel is "inimical" to their "legitimate business interests," two questions arise. First, what other expression do they deny because it runs counter to their business interests? Second, how can debate on this important issue ever flourish?

Through its efforts, the Media Foundation has forced both the media industries and the public to confront these questions. How do you defend your position on them?

definitions that serve the advertiser and not the culture. If love, for example, is something that can be bought rather than something that has to be nurtured, worked on, and maintained, how important can it be? If success is not something an individual values for the personal sense of accomplishment but rather is something chased for the material things associated with it, how does the culture evaluate success? Name the five most successful people you know. How many teachers did you name? How many social workers? How many wealthy or famous people did you name?

Critics contend that the consumer culture also demeans the individuals who live in it. A common commercial strategy for stimulating desire and provoking action is to suggest to people that they are inadequate, that they should not be satisfied with themselves as they are. We are too fat or too thin, our hair is in need of improvement, our eye color is all wrong, and our spouses don't respect us. Personal improvement is only a purchase away.

The box called The Right to Advertise Nothing presents the activities of a public interest group that works to counter what it sees as the hypercommercialism of our culture.

Chapter Review

Radio advertising was not unwelcomed by most people in its early days, and advertisers soon came to be radio's primary providers of programming. Then came television, which changed the nature of advertising content just as the commercials themselves changed the nature of the medium.

Because television employed moving pictures as well as sound, the nature of broadcast commercials changed. Television commercials often featured unique selling propositions to differentiate parity products. And when too many commercials packed into a break in programming produced clutter, commercials became more eye-catching and less informative to attract viewer attention.

Contemporary radio time is sold on three levels: retail sales, local sales, and national spot sales, the latter usually involving a national sales rep. Radio time is sold according to terms listed on a rate card that delineates the responsibilities and obligations of both station and advertiser. Radio has many strengths as a commercial medium: it is demographically targeted and intrusive, it has low production costs and rates for buying time, and it has good CPM. Several stations can be combined to air an advertiser's spots. Radio is flexible, mobile, emotional, has good reach into the suburbs, is timely, involves the imagination, and is well suited to reaching teens. But radio advertising also has a few drawbacks. It is ephemeral, fragmented, and listening is open to distraction.

Television time is sold in much the same way as radio time, but there are a few differences. Advertisers buy adjacencies, participations, and islands, usually within a pod or group of commercials. National television sales usually offer guaranteed CPMs, which, if not met, require the broadcaster to offer make-goods. Television has several advantages as an advertising medium. It is the most persuasive of the media, and it is intrusive. It has great reach, engages people's emotions, and its commercials are memorable. Television is credible, and its advertising conveys prestige and glamour to sponsors and their products. Television's drawbacks as a commercial medium are that it is expensive and spots often appear amid clutter. Advertisers often must pay for unwanted coverage, and television viewing is seasonal.

Cable advertising operates much the same as broadcast television advertising, but its audiences are much smaller, although demographically more narrow and young and upscale. Cable typically has a high inventory of available commercial time, leading to low costs to advertisers who can use it as a supplementary medium. Cable's major flaws as a commercial medium are its low ratings and lack of penetration.

Commercial time is not bought and sold only for cash. Stations and cablecasters often barter, and many local merchants rely on advertising allowances and co-op advertising to help them purchase broadcast time. Not all commercials are those 15- and 30-second spots with which we are familiar. Program-length commercials are 30-minute pitches, and public service announcements are free spots made available to nonprofit causes and organizations.

Despite its centrality to the U.S. economy as it now exists, advertising is often criticized for being intrusive and deceptive and for unethically targeting children, fostering monopolies, and debasing the culture. These criticisms are supported by pointing to the industry's common use of words that do not mean what the audience is led to believe they mean.

Questions for Review

1. What was the relationship between network radio sponsors and the programs on which their commercials aired prior to World War II?
2. What are unique selling proposition, parity products, and brand awareness? What do they have to do with contemporary television commercials?
3. What is clutter? Identify a few ways advertisers attempt to cut through clutter.
4. What are the three levels of radio time sales? Characterize each.
5. What information is likely to be included on a station's rate card?
6. What are the advantages and disadvantages of radio as an advertising medium?
7. What are the advantages and disadvantages of television as an advertising medium?
8. What are the advantages and disadvantages of cable television as an advertising medium?
9. What are the advantages and disadvantages of the World Wide Web as an advertising medium?
10. In television advertising, what are adjacencies, participations, and islands?
11. What is barter advertising? What two types of advertising are commonly referred to as co-op advertising?
12. What are demographic and psychographic segmentation?
13. In what ways can an ad be false?
14. Name and characterize as many different ways of testing the effectiveness of advertising as you can.
15. Explain five contemporary criticisms of advertising.

Questions for Critical Thinking

1. If you owned a television station, would you air commercials aimed at children? Why or why not?

2. What extent of puffery would you allow if you were an FTC regulator? Where would you draw the line between deception and puffery? Give examples.

3. Do you think the United States is overly materialistic? If you do, what role do you think commercials have had in creating this state of affairs?

4. What do you think of contemporary television advertising? Are its creativity and technological sophistication adequate substitutes for information about the product?

5. Have you encountered much online advertising? If yes, in what forms? Did you welcome its information or find much of it intrusive or annoying? What is your general reaction to online advertising?

Engaging Technology

1. We have outlined some of the complaints concerning advertising to children. Visit the Better Business Bureau's Children's Advertising Review Unit Web site at <http://www.bbb.org/advertising/childrensMonitor.html> to read the ad industry's take on this important issue. How convincing do you find its arguments?

2. Two of the most influential publications for those who make the commercials we see on radio, television, cable, and online are *Adweek* and *Adage*. Visit their Web sites at <http://www.adweek.com> and <http://adage.com> to determine what they have to say about the latest in advertising on these media. What are their perceptions of the state of that advertising? Can you identify trends they say are unfolding?

3. The people at the Media Foundation are in conflict with the major broadcast and cable networks over their refusal to air their uncommercials. Visit the Media Foundation's Web site at <http://www.adbusters.org> to get the latest news on this controversy. From what you read, can you determine if the Media Foundation will prevail? Do you think it should prevail? Why or why not?

4. Radio and cable television both count among their strengths as commercial media their ability to deliver demographically homogeneous audiences. Visit the Web sites of the primary advocates for these media at< http://www.rab.com/> and <http://www.cabletvadbureau.com/>. Read the arguments made in support of the commercial power of each medium, and build a case for where you would advertise your own narrowly targeted products, say, for example, a new type of in-line roller skates and a new cherry-flavored soft drink. Defend your choices.

Chapter Ten

Programming

Fred Silverman is famous as the "Man With the Golden Gut" because, as programming chief at each of the three major television networks, he was able to bring each to the top of the ratings. His secret, it seems, was that he could tell—in his gut—what people would watch.

The son of a television repairman in Queens, New York, 11-year-old Fred would watch everything there was on the 7-inch screen of the Transvision television set his dad built in 1948 from a kit. From this start, Silverman would go on to shape "a healthy chunk of modern TV tastes, beginning at CBS in 1970" (Rense, 1999). In the words of pioneering television executive Ethel Winant, who worked for him at CBS and NBC, "Fred has a real sense of what people like. He was the first executive that grew up on television, and really loved it and watched everything" (as quoted in Rense, 1999).

Silverman, at age 25, was hired to save CBS's failing daytime schedule. He replaced a mish-mash of repeats, soap operas, and game shows with hits like *The Joker's Wild* and *The Price Is Right.* He championed *Captain Kangaroo.* He turned CBS daytime programming into a money machine. But despite its top prime time ratings, the network was losing ad dollars to NBC and ABC. Silverman was asked to fix that problem. He totally revamped the schedule, dumping many popular but demographically older

The Man With the Golden Gut, Fred Silverman, programmed each of the three major television networks during his long and illustrious career.

programs including the *Red Skelton Show* and *The Beverly Hillbillies.* Many of the replacements he chose quickly disappeared, but one that was to change the face of the medium did not, *All in the Family.* "I put a bunch of bombs on the air," Silverman said, "It was very much a nervous time. But we did get lucky with *All in the Family.* It was actually an ABC pilot they passed on. We grabbed it because we thought it would be a big hit. When it moved to 8 p.m. Saturday, it exploded. It got a 60 share of the audience. You're talking about an over-30 rating! In today's terms, if you get a 20 rating, they'll put up a monument for you" (as quoted in Rense, 1999).

But Silverman was more than lucky. He soon scheduled hits-to-be *The Jeffersons, Maude,* and *Good Times.* He brought *M*A*S*H, The Mary Tyler Moore Show, The Bob Newhart Show, Rhoda, Cannon, Kojak, Barnaby Jones,* and the made-for-TV movie challenging lingering Southern racism, *The Autobiography of Miss Jane Pittman,* to the air. At the end of the 1973/74 season, CBS had 9 of the 10 top-rated shows.

How was he to follow this? Silverman moved to ABC in 1975, where he was equally successful. "There is no parallel in the history of broadcasting—

and few in any well-established industries—to ABC's sudden rise," reported *Time* magazine ("Man With," 1977, p. 46). At ABC Silverman was responsible for *Baretta, Starsky and Hutch, Barney Miller, Happy Days, Laverne and Shirley, The Love Boat, Charlie's Angels,* and *Welcome Back, Kotter.* He experimented with a new genre, the miniseries, airing what is now regarded as a television landmark, *Roots.* Broadcast over five nights in January 1977, the adaptation of Alex Haley's novel was, according to Silverman, "a risk. Who knew whether people were going to be interested? It was about slavery. I didn't know. . . . But we had the biggest ratings not only for *Roots* but also for our regular prime time schedule" (as quoted in Rense, 1999).

How was he to follow this? In 1978 Silverman moved to NBC, where he had to compete against himself; that is, he had to challenge the successful schedules he himself had built on CBS and ABC. He did so with *Real People,* a precursor to today's videotape reality shows, *St. Elsewhere, Cheers, Diff'rent Strokes, The Facts of Life, Gimme a Break, Hill Street Blues,* and another landmark miniseries, *Shogun.*

How was he to follow this? Silverman left NBC in 1981, and its programming was taken over by two people he had hired, Grant Tinker and then Brandon Tartikoff, both legendary programmers in their own right. Silverman moved into producing, where he brought to the networks the Raymond Burr *Perry Mason* made-for-TV movies, *Matlock, In the Heat of the Night, Jake and the Fatman,* and *Diagnosis: Murder.* "He should be in the Hall of Fame," says Winant, "because he changed television forever at all three networks, and he brought along a lot of really talented people, like Brandon Tartikoff. If he did nothing but *Hill Street Blues,* it would be an extraordinary accomplishment" (as quoted in Rense, 1999).

Preview

Fred Silverman may be telecommunication's most famous and successful programmer, but there are thousands of other radio, television, and cable professionals working to bring us what we see and hear, that is, programming. The programming they present is the product of a complex industrial system, shaped by a combination of technical, economic, and regulatory factors, audience demand and expectations, and flat-out imitation.

We visit these themes as we study, first, contemporary radio programming, differentiating between music formats, news and information formats, and noncommercial radio. In doing so, we examine the role of consultants, syndicators, and networks in shaping what we hear.

Our study of television programming focuses on programming network television, including daytime and late night as well as prime time, how shows make it onto network schedules, and how changes in the ownership of programming have influenced what we see on network schedules. We look in detail at a number of programming practices such as scheduling strategies, the use of reruns and repurposed content, and demographic targeting.

Not all television is network television, and we also look at local station programming, for both affiliates and independents. We see how syndication, in its different forms, influences the schedules of network and local stations alike and how syndicated fare is purchased. Finally, we examine national cable television programming to complement what we learned about programming local cable operations in Chapter 6.

Radio Programming

In its earliest days radio programming was haphazard and idiosyncratic (Chapter 2). Then, with the coming of the networks, sponsors took over the production of programming, using the national chains simply as carriers of their shows. With the coming of television, radio, now a local medium in search of inexpensive content, turned to formats and disc jockeys. Today, primarily because of advances in satellite technology and the relaxation of federal regulations limiting station ownership, radio programming has become a bit more sophisticated than a DJ spinning her favorite discs.

Programming Music Format Commercial Radio

Programming at local commercial radio stations is overseen by a program director, typically called the PD. This position came into universal use in the 1950s. When radio stations began to once again become profitable, their sound (and therefore their identity) became too important to be left to several different record-spinning personalities. Especially after the payola scandals of that decade, in which some disc jockeys were discovered taking payments from record companies in exchange for playing their music, the role of the PD in establishing a station's cohesive and recognizable sound not only became more important but evolved into a central management function at most stations. As such, the PD is often intimately involved in the work of station employees in other areas involved in what goes out over the air, promotions and production, for example, and at some stations, even news.

Ultimately, the function of a station's programming (and therefore of the PD) is to keep listeners tuned in. Music stations do this with music. To ensure a cohesive and identifiable sound, stations employ several devices, all of which leave control of music selection to the PD. Disc jockeys are given a playlist, the music available (within the station's format), arranged in order of play. Some stations allow a bit of discretion by their DJs, but most do not. The playlist is designed to structure the **rotation** or frequency with which different songs (or types of songs) are played. Songs that are in heavy rotation are played more frequently than those in light rotation. Rotation is typically tracked by computer.

But there is more to a station's sound than the music it plays. Not only does the PD ensure that station jingles and promotions, the style and sound of commercials, and the style of announcing fit the station's identity, she or

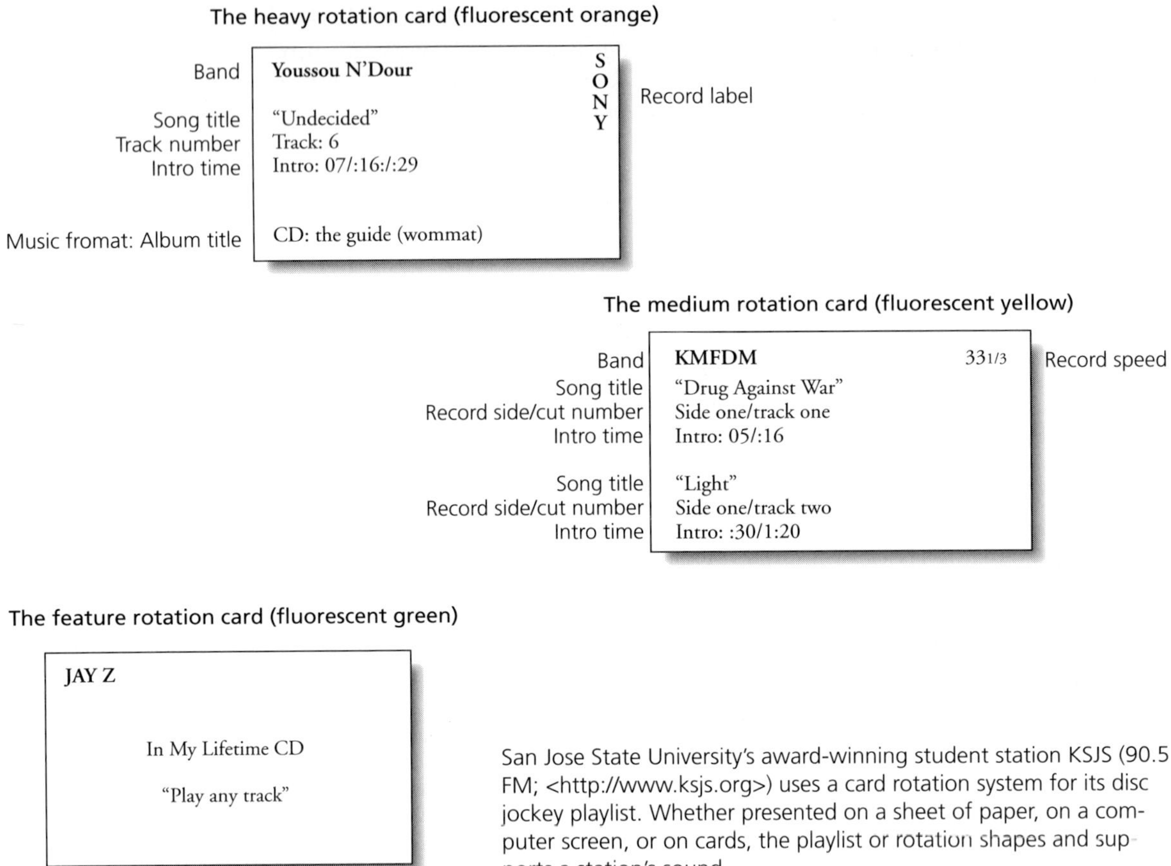

San Jose State University's award-winning student station KSJS (90.5 FM; <http://www.ksjs.org>) uses a card rotation system for its disc jockey playlist. Whether presented on a sheet of paper, on a computer screen, or on cards, the playlist or rotation shapes and supports a station's sound.

he must ensure that these elements of programming appear when and where they are most effective. To do this, stations employ a **program clock.** Sometimes called a **program wheel,** it is a visual representation of where the music, commercials, station identification, promotions, and news (if any) appear in a **sound hour**—literally a specific hour of broadcast time. Program clocks come in many different looks and styles, but all divide the hour into very specific segments with instructions on what is to be aired when. You can try your hand at being a PD using the box titled You Be the PD.

Programming News and Information Format Commercial Radio

Not all radio programming is music. Many stations adopt a news and information format, typically either All News (occasionally All Sports) or News/Talk. All News is characterized by the repetition of a **news cycle,** a rotation of news and information that might include national news, regional news, local news,

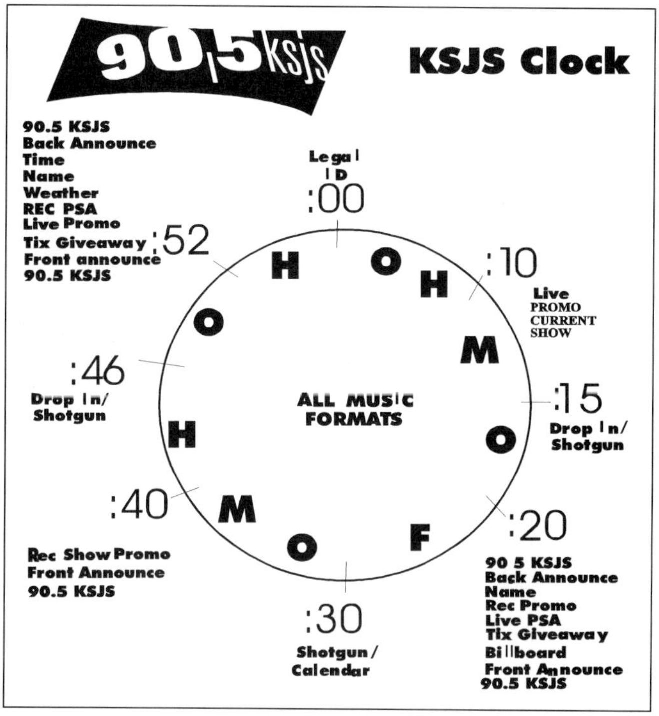

A typical program clock, also called a program wheel.

time updates, traffic reports, weather, sports, and a leisure or lifestyle feature. All News stations will frequently interrupt their news cycles to present a specific program such as a sporting event or a nationally syndicated talk show. But All News stations' primary format is airing cycles of headlines and short reports designed to attract listeners and briefly hold them.

Talk stations, conversely, attempt to build a loyal following of listeners who will stay with the station for longer periods of time. Talk stations do present news, typically during morning and afternoon drive time (Chapter 9), but their primary format is to broadcast lengthy talk and phone-in shows designed to attract and keep listeners. This fact accounts for the biggest difference in programming the two information formats. "The All-news programmer oversees the equivalent of a single program airing twenty-four hours a day. Similarly, a Music station program director must follow a common thread throughout the format day—a sound that defines a station. The Talk programmer, on the other hand, deals with diverse blocks of programming lasting from one to four hours. Thus, a Talk radio program director is much like a television programmer: Both must consider how one program leads into the next, strategically plotting **audience flow,**" the carry-over or movement of audience members from one program to the next (Barr, Brewer, & Eastman, 1993, p. 459). It is common, then, for an All Talk PD to schedule blocks of two or three politically conservative phone-in shows or back-to-back how-to call-in programs.

The Real World *You Be the PD*

After all those years working your way up the ladder—from small market AM easy listening station to medium size FM rocker; from music librarian to traffic reporter to DJ—you've arrived at the destination you targeted when you graduated, PD of a major market country station. And best of all, your new station is already Number One in the market and completely dominant in morning and afternoon drive times. All you have to do is maintain the winning formula set up by your predecessor.

But you're not on the job 2 weeks when controversy strikes. The Dixie Chicks, Grammy Award winners and favorites of young country music fans, especially preteen girls, have released the single "Goodbye Earl." It's an instant hit around the nation. All the major country stations have put it into heavy rotation. But its subject matter troubles you. The tune is about a woman who murders her abusive husband with poisoned black-eyed peas and, with the help of a friend, wraps him in a tarp, stuffs him in the trunk of her car, and dumps him in a lake.

Columbine (and Paducah and the rest) are on your mind. There's simply too much violence in popular culture and in American schools. Just last week a 6-year-old child in Michigan brought a gun to school and shot a 6-year-old classmate. So you refuse to air the chart-topping song—but your DJs protest. They want to sound hip and with it. Ignoring a smash hit that the band played before a national television Grammy Award audience just days before makes them look silly.

You're the PD; this is your decision. You ask yourself about the wisdom of programming music that glamorizes, and even justifies, violence to attract an audience. What is your responsibility to your listeners? To your DJs? To your employers? To your career? To your own set of values? You didn't get where you are as fast as you did by running away from tough decisions, but you know that whatever you decide you will anger thousands of listeners. So you choose to go public. You announce an on-air "town meeting" in which you will listen to and answer listeners' opinions and concerns.

The first caller argues that the song might make some abusing men think twice before they strike their partners. But the second responds that it will encourage abused women to take vigilante action rather than work within the system to solve the problem. Several callers say that the song is "only a song," "just fantasy," and "make believe." But you respond that popular culture can and does have important effects, whether its content is fantasy or not. If music didn't matter, you tell the listener, you wouldn't be a radio PD. And you ask one caller what he thinks of Ice-T's controversial "Cop Killer." Ban it, he replies, it's offensive. It's only a song, you counter, so what's the difference? Why is "Goodbye Earl" OK and "Cop Killer" not? But, argues the next listener, your station frequently played Garth Brooks' "Papa Loved Mama," a song about a man who kills his cheating wife, so what's the difference there? The difference, you say, is that where Earl's killers get away with and even celebrate murder, Papa ended up "in the pen." Another caller offers a compromise. Why not, she suggests, give the number of the National Domestic Violence Hotline every time the song plays. She even helpfully gives it to you—1-800-799-SAFE (or 7233). Good idea, you respond, but you don't want your DJs (or your station) to come off as too preachy. Finally the give and take ends. You promise to announce your decision during drive time the very next day.

What do you do in this situation? Defend your position. After you read the discussion of ethics in Chapter 13 and the examination of the effects of media violence in Chapter 14, revisit this scenario to see if your decision or your reasoning have changed.

All Talk and News/Talk do have several things in common. Because news and talk are more expensive to produce than music, these stations tend to be clustered in larger markets. (Record companies supply discs to music stations for free; content is rarely provided for free to information formatted stations.) All Talk and News/Talk stations both employ a program clock in the same manner as do music stations, and again like music stations, they rely increasingly on consultants, syndicators, and networks.

Consultants, Syndicators, and Networks

Local radio is programmed locally. However, because of increased audience fragmentation, concentration of ownership, and the ready and relatively inexpensive availability of satellites, stations are turning more and more to outside sources for programming.

In an effort to attract listeners of ever-narrower demographic specificity, many stations employ the services of consultants, sometimes called **format doctors.** Some consultants only provide guidance to the PD. Because of their position outside the station's management, they can be more objective than can those working on the inside. They examine all aspects of the station's sound and talent, conduct surveys and focus groups to solicit listener input, and render a market analysis to be used by the PD to sharpen or even sometimes change a format. Most commonly, after the initial evaluation is complete, the consultant remains under contract with the station for periodic examination and fine-tuning over time.

Other consultants provide programming in addition to advice. Typically, this type of consultant specializes in a particular format. A station that wishes to strengthen or change to a country format, for example, might contract with a company such as Toby Arnold and Associates, which will help the station perfect its sound and provide it with national quality programming as well.

When television displaced radio in the 1950s, the national radio networks went into decline, leading to the expansion of the radio syndication business. Stations had to get their features and specials from somewhere. Today, as individual stations continue to become parts of larger and larger station groups, there is less incentive for them to develop, nurture, and perfect distinctive programming. As a result, syndicators have become central to the business of providing well-produced, national radio programming. Radio syndicators come in two primary forms. **Format syndicators** such as Bonneville (easy listening, beautiful music, adult contemporary) and Drake-Chenault (country, classic rock, urban, contemporary hits, adult contemporary, beautiful music) provide **long-form programming.** Long-form programming refers to complete shows, typically 4 hours or longer, that are provided via satellite to local, usually automated (Chapter 2), stations. These might be a DJ-driven show or something like Creative Radio Networks' history of rock 'n' roll program, *Rock Files.*

A second form of syndicator provides **short-form programming,** that is, individual programs or series. These **feature syndicators** use satellites to deliver brief, 1- to 30-minute pieces of content such as Westwood One's 90-second celebrity interview bits, *Off-the-Record With Mary Turner,* AMFM Radio Networks' *Rick Dee's Weekly Top 40,* and All Star Radio's comedy shows.

Stations pay for syndicated programming in one of two ways. The first is **outright cash purchase** in which a station simply pays a **licensing fee** for a given piece of content. It then sells advertising time within the show, keeping the income. Cash payments typically ensure **market exclusivity**—no other station in its broadcast area can carry the syndicated material in question. The second way syndicated programming is acquired is through barter. In **barter syndication** the local station receives the content in exchange for its promise that it will air the commercials (sold by the syndicator to national advertisers) contained in it. Sometimes a program is all barter; that is, the local station gets the show completely for free, making its money by selling the few vacant commercial minutes also included in the program. *Earth News Radio* is an all barter program from Westwood One. Sometimes local stations combine cash payments and barter. In either case the syndicator sells its commercial time to national advertisers, and the local station sells its avails as it wishes, keeping the proceeds.

Before satellites the distinction between syndicators and networks was clear. Networks used telephone lines and microwave relays to distribute content; syndicators used physically delivered tapes and discs (referred to as *bicycled*). Because it was live and simultaneously delivered, network programming was time specific, whereas bicycled syndicated material was less so. But satellites, which allow syndicators to reach as many stations with the same immediacy as do the networks, have made this distinction almost meaningless (at least in terms of the quantity and quality of programming available to stations and listeners). The primary difference between the two (at least for radio) is that networks pay compensation to affiliates to encourage them to carry their programming and commercials, whereas syndicators are paid (either in cash or barter) by licensees for the right to air the syndicators' material. Satellites have not only reduced the distinction between national radio networks and syndicators, but they have also led to an explosion in the number of radio networks. Networks such as CNN (news), Disney (children's programming), and ESPN (sports) now operate alongside syndicators-turned-networks such as Westwood and Unistar and traditional networks such as CBS and the five distinct services run by ABC.

Programming Noncommercial Radio

As we saw in Chapter 4, there are several different types of noncommercial radio stations—college, community, religious, and public—and within each category there is great variation and some overlap (some community and college operations, for example, are public broadcasting stations as well). This makes it difficult to discuss programming of noncommercial stations in any general way. College stations that play music are programmed much like their commercial counterparts, albeit typically with more leeway in format, playlist, and adherence to the program clock. Community stations that mix talk, news, and music combine program and programming elements usually found in commercial stations using these three labels.

But this does not mean that there are no programming concerns specific to noncommercial radio broadcasters. Noncommercial stations typically are programmed to be alternative; that is, they intentionally strive to be unlike the other stations in their broadcast area. This is why Celtic music, lessons in holistic healing, and discussions of the mystical power of crystals are at home on noncommercial radio. Noncommercial stations also tend to have a greater commitment to localism than do their commercial brethren. Even national network noncommercial stations, for example, those affiliated with PBS, PRI, and Pacifica, program material of interest to their local communities to a far greater degree than do the vast majority of commercial stations.

Broadcast Television Programming

Once upon a time, programming network and local television was a relatively straightforward matter. The Big Three—ABC, CBS, and NBC—simply aired content that would bring them each about a third of the audience. If each network averaged more or less that proportion of the national audience over the span of a season, all would prosper.

In local affiliate television, the programming strategy was straightforward—fill the time around the networks' offerings with syndicated fare, most typically shows that once ran on the networks. On weekends the networks provided Saturday morning children's shows, so the local programmer could run her FCC-satisfying public affairs programs on Sunday, waiting for a sporting event, usually provided by the network, to begin. The strategy for local independents was even simpler: run whatever syndicated material the station could afford (first-run game shows, old movies, off-network golden oldies) and hope to attract enough viewers to keep the station on the air. Then came cable.

Cable changed almost all aspects of the television business, including programming and programming strategies (see Chapter 3). For example, independent stations (mostly UHFs) became increasingly successful because cable erased the technical difference between them and the affiliates clustered primarily on the VHF channels. This created new markets for high-quality **first-run syndicated** fare such as *Baywatch* and *Star Trek: The Next Generation*. This, in turn, made the independent stations more attractive to entrepreneurs wanting to start their own broadcast networks. So where there was once ABC, CBS, and NBC, there is now ABC, CBS, NBC, Fox, the WB, UPN, and Pax. Scores of independent cable networks resemble the traditional networks in all ways other than the manner in which they enter our homes, but because a majority of U.S. homes have cable service, even this difference matters little.

Xena: The Warrior Princess has been extremely successful in first-run syndication.

Today's world of broadcast television programming is not as straightforward as it once was, but to understand it is to understand the dynamics that are shaping and reshaping the content on our sets, regardless of the source.

Programming Prime Time Network Television

In 1975 HBO became a national cable pay channel, beginning the networks' loss of dominance. The 25th highest rated program in the United States was CBS's *Paul Sands in Friends and Lovers.* It had a rating of 20.7, which today would make it the runaway Number One show in the United States. (The ratings champ for 1999 was *ER,* with its 16.8 rating.) In 1975 the average prime time, 1-hour program cost a network $227,075 (Adams, Eastman, & Lewine, 1993). In 2000 NBC paid $13 million per episode, or $286 million for a 22-episode season for *ER.* It is against this backdrop and a host of video and other media that the networks must compete for people's attention.

It is important to keep in mind that although its overall share of the television audience has declined in the last 25 years, broadcast network television is still the most watched form of programming. The Big Four—NBC, CBS, ABC, and Fox—drew 64% of all prime time viewers who tuned in in 1998/99 (see Figure 10.1). The networks and the **netlets** (WB and UPN) continue to program content that millions of people watch and enjoy.

How a Program Gets on the Air

The national broadcast networks look at about 4,000 proposals a year for new television series in a process characterized by former NBC programming chief Brandon Tartikoff as a dance between beggars (the producers) and choosers

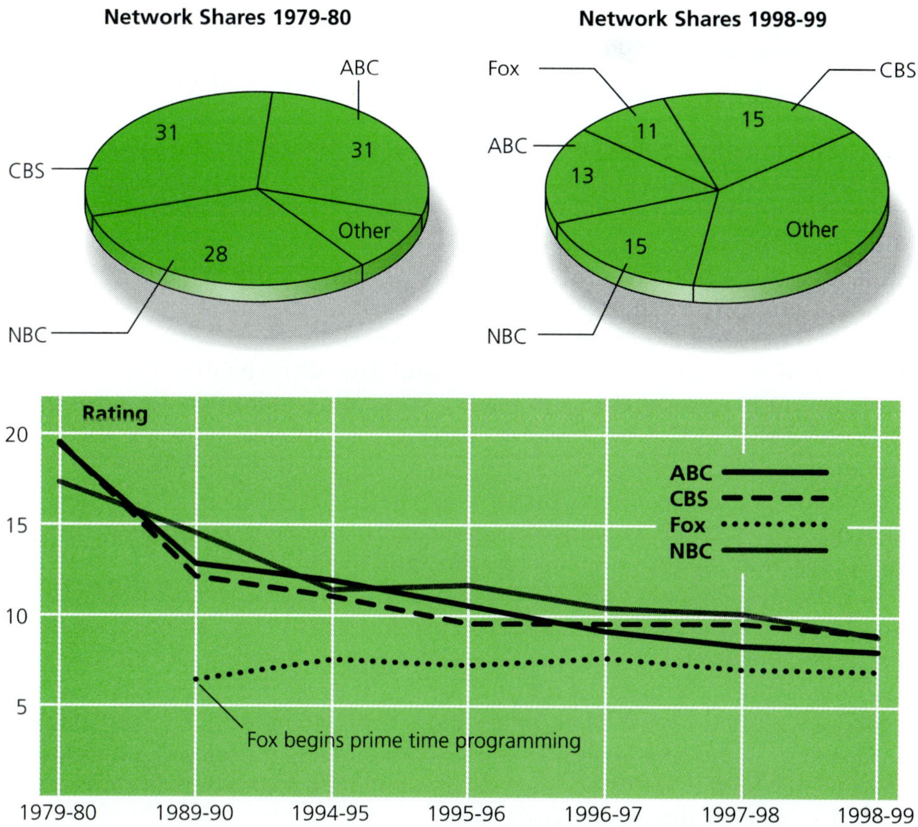

Figure 10.1 Declining Network Ratings
Prior to cable's introduction, the national television networks would each routinely draw an annual cumulative share of near 30. But times have changed. For the 1998/99 season, not only did none of the Big Four networks score anywhere near a 30 share, but it was the first year in their history that none had an annual rating in the double digits. We can see the networks' steady decline in ratings and shares from this table of their annual Nielsen numbers. *Source:* Various issues of *Broadcasting & Cable* and *Variety.*

(the networks) (Franklin, 1999). Many, if not most, proposals are submitted at the networks' invitation or instigation. Of the 4,000, about 100 will be filmed as **pilots,** or trial programs. Perhaps 20 to 30 will make it onto the air, and only half of these will last a full broadcast season. In a particularly good year 3 or 4 will succeed well enough to be called hits. For this reason the networks prefer to see ideas from producers with established track records and financial and organizational stability. For example, David E. Kelley is the source of *Ally McBeal, Chicago Hope,* and *The Practice,* and Twentieth Century Fox produced 22 different prime time shows on the 1999 schedules of the six national networks.

The way a program typically makes it onto the air differs somewhat for those who have been asked to submit an idea and for producers who bring their concepts to the networks. First, a producer has an *idea,* or a network has an idea and asks a proven producer to propose a show using this idea. The network may offer a **put,** which is a deal that guarantees the producer that the network will order at least a pilot or it has to pay a hefty penalty. The producer must *shop* its idea to one of the networks. (An invitee only presents its idea to the network that asked for it.) In either case, if the network is persuaded, it *buys the option* and asks for a written *outline* in which the original idea is refined. If still interested, the network will order a *full script.*

If the network approves that script, it will order the production of a *pilot.* Because of increased production costs and dwindling profits, the networks have begun penny-pinching on pilots, demanding that pilots for 60-minute dramas be shot as shorter, less-expensive presentations. Pilots are then subjected to rigorous testing by the networks' own and independent audience research organizations. Based on this research, networks will often demand changes, for example, writing out characters who tested poorly or beefing up story lines that the test audiences particularly liked.

If the network believes the show will be a hit, it orders a set number of episodes and schedules the show. In television's early days an order might be for 26 or 39 episodes. Today, however, because of escalating production costs, the convention is to at first order 6 episodes. If these are successful, a second order of 9 more is placed. Then, if the show is still doing well, a final 9 episodes (referred to as the *back nine*) will be commissioned. Untested program ideas or producers who are not fully established might get initial orders for only 2 or 3 episodes. This is called **short ordering.**

At any point in this process the network can decline interest. Moreover, it invests very little of its own money during the developmental stages of a program. Even when a network orders a package of episodes, including those for an established hit that has been on for years, it typically pays producers only one half of the show's entire production costs. In other words, producers engage in deficit financing. They lose money throughout the development process and continue to lose even more the longer their show stays on the network's schedule. There are exceptions, however. In its effort to save the Thursday night lineup that made it the top-rated network in the mid-1990s, NBC paid the producers of *Cheers* 100% of its production costs and, as already mentioned, entered into the $13 million per episode deal with the makers of *ER,* more than covering their production costs.

The reason television program producers participate in this expensive enterprise is that they can make vast amounts of money in **off-network syndication**—that is, selling their one-time network programs to stations on a market-by-market basis. Even though the networks control the process from idea to scheduling and decide how long a show stays in their lineups, producers continue to own the rights to their programs. Once enough episodes are made (generally about 50 or 60, the product of 3 or 4 years on a network), producers can sell the syndicated package to the highest bidder in each of the country's 211 television markets, keeping all the revenues for themselves.

The price of a syndicated program depends on the market size, the level of competition between the stations in the market, and the age and popularity of the program itself. The station buys the right to a specified number of plays, or airings. After that, the rights return to the producer to be sold again and again.

A program that has survived 3 or 4 years on one of the networks has proven its popularity, has attracted a following, and has accumulated enough individual episodes so that local stations can offer weeks of daily scheduling without too many reruns. That means it's a money maker. *The Bill Cosby Show,* for example, will ultimately make a half billion dollars in syndication. *Seinfeld* will generate over $1 billion. Eyemark Entertainment, CBS's syndication unit, predicts that the comedy *Everybody Loves Raymond* will earn $300 million in its first 4 years of syndication (Schlosser, 1998), and in July 1999 cable network Lifetime paid $58 million for *cable only* syndication rights to *Mad About You,* freeing that show's producers to reap hundreds of millions of dollars more on the local television syndication market.

Many critics of television argue that it is this deficit financing system that keeps the quality of content lower than it might otherwise be. Because a producer must attract the interest of a network with an idea that is salable on the network's schedule today, while incurring years of financial loss in hopes of having content that will be of interest to syndication viewers 4, 5, or 10 years in the future, there is very little incentive to gamble with characters or story lines. There is little profit in pushing the aesthetic boundaries of the medium.

Changes in Program Ownership

Concerned that ABC, CBS, and NBC held a virtual monopoly over the production and distribution of television content, the FCC issued **Financial Interest and Syndication Rules (Fin-Syn)** in 1970. These rules denied the networks ownership of the entertainment shows they aired, hoping that many independent producers bringing shows to the Big Three would result in a more diverse array of programming. Fox, however, was not bound by these limits when it began operation in 1985 because the FCC wanted to give the new network every chance to survive. The traditional networks were losing audiences to cable and other alternatives, and they lobbied the commission for change, pointing to Fox's success. Finally, in the mid-1990s the FCC amended Fin-Syn, allowing networks to own up to 50% of their prime time entertainment fare. Eventually, the rules were simply eliminated. ABC owns all or part of *Ellen, Home Improvement, Boy Meets World, The Wonderful World of Disney,* and *Sports Night.* NBC owns all or part of *Veronica's Closet, Stark Raving Mad, Suddenly Susan, NewsRadio, Profiler, Will & Grace,* and *Homicide: Life on the Street.* CBS has interests in *Dr. Quinn, Medicine Woman, Now and Again, Touched by an Angel, Everybody Loves Raymond, L.A. Doctors, Marshal Law,* and *Walker, Texas Ranger.* Fox owns or has a stake in *The X-Files, The Simpsons, Ally McBeal, King of the Hill,* and *Millennium.* UPN has *Star Trek: Voyager, Seven Days,* and *Love Boat: The Next Wave.* The WB owns *The Parent Hood, Wayans Bros.,* and *Jamie Foxx.* In the 2000–2001 season Fox owned outright or held a financial interest in 79% of its prime time schedule. CBS and UPN held 67%, ABC 59%, NBC 45%, and

Everybody Loves Raymond (CBS) and *Seven Days* (Fox) are among the recent crop of network-owned prime time shows.

the WB held 39% (Schneider & Adalian, 2000). In fact, the major networks now routinely expect to hold some degree of ownership in all the prime time programs they air.

The change in ownership rules is not without controversy. The networks claim that content gets better, because with their production money and potential syndication profits at risk, they are more likely to stay with a high-quality or innovative show until it finds an audience. Producers argue that quality suffers for two reasons. First, independent producers with less of a stake in tomorrow's syndication money do not take innovative chances today. Second, because a longer network run makes syndication success more likely, and the networks want to guarantee their own syndication income, they program and keep the shows they own on the air longer, not only with minimal concern for quality but at the expense of programs from independent producers. In fact, however, ownership did not stop ABC from canceling *Ellen,* CBS from dropping *Dr. Quinn, Medicine Woman,* NBC from axing *Wind on Water,* or Fox from canceling *Getting Personal.*

Programming Practices

Once a network decides to broadcast a program, it must determine when and how it will do so. As is the case with almost every other aspect of telecommunications, developments in technology and convergence have altered network television's programming practices.

The Programming Year Network television once scheduled for the **viewing year,** traditionally thought of as the 30 weeks running from late September to mid-May, and the **off-season** (all other time). This is no longer necessarily the case. When Fox was a netlet, it recognized that it could not compete with the promotional blitz mounted by the Big Three to announce their new fall seasons. Fox began debuting its new shows in late summer, successfully pitting them against network reruns. Now it is not uncommon to see all six national broadcast networks premiering new shows throughout the year. In January 1999, for example, NBC launched *Providence,* Fox began *The PJs,* and CBS premiered *60 Minutes II. Becker* (CBS), *King of the Hill* (Fox), and *Dawson's Creek* (WB) all enjoyed midyear viewing debuts. Although ABC, CBS, and NBC rely more than the others on the traditional fall opening, this new **year-round scheduling** has become so pervasive that the networks and their advertisers are openly questioning the validity and worth of that long-time, traditional seasonal audience measurement—the sweeps (Chapter 8).

Scheduling Strategies Regardless of when a show first appears on a network's schedule, it must have a specific place in that lineup. Different locations on the schedule bring different benefits. Programmers always want a strong show, such as *The Simpsons* or *Friends,* in the **lead-off position.** This is the first show of an evening's prime time lineup. Because the network that wins the first hour of prime time is in a good position to win the ratings race for the entire evening, this is a crucial position. A good lead-off provides **lead-in** for a second, presumably weaker show, which inherits some of the audience of its lead-in. CBS used *M*A*S*H* for several years as something of a universal lead-in, moving it from its original Sunday 8 p.m. (Eastern time zone) spot to Saturday, Tuesday, Friday, back to Tuesday, and then to Monday, sometimes at 8:30 p.m. and sometimes at 9 p.m.—all in an effort to help new or struggling shows.

Programmers sometimes place a new or struggling show between two proven programs, a practice known as **hammocking.** The goal is to reduce audience loss because of the presence of viewer favorites before and after. Hammocking can be dangerous, however. *Paul Sands in Friends and Lovers,* a Top-25 show in 1975 with a 20.7 rating, lasted only half a season because it was hammocked between two giant hits, *All in the Family* (with a 30.2 rating) and *The Mary Tyler Moore Show* (with a 24.0 rating). CBS programmers believed the 6-point drop between the two classics was due to the weakness of the *Paul Sands* show and that the success *Sands* did have was due to the *All in the Family* lead-in. As a result, it was canceled not for poor ratings but for **forfeiting audience.**

When a network has a strong lineup of similar programs, it combines lead-off, lead-in, and hammocking in a practice called **blocking** (or **stacking**). That is, it puts several winners together to protect the weakest among them and to catch and hold an audience for an entire evening. The 1970s CBS Saturday night lineup of *All in the Family, M*A*S*H, The Mary Tyler Moore Show,* and *The Bob Newhart Show* is often credited by programmers as establishing

As early as its second season (1973–74), *M*A*S*H* was shifted around in CBS's prime time schedule in an effort to help the network's new or struggling shows.

this practice. Throughout the 1980s and 1990s, NBC blocked Thursday night with a string of powerful comedies including *Cheers, Seinfeld, Frasier, Friends, 3rd Rock From the Sun, Just Shoot Me, Suddenly Susan, Veronica's Closet,* and *Mad About You.* UPN found success in its early days by programming evenings of science fiction and fantasy, anchored by *Star Trek: Voyager.*

An alternative to blocking is **tent-poling,** which is placing a proven winner in the midway spot of an evening's schedule (typically 9 p.m.) in hopes of having it benefit the programs surrounding it. *The X-Files* was a long-standing Fox tent-pole. Figure 10.2 gives a clear look at how NBC employed several of these strategies to win not only Thursday nights but the 1999 year-long ratings race.

Two scheduling strategies are designed not so much to build the strongest lineup for a programmer's own network but to directly challenge the opposition's. **Counterprogramming** involves scheduling content that is unlike that on the other networks at a given time. For example, a network might counter a lineup of comedies with action shows or place demographically younger shows like *The Simpsons* and *The Family Guy* against the demographically older *Touched by an Angel.* **Stunting** is the practice of programming special events such as hour versions of 30-minute shows, having characters from one series appear on another, or having every show on an evening's lineup deal with a

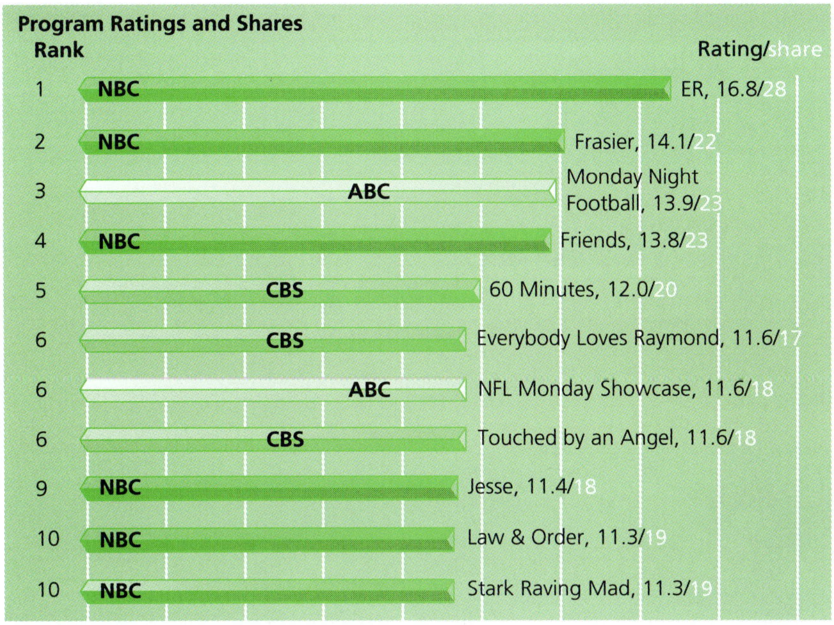

Here is NBC Thursday evening lineup for that year

8:00 p.m. *Friends*

8:30 p.m. *Jesse*

9:00 p.m. *Frasier*

9:30 p.m. *Stark Raving Mad* (premiering that year)

10:00 p.m. *ER*

Can you find examples of programming strategies for lead-off, lead-in, hammocking, and blocking?

Figure 10.2 1999 Network Television Ratings Leaders
Source: "Prime Time Champs," 2000.

single topic. Recently, *Ally McBeal* and *The Practice* have shared characters, and *Homicide: Life on the Streets* and *Law & Order* have shared plots.

Bridging is a strategy that involves the creative use of a show's start time. Sometimes networks (especially national cable channels) will start a movie or special in the half hour just before their traditional prime time. The goal is to catch viewers and keep them when the opposition starts its nightly lineup. A second form of bridging, used extensively by cable channel WTBS and sometimes by large market network affiliates' news shows, is to begin a program at 5 or 10 minutes past the hour and run it to 5 or 10 minutes past the next hour. Here the goal is to cause viewers to miss the start of other programming, leaving them with only one option for a show that can be watched in its entirety—the bridging program.

The Use of Reruns The networks once used **reruns,** rebroadcasts of fare aired earlier in a season, as filler programming for periods when there was reduced viewership, for example, around major holidays and summers. But reruns now occupy a more central role in how network programmers fill their broadcast year. NBC, for example, aggressively promotes its rerun schedule by encouraging audiences to "view it again for the first time." In fact, given the fees the networks must pay for quality prime time fare, they do not earn profit on much of this material until the second or third broadcast. Additionally, because the audience is so fragmented—most fans see only one of every

ESPN's *SportsCenter* is aired six times a day. Two installments are original, the other re-runs. Rather than complaining, viewers welcomed the expanded coverage of this popular sports news show.

three episodes of their favorite shows when they first run—what a programmer schedules as a rerun is often seen for the first time by a majority of viewers who do tune in.

It is cable that has helped wash away the stigma once attached to reruns. Virtually all the cable networks rerun their fare multiple times a day (ESPN's *SportsCenter*), a week (Comedy Central's *South Park* and HBO's *The Sopranos*), or a month (all the movie channels employ multiple airings of hit films). Audiences have come to expect and even welcome them.

Repurposed Content A close relative to reruns is **repurposing,** in which a piece of content is reused or reshaped for use as original programming. Repurposing is increasingly a part of network programmers' available strategies. Repurposing occurs in a number of ways, some more controversial than others. One noncontroversial form is re-editing a popular show for airing as a new entity. For example, in 1999 Fox programmed the 60-minute *Ally McBeal* on Monday night and an edited 30-minute version highlighting the show's comedic aspects on Tuesday. Another version of repurposing involves producing an entirely new version of a familiar and popular program, maintaining much of the original's identity (and hopefully following). NBC, for example, began in 1999 to air its hit *Law & Order* on Wednesday evenings and *Law & Order: Special Victims Unit,* featuring many of the same cast members and focusing on particularly heinous crimes, on Mondays.

Program producers do not mind repurposing in any form; it means more of their material makes its way onto our screens. And affiliates are not troubled by the type of repurposing just described; it means more network-quality fare for their schedules.

But one aspect of the NBC *Law & Order* arrangement represents another form of repurposing that is controversial with local affiliates. To reduce the expense of licensing the two *Law & Orders,* NBC struck a deal with cable channel USA Networks that allows USA to schedule the *Special Victims Unit* version within one week of its NBC airing. This saves NBC $50 million a year on the price of the two shows, but the affiliates chafe at the further erosion of the distinction between them and what were once considered the second-class cable channels. It doesn't matter to viewers if they catch the show on USA or an NBC-affiliated station. NBC affiliates also complain about the same-day afternoon rebroadcast of the *Today* show on NBC's cable channel MSNBC. ABC repurposes its daytime soaps on the cable soap opera channel it owns and its prime time *Once & Again* on Lifetime, putting them in direct competition with programming on its own affiliates.

Programming Demographically As advertisers began to buy network time with an eye toward the demographic composition of the audience, networks began to make programming decisions on demographic grounds (Chapter 8). Especially since the coming of cable, demographic scheduling has become a matter of survival for the networks. The many cable channels offer advertisers smaller, demographically concentrated audiences at a fraction of the cost they would pay to reach network viewers.

In fact, contemporary practice is to brand or otherwise demographically identify a network: CBS is for older, financially better off and better educated viewers (*60 Minutes, Touched by an Angel*), NBC is the network for hip urban professionals (*Friends, ER, Law & Order*), ABC attracts younger and especially male urbanites (*Sabrina the Teenage Witch, The Hugleys, Sports Night*), Fox aims

In the late 1990s ABC began to brand itself as the fun-place-to-be network.

Critical Issues *Programming Spanish-Language Television Networks*

There are two important and influential national broadcast networks beyond those we discuss in this chapter. Worthy of our attention are Spanish-language Telemundo and Univisión. They represent demographic programming at its highest.

Univisión is the older and more successful of the two. Founded more than 30 years ago, it has only 15 affiliates (compared to Telemundo's 63) but 80% market share among Spanish-speaking viewers. Its programming looks much like that of the English-language networks. Univisión strips morning talk shows (*Maite*) and soap operas, called **telenovelas** (*Luz Maria*), afternoon game (*El Blablazo*) and magazine shows (*El Gordo y La Flaca*), and early evening news (*Noticiero Univisión*). Prime time is somewhat different from typical network television in that it consists primarily of telenovelas. Weekend programming, however, looks like the traditional English-language network model: kids' shows, dance shows, and sports.

What differentiates the programming on Univisión from its English-language competitors is that all the programming is in Spanish, with very few shows—only four in 1999—presented bilingually. Univisión targets older viewers with its primarily Mexican- and Venezuelan-produced programming.

One of Telemundo's most successful daytime programs is its late afternoon talk show, *Padre Alberto,* hosted by a Catholic priest, Father Alberto Cutie. The show is broadcast in Spanish and targets an older audience.

But this also differentiates Univisión from its more direct competitor, upstart Telemundo.

Telemundo was founded in 1988 and targets a different niche in the Spanish-language audience, bilingual and bicultural young adults, with story lines, characters, and settings that are representative of Latino lifestyle in the United States. To do this, it programs fare that would be very much at home on

for young urbanites with attitude (*The Simpsons, King of the Hill, Futurama, Ally McBeal*), and UPN (*Moesha, Clueless, WWF Smackdown!*) and the WB (*Buffy the Vampire Slayer, Felicity*) both identify themselves as the networks of teen urbanites, males for UPN, females for the WB. Two networks branding themselves differently for the Latino audience are the focus of the box titled Programming Spanish-Language Television Networks.

Programming Non–Prime Time Network Television

We may know the networks best by their prime time fare, but their programmers must also fill daytime, late night, and weekend schedules. The major networks employ similar programming strategies: early morning talk shows (*The Today Show, Good Morning America*), morning and afternoon soap operas (*Passions, The Bold and the Beautiful*), early evening national news (*ABC World*

NBC or Fox. For example, there is the comedy *Solo en América,* about a recently divorced, 34-year-old Venezuelan mother trying to raise her two bilingual daughters in New York. Action show *Reyes y Rey,* the first police action program produced for Spanish-language television in the United States, presents two detectives, one Mexican, the other Mexican American, who share the same last name but not much more, as they police the Mexico-Texas border from Tijuana. As you might guess from these brief descriptions, the actors use Spanish and English interchangeably, and English subtitles appear when they speak Spanish.

There has been a generally positive response from Latino advocacy groups, academics, and, especially important, advertisers. "We know that this bilingualism and biculturalism exists, and it's very much the trait of many in our community," said Lisa Navarrete of the National Council of La Raza. "Most Americans see us as a predominantly immigrant community, but that's not the case. Most of us were born here" (as quoted in Garcia, 1999, p. 18A).

Telemundo programmers recognize this, pointing to data that indicate that 60% of the country's 30 million Latinos consider themselves bilingual. Among all Spanish-language television viewers, 53% of the U.S.-born audience watches English-language programming exclusively or primarily, and another 35.6% tune in to both languages equally (Littleton, 1999c).

Criticism does exist, however. Some members of the Latino community see Telemundo's programming strategy as "culture-lite," a desperate attempt to cut into Univisión's dominance. Others see it as a rejection of traditional Latino culture in an attempt to satisfy advertiser demands for ever-narrower demographics. Still others object to the use of the tired American television cops and robbers genre to present modern Latino culture. But according to the network's Nely Galán, president of entertainment and programming, "Telemundo is trying to create cultural pride by casting Latinos in positive roles, which is something the mainstream networks have rarely done. We want Latinos to feel that being Latino in the U.S. is a wonderful thing" (as quoted in Garcia, 1999, p. 18A).

What do you think? Put yourself in the position of a Telemundo programmer as the network debated this new programming strategy. What would you have argued when the idea for bilingual and bicultural Spanish-language television was raised? Is this a step forward or a cultural sellout?

News Tonight, CBS Evening News With Dan Rather), late night talk (*Nightline, The Tonight Show With Jay Leno*), weekend sports, and Sunday morning public affairs (*Meet the Press, Face the Nation*).

There is little programming diversity during these non–prime time periods for several reasons. First is that the audience is relatively small, at least when compared to prime time. This makes expending large amounts of money on more sophisticated or innovative programming unwise. A second reason resides in the nature of the audience itself, which is highly demographically homogeneous during these viewing periods. Women watch soaps, kids watch kids' shows, and men watch weekend public affairs and sports. It would be illogical for a network to counterprogram to win a small portion of an already small audience.

There is a somewhat different reason ABC, CBS, and NBC program their national news shows at the same time every evening (although not all affiliates air them at those times). Since television's earliest days, a network was defined by its news operation. CBS was the nation's "Tiffany Network." It was the top-rated network for 30 years, largely on the strength of its news show and its links to Edward R. Murrow and Walter Cronkite. To be a major national television network, it is important to demonstrate service and commitment, and the early evening national news shows do this. They are a must. Fox, with no national evening news show, may disprove this long-held assumption, but many observers argue the opposite. That is, Fox will never join the "big boys" until it can mount a respectable evening news program.

Programming Local Commercial Television

Local network affiliate and independent station program directors face somewhat different challenges. Affiliates receive, among other benefits, a great deal of high-quality programming, up to 60% of the material they air (Chapter 3). The remainder is composed of local programming (10%), first-run syndicated programming such as *Wheel of Fortune* and *The Oprah Winfrey Show* (15%), and off-network syndicated shows (15%) such as *Cheers, Frasier,* and *Mayberry RFD* (Haldi & Eastman, 1993). Local affiliate programmers make different use of these program types in different dayparts.

On weekdays the **early morning daypart,** typically 7 a.m. to 9 a.m., might see syndicated cartoon packages for children, off-network programs, women-oriented syndicated talk shows such as *Ricki Lake* or *Donnie & Marie,* and paid programming or possibly a paid religious broadcast. Many stations in larger markets produce their own morning talk and information programming to follow the networks' offering. *Morning Exchange* on WEWS in Cleveland, long the highest rated locally produced television show in the country, was so successful that it served as the model for ABC's *Good Morning America.*

The **morning daypart,** 9 a.m. to noon, when network material is not scheduled, is generally programmed with first-run syndicated game and talk shows, off-network fare, movies, and in some larger markets local live talk and information programming. The **afternoon daypart,** noon to 4 p.m., is dominated by network soaps and is normally filled out with syndicated talk shows, movies, and in many larger markets, a locally produced, noontime news program. **Early fringe,** 4 p.m. to 6 p.m., typically follows one of two models. In many large markets, affiliates have expanded their local news programs to 60, 90, and even 120 minutes, beginning in early fringe and most often wrapping around the network news. Other stations find that this period is perfect for **stripping** (running the same show at the same time five, six, or even seven times a week) either first-run or off-network syndicated programming (sometimes for children, sometimes for adults). People's routines bring them home and to the television set at about the same time every day, and habit viewing results from and encourages stripping. The opposite of stripping is the less common **checkerboarding,** in which different programs are rotated through the same time slot on different days.

The ultimate evergreen: *The Andy Griffith Show.*

The **early evening daypart,** 6 p.m. to 7 p.m. or 7:30 p.m., is dominated by local news in markets of all sizes. In fact, news-producing stations in larger markets can earn 50% to 70% of their annual profits from their local news shows alone. **Prime access,** 7 p.m. or 7:30 p.m. to 8 p.m., is the half to full hour just before network prime time. Here, most stations employ highly rated first-run syndicated material such as *Jeopardy* or a guaranteed off-network winner such as *Cosby, Friends,* or *Cheers.* **Late fringe,** the 30 or 60 minutes immediately after network prime time, is reserved for news on most stations, and the **late night** and **overnight dayparts** (those hours after the networks' late night entries such as *The Tonight Show With Jay Leno* or *Late Night With David Letterman)* either find stations off the air or showing program-length commercials (Chapter 9) or old movies.

Programming weekends is made somewhat simple by long years of television tradition. When there is no network programming available, Saturday mornings are filled with syndicated children's programming and Sunday mornings with public affairs and religious programming. Weekend afternoons are almost always programmed with local or regional sports (either self-produced or delivered by a regional network) and old movies.

Independent stations face a tougher task. They must program 100% of their schedules. As such, they rely on all forms of syndicated material and movies. Most independents choose their programming with an eye to overall station identity as well as to ratings. For example, a station might schedule afternoon and early fringe with **evergreens,** which are classic off-network programs and guaranteed syndication winners such as the *Andy Griffith Show, Gomer Pyle,* and *Petticoat Junction* and bill itself as "Your Television Home at Home."

It is important to note, however, that what constitutes an independent station is less clear today than it was in the days when ABC, CBS, and NBC dominated broadcast television. Stations carrying the limited programming of the new weblets are affiliates, but they operate much of the time as independents, and independents sometimes join **ad hoc networks.** These occasional networks are created by the union of several stations for the purpose of broadcasting some specific piece of content, for example, satellite-fed Spanish-language programming, religious shows, shopping networks, or sports.

Syndication

Some industry people consider syndication the force that drives all television programming. Producers produce either for syndication or for network runs that will lead to syndication. Networks, now allowed to produce their own fare, attempt to balance their immediate network programming needs with the promise of future syndication earnings. Local stations, affiliates and independents alike, rely heavily on syndicated material to complete their schedules. As such, the syndication business has expanded beyond the sale of off-network programs, inexpensive game and talk shows, and movie packages.

In the era of network domination first-run syndication (except for game and talk shows) was rarely attempted. There was no need; a reservoir of already proven off-network fare was there for the choosing. But the increase in the number of independent stations hungry for quality fare, combined with their new financial clout, has made the practice common. For several years, Paramount's *Star Trek: The Next Generation,* airing on more than 200 stations, would frequently beat network programming in the ratings. *Baywatch,* dubbed into 32 different languages, has a billion worldwide viewers in 141 countries. *Baywatch, Babylon 5,* and *Journeys of Hercules* frequently rank in the top 20 most-watched syndicated programs, right next to *Seinfeld* and *Wheel of Fortune.* First-run syndicated fare is especially attractive to independent station programmers because its high quality helps further blur the distinction between their stations and the usually richer affiliate competition, and it often serves as a tent-pole for an evening of material.

Satellites have helped boost the number and variety of programs in first-run syndication. Game and talk shows, standard first-run fare in the past, have proliferated and been joined by *Judge Judy* and *Judge Joe Brown,* court shows distributed daily by satellite to hundreds of stations. They are inexpensive to make, inexpensive to distribute, and easily stripped. They allow an inexhaustible number of episodes with no repeats and are easy to promote ("Watch the case of the dirty dancing dentist").

Buying Syndicated Programming

Syndicators offer station programmers three ways to acquire the license fee for their programming. Cash payments require a relatively small down payment, 10% to 15%, and then regular payments to the syndicator once the program is actually on the air and earning revenue. Paramount, for example, sells *Star Trek: Voyager* for cash.

Barter involves no exchange of cash. Syndicators provide programming to stations with several minutes of commercials already included (for which the syndicator is paid by the advertisers). A specified number of commercial spots are left empty, and the local station sells that time, keeping the proceeds. Stations like barter because they do not have to spend (or gamble) their own money. Syndicators like to use barter, especially for first-run programming, because production costs are typically so high that it takes a combination of several national advertisers to underwrite that production. The syndicator, however, must guarantee that the programming will reach 70% to 80% of all U.S. television homes (von Soosten, 1993).

Barter deals are typically presented to stations in terms of their **time splits.** This is the number of national commercial minutes that must be run plus the number of local minutes available for sale. For example, in 1999 MGM's *National Enquirer* was offered as 3.5/3.5, meaning that each half hour comes with 3.5 minutes of time for the local station to sell and another 3.5 minutes of commercials built in. Twentieth Century Fox offered *Buffy the Vampire Slayer* for a 7/7 split. The value of these splits to the distributor can be seen in the amount of money generated by the national advertising that fills this time. *Drew Carey* earns Warner Brothers $143,300 for each 30-second national spot; *Friends* earns that same company $119,800 per 30-second spot. Even an inexpensive-to-produce daily first-run syndicated program such as *Entertainment Tonight* makes $84,900 for each 30-second spot for its distributor, Paramount. The average split for all syndicated programming hovers around 4 minutes of national and 10 minutes of local advertising per hour (Accas, 2000).

The final way stations acquire syndicated material is through **cash-plus-barter,** a combination of cash and barter. These deals are characterized by reduced amounts of built-in commercial time. For example, Paramount offered *Spin City* in 1999 on a cash-plus-barter basis with a time split of 5.5/1.5 (Seltel, 1999). Figure 10.3 presents the ratings for a sample week of syndicated programming.

Concentration of ownership is reshaping the syndication business. A frequent lament among contemporary programmers is that "it's getting hard to tell the buyers from the sellers in syndication these days." This is because, in a reversal of the normal way of buying and selling syndicated programming, owners of large groups of stations are now demanding cash payments—in addition to time splits—from syndicators. "In exchange for giving syndicators

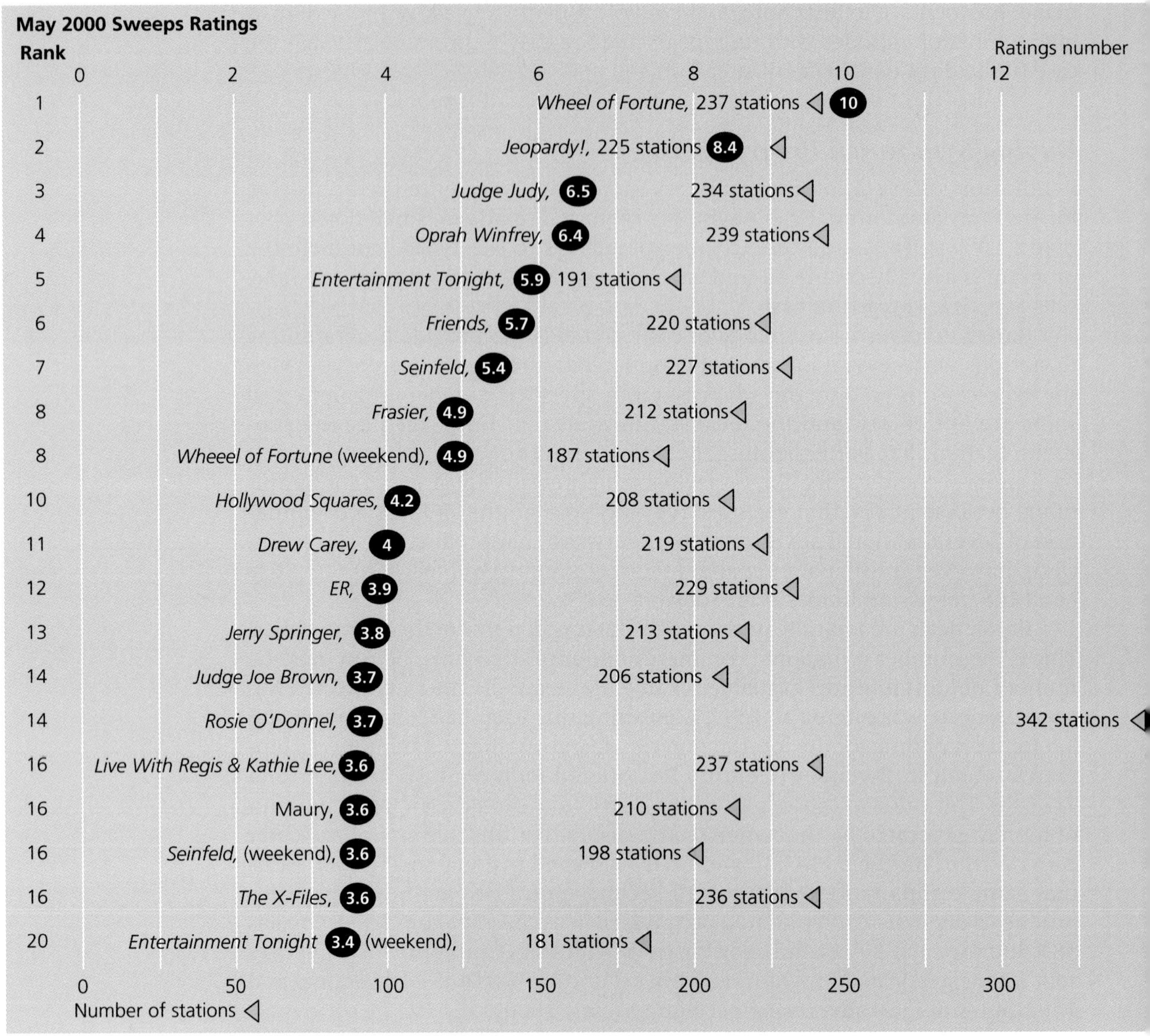

Figure 10.3 Syndication's Ratings Leaders
Here are the top 20 rated programs in national syndication for the 2000 May sweeps (April 27 to May 24). These ratings are computed by totaling all viewers in all the markets who tune in. Can you identify the first-run and off-network programs among them? *Source:* Nielsen Syndication Ratings, *Variety,* March 20–26, 2000, p. 11.

Careers *Working in Programming*

Regardless of medium—radio, local television, network television, or cable—some aspects of working in programming are the same. Titles may differ (program director in radio, program manager in the video media) but the responsibilities of the programming chief do not. They include these jobs:

Program planning. Developing the operation's short-, medium-, and long-range strategy for meeting its programming and financial goals and objectives. In television this planning is geared to choosing and scheduling programs that appeal to the greatest number of viewers among the available audience. For radio planning centers on the selection of a format and other content consistent with that station's sound and character to attract and hold the interest of a specific demographic group. This planning is typically conducted with the general manager, who has overall responsibility for the success of the operation, and the sales manager, who has the responsibility of selling audiences attracted by programming to advertisers.

Program acquisition. Overseeing the production of programming by the operation itself and obtaining content from other sources such as networks and feature and program syndicators.

Program execution. Scheduling or otherwise using programming to meet the goals established in program planning.

Program control. An operation's programming head "is often called the 'protector' of the station's license because of the responsibility for ensuring that the station's programming complies with the terms of its license" (Pringle et al., 1991, p. 105). This responsibility requires the development of program standards, supervision of all content for adherence to station, FCC, and other standards, and maintenance of the appropriate program records.

Successful programmers know their operation and management goals and objectives, their market, their competition, how to manage programming and people, their audiences, and their content. Moreover, successful programmers must be able to work as part of a team because "in most cases management teams headed by vice presidents, not just program directors, make programming decisions of any consequence. . . . Programmers do not enjoy the luxury of autonomous decision making" (Head, Eastman, & Klein, 1993, p. 36). As such, the two most important skills a successful programmer can possess are a thorough understanding of his or her medium and its relationship with its audience and excellent speaking and writing skills. As broadcast programming executives Edward Aiken and Robert Affee (1993) argue, "If you cannot write, speak, and comprehend (language) masterfully, how can you express yourself and convince others of the merits of your ideas?" (p. 236).

premium clearances in 21% of the country, we think it's appropriate that we get more than a 50–50 split," argued Bill Butler, vice president and group program director for Sinclair Television, owners of 55 stations that reach nearly a quarter of the nation's television homes (as quoted in Littleton, 1999a, p. 53). For now, the practice of local stations (even big chains of stations like Sinclair) being paid by syndicators to air their fare is only a trend, but given the rapid rate of concentration in television ownership (Chapter 3), it deserves programmers' attention.

If you're interested in a career in programming, you may find the box titled Working in Programming of interest.

National Cable Television Programming

We examined cable television programming in some detail in Chapter 6 in our discussion of the structure of its various tiers and the various types of channels available to local cable operators. Here we will describe several trends in national cable television programming.

The national reach and financial clout of many basic cable channels such as USA Network and Lifetime have made them major players in buying and exhibiting off-network and first-run syndicated programming. *Miami Vice, Murder, She Wrote,* the lineups of Nick at Night and TV Land, *ER, The Golden Girls,* and *NYPD Blue* are examples of off-network fare drawing sizable cable audiences. *The Journeys of Hercules, Xena, Warrior Princess,* and *Silk Stalkings* are first-run shows that also enjoy success on national cable channels.

Cable channels are increasingly being recognized by producers and audiences alike as the home of innovative and challenging programming. For example, *The Man Show* was rejected by ABC only to become a hit on Comedy Central. Similarly, the critically acclaimed audience favorite *The Sopranos* was declined by NBC, CBS, and Fox. But in its first year on HBO it became a much-talked-about national sensation, drawing more than 10 million viewers an episode. It also garnered 16 Emmy nominations and 4 of the 10 prizes awarded by the Television Critics Association (outstanding achievement in drama, program of the year, best new program, and for its star James Gandolfini, individual achievement in drama), making it television's most honored program in many years. The *New York Times* went as far as to call it "the greatest work of American pop culture in the past quarter century" (McCollum, 2000, p. 1G). At a time when traditional broadcast networks seem to be retreating from their commitment to quality miniseries (airing audience and critical failures such as NBC's *The Magical Legend of the Leprechauns* and CBS's *Shake, Rattle & Roll*), the cable networks are moving in the other direction. HBO recently invested $68 million in production and $10 million in promotion of its 12-part *From the Earth to the Moon.* In the last few years TNT produced George Orwell's *Animal Farm,* Stephen E. Ambrose's story of the Lewis and Clark expeditions, *Undaunted Courage,* and an Emmy-nominated miniseries on former Alabama governor and presidential candidate George Wallace. Showtime undertook productions of stories about Fidel Castro and the Middle East peace process, and USA had a ratings and critical hit with its miniseries *Moby Dick.* The cable networks are programming this challenging long-form content for several reasons. First, it further sets them apart from the broadcast networks, justifying the additional expense incurred by viewers of cable. Second, the **aftermarket** for high-quality miniseries, both in videotape rental and overseas distribution, is lucrative.

Time magazine television writer Jeanne McDowell (1999) explained why creative, challenging, and innovative programming appears increasingly on cable channels.

> Network chiefs, having watched their prime-time audience share erode from 91% for the big-three networks 20 years ago to 60% shared by six of them today,

USA's *Moby Dick,* starring Patrick Stewart, was not only a ratings and critical success, its aftermarket income (both in videocassette and international distribution) was substantial.

> seem too paralyzed to make real changes. "Networks are locked in a box like the rest of corporate America," says Normal Lear, who created *All in the Family.* "In TV terms that translates into *Gimmee an instant hit* at the expense of every other value, like creativity." Instead of looking beyond Burbank for people with fresh ideas, the networks return to the same talent pool over and over. (p. 71)

Producers interested in creativity gravitated to cable. "I just think working in cable I died and went to heaven," said Ethel Winant, producer of TNT's *Wallace* (as quoted in Sterngold, 1998, p. 87).

There is another more practical reason cable appears to be able to produce better quality original programming than the traditional over-the-air networks—they don't have to produce as much. Cable channels fill the majority of their schedules with movies and off-network fare.

Chapter Review

Radio station programming follows a program clock developed by the program director to ensure that the DJ adheres, in the case of music radio, to the playlist and rotation, and in the case of news and information radio, to the station's desired sound. Often, local radio stations depend on consultants, syndicators, and networks for some or much of their content, and that content is acquired through affiliation, cash purchases, barter, and cash-plus-barter. Because there is so much variety in the types and sound of noncommercial radio, its programming is more idiosyncratic than that of commercial radio, but most noncommercial stations strive for localism and program appropriately.

Programming broadcast television, once a relatively straightforward matter, has developed into a more complex enterprise since the advent of cable. With declining audiences and scores of video competitors, the networks have had to become more sophisticated in their approach to programming. Most programs, even those requested by the networks, undergo a rigorous route before they can be scheduled, and most often the financial risk is borne by the producer rather than by the network. Producers accept this state of affairs in hopes of earning profits when their shows enter syndication. Changes in FCC rules of content ownership have put much more programming under the networks' own ownership, and there is disagreement about whether this is good or bad for the quality of television fare.

Once a show makes it onto a network schedule, programmers must determine its position. Programming strategies now include premiering shows throughout the broadcast season and using scheduling strategies such as lead-off and lead-in, hammocking, blocking, tent-poling, counterprogramming, stunting, and bridging. Networks, too, are making greater use of reruns and repurposed content in their quest for ratings.

Network programmers must also consider their daytime, late night, and weekend schedules, and most follow a similar pattern. They program early morning talk shows, morning and afternoon soap operas, early evening national news, late night talk, weekend sports, and Sunday morning public affairs.

Local television programmers face different challenges than do their network counterparts. Although affiliate stations have the benefit of network-provided content to fill much of their broadcast day, local programmers must find material for morning and afternoon dayparts, early fringe, early evening daypart, prime access, late fringe, and the late night and overnight dayparts. To do so, they rely primarily on syndicated material, typically either first-run or off-network. They secure this programming through cash payments, barter (with varying time splits), and cash-plus-barter. The growth of large groups of television stations may upset the traditional syndication business, with big chains such as Sinclair asking syndicators to pay *them* to run their shows.

Central to the decline of the networks' share of audience has been the ascendance of numerous national cable networks, themselves programming high-quality fare, oftentimes off-network hits, first-run material, and increasingly innovative and challenging shows considered too controversial for the over-the-air networks.

Questions for Review

1. How do the playlist, rotation, and program clock influence what we hear on music formatted radio?
2. What is a sound hour? A news cycle? Audience flow?
3. Differentiate between a radio consultant, syndicator, and network.
4. What are format and feature syndicators?
5. What is long-form and short-form programming in radio?
6. What are licensing fees and market exclusivity? What are outright cash purchase and barter?
7. Describe the steps that must be taken for a television show to make it onto a network schedule.
8. How have the rules governing ownership of network television shows changed over the years? What has been the effect?
9. Define viewing year, year-round scheduling, lead-off position, lead-in, hammocking, forfeiting audience, blocking, tent-poling, counter-programming, stunting, and bridging.
10. What is repurposed content?
11. What are the different dayparts that local television station programmers must program?
12. What are stripping and checkerboarding?
13. What is a time split in television syndication?

Questions for Critical Thinking

1. Although radio stations are licensed locally and must promise to serve their local broadcast areas, material from consultants, syndicators, and networks fill their schedules with increased amounts of non-local fare. Does this concern you? Do you think it is important for local radio to be local in the sense that it is programmed locally for a specific local audience?

2. Many critics believe the deficit financing involved in the production of network television content limits producers' willingness to take chances or make innovations in the programs they offer the networks. Do you agree? If you do, how do you explain innovative, groundbreaking network programs such as *All in the Family* and *Hill Street Blues*? If you do not, how do you explain the apparent difference between most network fare and cable innovations such as *The Sopranos* and *Win Ben Stein's Money*?

3. Fred Silverman seemed to succeed as a programmer because he had "the feel" or a sharp sense of what audiences wanted. What about you? Do you think you would make a good programming executive? What strengths would you bring to the job? What deficiencies might you have?

4. Have you ever seen the cable series *The Larry Sanders Show, Arli$$, Oz, The Man Show, The Daily Show, Sex in the City,* or *The Sopranos*? If yes, how different might they look if they were to find a place on the schedule of one of the over-the-air networks? Would you, as the producer of one of these shows, exchange

the artistic freedom of cable for the larger audience and presumably greater riches of the networks?

5. In light of contemporary concerns about the degradation of American culture, with critics pointing a finger at television's contribution to falling standards of civility and courtesy and to increases in teen violence, should we long for the days of the Big Three? Were we better off as a nation when there were three networks programming more or less predictable similar fare designed for a large national audience?

Engaging Technology

1. The debate surrounding Telemundo's move to bilingual and bicultural fare involves its primary Spanish-language competitor, Univisión. Visit Univisión at <www.univision.net> and Telemundo at <www.telemundo.com> to determine for yourself which network's programs and programming strategy seem most appropriate to you. Defend your position.

2. Here are the URLs of a few television program syndicators. Visit as many as you can and report on the titles they offer and the financial arrangements (cash, barter, cash-plus-barter) they require.

Alfred Haber Distribution	www.haberinc.com@haberinc.com
BBC Worldwide	www.bbc.co.uk
Buena Vista Television	www.tvplex.com
Cable Ready Corporation	www.cableready.net
Paramount Television	www.paramount.com
Planet Pictures	www.planetpictures.com
Radio Netherlands Television	www.rnw.nl
Studios USA	www.studiosusa.com
Warner Brothers Domestic Television Distribution	www.warnerbros.com
Zia Film Distribution	www.ziafilm.com

Chapter Eleven

Global Telecommunications

First-run syndication hit *Baywatch*, dubbed in 32 different languages, has a billion worldwide viewers in 141 countries. The World Wrestling Federation is aired in 120 countries in 11 languages. Discovery Communications, parent company of the Discovery Channel, the Learning Channel, Animal Planet, and the Travel Channel, reaches 1.2 billion cable and satellite subscribers in 149 countries. Blockbuster Video has more than 6,000 outlets in 26 countries. The Number One rated television show for 1999 in Australia features six New York City twentysomethings—*Friends. South Park* appeared on German television screens in 1999, quickly making "Stanley" one of the most popular names for Deutschland's newborn baby boys. England's BBC has a channel on hundreds of U.S. cable systems and its storied *BBC Television News* can be watched by some 36 million American PBS viewers. If you're in the right place (and in the mood), you can watch MTV Latin America, MTV Asia, MTV Australia, MTV Brazil, MTV Japan, and MTV Europe. TV France International, that country's umbrella distribution organization, has partnership agreements with Fox, Warner Brothers, the Discovery and Sundance Channels, and Bravo. Its all-French channel is available to viewers in the United States on satellite provider DirecTV. Sixty nations receive CNN by satellite. Radio Beijing broadcasts to a worldwide audience in 40 languages. Hundreds of millions of

Internet users spread across scores of countries can tune in to the more than 7,500 Web radio stations originating from every continent except Antarctica. AT&T, the largest U.S. telecommunications company and multiple systems operator, and British Telecom, Britain's biggest telecommunications provider, have merged their international operations into a single $10 billion unit. Telecommunications knows few national borders.

But the global flow of expression and entertainment is not welcomed by everyone. French law requires that 40% of all music broadcast by its radio stations be in French. The Israeli government is considering a law requiring all "radio stations to devote half their airtime to songs sung in Hebrew in order to slow down Israel's cultural shift toward Americanization" (Zachary, 1999, p. 56). In the kingdom of Bhutan, near India and Tibet, television was illegal until June 1999. To ensure that its people do not access "foreign" or otherwise "counterrevolutionary" Internet content, the Chinese government requires that all Internet accounts be registered with the police. Telecommunications may know few national borders, but there is growing concern that it at least respect the cultures within them.

Preview

Throughout this text we have discussed globalization of telecommunications as a significant trend in reshaping media industries and their relationship with their audiences. Here, we look more deeply at that globalization and its impact on world media and on the media of individual nations. We begin with an examination of international telecommunications in the days before communication satellites when pirate and clandestine stations were the forerunners of today's global media. We then look at how satellites transformed international media, making them truly global.

We use comparative analysis to study today's global telecommunications, focusing on the normative theories that shape media operations in different countries. This naturally leads us to an examination of the programming offered by the media in different lands.

Finally, we visit in some detail the issue of cultural imperialism. Powerful arguments exist on both sides of the debate over whether one culture can impose its values on another through mass media.

The Roots of Global Telecommunications

Radio and television were, in effect, international in their earliest days. Guglielmo Marconi was the British son of an Italian diplomat, and among his earliest successes was the 1901 transmission of a wireless signal from England to Newfoundland. U.S. inventors, in the persons of Philo Farnsworth and Russian immigrant Vladimir Zworkyin, met and eventually overcame the challenge posed by Scottsman John Logie Baird, who counts among his greatest achievements the successful transmission of a television picture from London to New York in 1928.

Both the Marconi and Baird transmissions were experimental, designed to attract attention and money to their infant technologies. But it was not much later in the development of radio and television that these media did indeed become, if not truly global, at least international. To best understand how these and other telecommunications technologies became global, it is necessary to divide our discussion into two eras: before satellites and after satellites.

International Telecommunications Before Satellites

Almost from the very beginning, radio signals were broadcast internationally. By the mid-1920s the major European colonial powers—the Netherlands, Great Britain, and Germany—were using **shortwave radio** to connect the Motherlands with their various colonies, primarily in Africa, Asia, and the Middle East, and in the case of the British, North America (Canada) and the South Pacific (Australia). Shortwave was (and still is) well suited for transmission over very long distances because its high frequencies (3,000 to 30,000 kHz) easily and efficiently reflect, or **skip,** off the ionosphere, producing **sky waves** that can travel vast distances. The box titled Sky Waves and Shortwave Radio graphically represents this process.

Clandestine Stations Colonial powers were not the only ones using international radio. Antigovernment or antiregime radio also comprised an important segment of international broadcasting. These were typically **clandestine stations.** These stations emerged "from the darkest shadows of political conflict. They frequently are operated by revolutionary groups or intelligence agencies" (Soley & Nichols, 1987, p. vii). In World War II, for example, stations operating from England and other Allied nations encouraged German soldiers and sailors to sabotage their vehicles and vessels rather than be killed in battle. Allied stations also intentionally broadcast misleading reports. For example, Atlantic Station and Soldiers' Radio Calais, posing as two of the many official stations operated by the German army, frequently transmitted false reports to confuse the enemy or to force official Nazi radio to counter with rebuttals, providing the Allies with exactly the information they sought.

But it was in the Cold War that clandestine broadcasting truly flowered. In the years between the end of World War II and the fall of European communism in 1989, thousands of radio, and sometimes television, pirates took up the cause of revolutionary (pro-communist) or counterrevolutionary (anticommunist) movements. In addition, other movements tangentially related to this global struggle, especially the growing anticolonial movements in South and Central America and Africa, made use of clandestine broadcasting.

The Cold War has ended, but clandestine stations still exist. As in the Cold War, they typically operate outside the nations or regions to which they broadcast. These stations are unauthorized and opposed to the regimes or govern-

Tech Talk *Sky Waves and Shortwave Radio*

The ionosphere does not allow all electromagnetic waves to dissipate into space. It bends back toward earth certain medium- and high-frequency waves. Because of its high frequencies, shortwave radio intentionally takes advantage of this phenomenon to extend its reach. A shortwave signal is sent upward from a transmitter, strikes the ionosphere, and bounces back toward earth several hundred miles away. The product of this effort is sky waves that skip. The process looks like this:

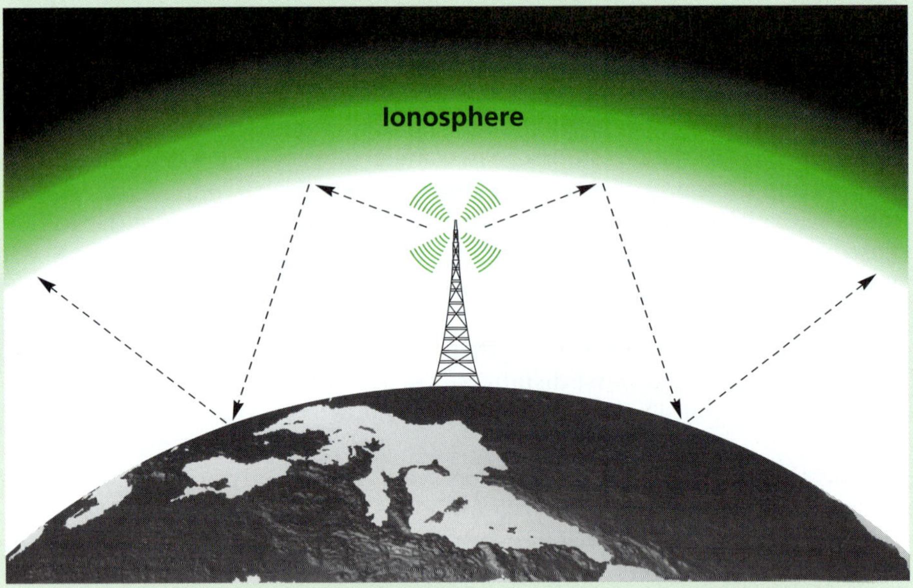

But because the ionosphere's ability to bend signals back to earth is affected by its temperature, engineers face an interesting problem in adjusting to atmospheric conditions. They do this by transmitting on a number of successive frequencies, allowing for different times of day and season. AM waves also experience skip, but because of their lower frequencies, they skip only at night when the ionosphere cools off. This is one reason we experience more interference when listening to AM in the evening.

ments in power. If they operate within those borders, they risk discovery, capture, and imprisonment or death. Those relatively few clandestine operations that operate within the regions to which they transmit can be classified as **indigenous stations,** and those operating from outside are labeled **exogenous stations.**

Opponents of military rule in Haiti utilize indigenous radio to challenge those in power. Said Jean Dominique, that country's leading radio journalist, "Radio, all by itself, is a very dangerous medium in an illiterate country like Haiti. People find, through the media, a way to amplify their voices, to express their hunger, their frustration, and their strategy to fight. . . . News can mobilize power. This, and the truth, are the fears of the military" (as quoted in "Soundbites," 2000, p. 2). Dominique was assassinated in Port-au-Prince, Haiti's capital city, on April 3, 2000.

Hezbollah TV is an example of an exogenous (or international) operation. In the months leading up to Israel's May 2000 departure from southern Lebanon, ending 18 years of occupation, the Iranian-backed guerrilla organization Hezbollah routinely produced video reports of its military actions against Israeli forces. Hezbollah would broadcast these sophisticated, multicamera minidocumentaries in both southern Lebanon and Israel from its transmitters deep inside Lebanon. The goal of these cross-border telecasts was to boost the morale of Muslims in occupied South Lebanon and to discredit official government accounts of success against the guerrillas in the eyes of Israeli viewers.

Pirate Broadcasters Another type of broadcast operation transmitting from outside its desired audience's geographic location involved something a bit more benign than war and revolution. Broadcasting into Great Britain beginning in the 1960s, these **pirate broadcasters** were illegally operated stations broadcasting to English audiences from offshore or foreign facilities. Among the more notable were Radio Caroline, reaching a daily audience of a million listeners with its signal broadcast from the *M. V. Frederika* anchored 3.5 miles off the Isle of Man, and Radio Veronica, broadcasting from a ship off the coast of the Netherlands.

Unlike their politically motivated clandestine cousins, these pirates were powerful and well subsidized by advertisers and record companies. Moreover, much like the commercial radio stations with which we are now familiar, they broadcast 24 hours a day, every day of the year. These pirates offered listeners an alternative to the controlled and low-key programming of the British Broadcasting Corporation's (BBC) noncommercial stations. Because the BBC was noncommercial, advertisers who wanted to reach England's consumers were attracted to the pirates. And record companies intent on introducing Britain's youth to their artists and to rock 'n' roll saw the pirates as the only way to reach this audience ignored by the staid BBC.

Enterprising broadcasters also made use of foreign locales to bring commercial television to audiences otherwise denied. The top-rated network in Germany today, for example, is RTL. Now broadcasting from the German city of Cologne, it began operation in January of 1984 in Luxembourg, transmitting an American-style mix of children's programming, sports, talk shows, and action-adventure programming into Germany to compete with that country's two dominant public broadcasters, ARD and ZDF.

The United States as International Broadcaster World War II brought the United States into the business of international broadcasting. Following the model of England, which had recently augmented its colonial broadcast system with an **external service,** the United States established in 1940 what

The Voice of America home page

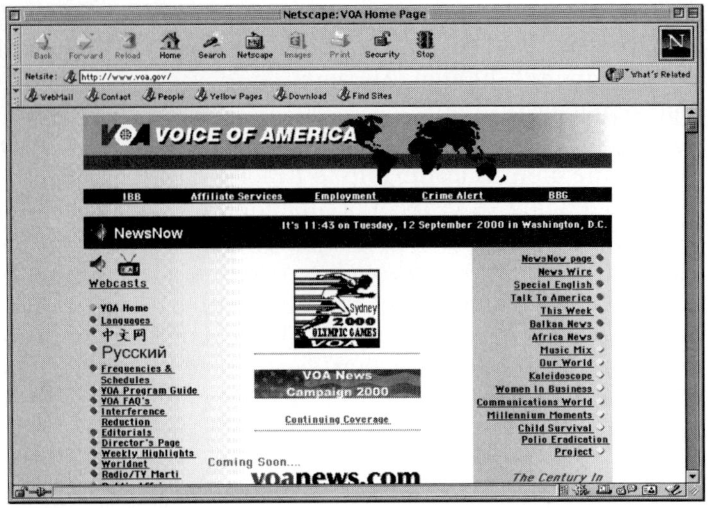

would eventually be known as the Voice of America (VOA) to counter enemy propaganda and disseminate information about the United States. The VOA originally targeted countries in Central and South America friendly to Germany. But as war engulfed the world, it quickly began transmission to scores of other nations, attracting, along with England's BBC World Service, a large and admiring listenership, first in countries occupied by the Axis powers and later by those in the Soviet sphere of influence.

It was this Cold War with the Soviets that moved the United States into the forefront of international broadcasting, a position it holds today. To counter the efforts of the Soviet Union's external service, Radio Moscow, three additional services were created. Radio Free Europe (RFE) broadcast to all of the communist bloc Eastern European countries in their native languages. Radio Liberty (RL) was aimed at listeners in the Soviet Union itself, and Radio in the American Sector (RIAS) served people inside East Berlin and East Germany. When these services were initiated, people in both the United States and abroad were told that they were funded by contributions from U.S. citizens. However, in 1971 it was revealed that they were, in fact, funded by the Central Intelligence Agency. The resulting furor brought them into open government funding and under the administration of the Board of International Broadcasting whose members were appointed by the president.

Communist nations targeted by these services attempted to jam their signals by broadcasting on the same frequencies at higher powers, but they were only minimally successful in keeping their people from these broadcasts. It was the success of these **surrogate services**—broadcast operations established by one country to substitute for another's own domestic service—that prompted President Ronald Reagan in 1985 to establish a special division of

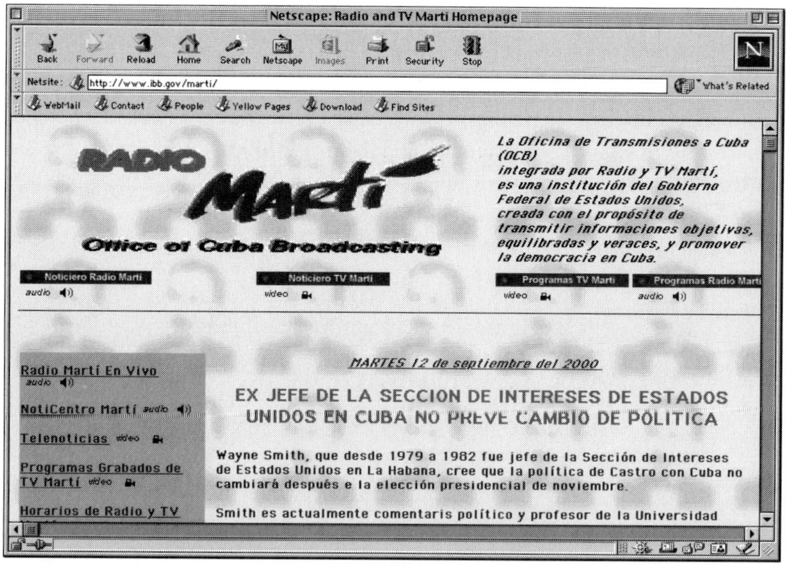

The Radio Marti home page

the VOA, Radio Marti, to broadcast into Communist Cuba. Radio Marti, still in operation, was joined by TV Marti in 1990. Beginning in October 1999, the administration of these surrogates has been supervised by the International Broadcasting Bureau, an autonomous civilian broadcast service associated with the U.S. State Department.

A final United States external service established during World War II and the Cold War, Armed Forces Radio and Television Service (AFRTS), remains active today. Maintained by the U.S. military, its stated mission is "to deliver Department of Defense internal information and radio and television programming services which provide 'a touch of home' to Department of Defense personnel and their families serving overseas" (AFRTS, 2000). With more than 100 stations, AFRTS reaches listeners in 70 countries with commercial-free fare.

The VOA Today Today, 83 million listeners a day tune in to VOA broadcasts in 52 languages, and another 20 million people in 23 developing countries listen to its surrogate operations, RFE, RL, Radio Marti, and the recently added Radio Free Asia (Barber, 1998). Throughout its history, the VOA has frequently vacillated between two roles in response to world events and political pressures at home, sometimes disseminating Western propaganda and at other times providing objective information. With the threat of communist world domination virtually nonexistent, the VOA now focuses on the less conflicting goals of spreading U.S. culture and disseminating health and social information.

Its commitment to the spread of U.S. culture can be seen in the establishment in 1992 of a 24-hour-a-day, Top-40-style service, VOA Europe, and in

The Real World *The United States as International Propagandist*

When the Voice of America went on the air in the 1940s, its pledge to foreign listeners was straightforward: The news may be good. The news may be bad. We shall tell you the truth. Ever since, and especially since the fall of Soviet communism in 1989, critics have attacked the VOA for placing "the truth" second to what former VOA Moscow bureau chief Mark Hopkins (1999) calls "political programming with clear ideological agendas" (p. 44).

In addition to the VOA and its 767 hours a week of radio broadcasts in 51 languages and 20 hours a week of television broadcasts in 9 languages, the United States maintains a number of other external services:

- Radio Free Europe/Radio Liberty (RFE/RL) broadcasts to audiences in the former Soviet Union and Eastern Bloc nations in 23 languages for 715 hours a week.
- Radio Free Asia broadcasts cover most of Asia, but primarily targets China. It offers 238 hours of programming a week in 10 languages.
- Worldnet Television, a satellite-delivered international television service, blankets the globe with 70 hours a week of English-language content.
- Radio Marti and TV Marti transmit 163 hours of Spanish-language radio and 31 hours of Spanish-language television to Cuba.
- Radio Free Iraq uses its $3 million annual budget to reach listeners in that country.
- The Persian Service uses its $3.2 million to reach those in Iran.

Additional service may be established at any time, given world events. For example, TV Bosnian, an offshoot of RFE/RL, recently targeted Bosnia with 30 minutes of television a week, and the South Slavic Service was aimed at the Balkan nations.

What concerns critics is that other than Worldnet and the VOA itself, each of these services specifically targets countries where the United States has strategic interests, in other words, where it would like to influence political and economic matters. Even the VOA and Worldnet are not above criticism. The argument of a VOA language service director that "as a government broadcaster, you can't be neutral" (Hopkins, 1999, p. 46) is unconvincing to those who want U.S. surrogate services to be more like the highly regarded BBC World Service or the smaller but respected German Deutsche

1998 of a 24-hour, All News English-language worldwide radio service characterized by a snappy, commercial station–style sound. The VOA's commitment to the dissemination of health and other practical information can be seen in its increased programming of AIDS prevention, nutrition, and vaccination information aimed at developing nations. In pursuit of this humanitarian goal, the VOA now frequently strikes agreements in these countries with local stations to broadcast its programming over their AM and FM stations, making it accessible to those people who listen on these rather than on the shortwave band. Still, not everyone is comfortable with the role of the VOA and other U.S. surrogate services, as you can see in the box titled The United States as International Propagandist.

Satellites and Global Telecommunications

We saw in Chapter 6 how satellites turned cable television from an importer of distant signals to a provider of original programming, setting off a revolu-

Welle. That is, they want a unified, objective international U.S. radio and television service free of political objectives—one listened to and watched because of its reliability and credibility.

Their arguments are several. First, a worldwide propaganda system seems unnecessary in an era of global telecommunications. Why would anyone tune in to a propaganda channel when they can receive objective, high-quality content not only from the BBC World Service but from other U.S. and Western sources? Critics point to Radio and TV Marti, both of which are ignored by a Cuban audience within easy reach of U.S. commercial broadcasters. Second, propagandistic broadcasts can alienate the leadership of targeted countries. President of RFE/RL Thomas Dine says the goal of his service is to "foster democracy, promote free market reforms," and Radio Free Asia focuses its news reports on dissidents who challenge regimes considered unfriendly to the United States (Hopkins, 1999, p. 46). Neither RFE/RL's goals nor Radio Free Asia's practice endear the United States to leaders in Eastern Europe and China who, regardless of their political orientations, are our global neighbors. Third, there is too much waste and duplication among what John Lennon, Worldnet's director, calls "Radio Free This or That." Twenty-seven different languages services are duplicated among the various surrogates (Hopkins, 1999, p. 46). Finally, the cost—at nearly $400 million a year—is simply too high to engage in practices that not only would be considered offensive at home but make a lie the American ideal of a free and open media embracing all shades of opinion.

What do you think? Is $400 million not a particularly steep price to ensure that people around the world have access to the American point of view? Do you think any international service, commercial or surrogate, should produce and distribute content specifically designed to transmit any government's "official" viewpoint? Does it concern you that Congress, in exchange for granting funding, requires the VOA to broadcast U.S. State Department approved editorials? How free of government influence should any taxpayer-supported international telecommunications service be?

tion in television. The impact of satellites on international telecommunications was no less profound. With the coming of satellites signals could now be distributed internationally, between two countries, as had been the case, and globally—that is, everywhere.

The satellite revolution began in 1957 with the successful launch and orbiting of the Soviet Union's Sputnik. Sputnik had no real function other than to prove (especially to the United States) that the Soviets could indeed produce the world's first artificial satellite. But it had a significant impact nonetheless. It shook the confidence of the Western nations, leading them to redouble their efforts to conquer space.

The United States placed the second satellite in orbit in 1962, AT&T's Telstar I. In that same year Congress established the Communications Satellite Corporation (COMSAT) to coordinate ownership and operation of the U.S. communications satellite system. Even at the time, however, it was clear that the greatest potential for satellites was in their ability to globalize telecommunications. In 1964 President John F. Kennedy convened a consortium of Western

The 1965 launch of INTELSAT I, better known as Early Bird, made possible the first regularly scheduled transmission of live television between Europe and North America and established the United States' leadership in global satellite.

and nonaligned countries to establish the International Telecommunications Satellite Organization (INTELSAT). Kennedy had two goals in mind: First, to effect the creation and maintenance of a global communication satellite system serving its many member nations; and second, to ensure that its leadership remained with the United States.

U.S. leadership in global satellite matters was assured when COMSAT was declared managing agent of INTELSAT's system, begun on April 6, 1965, with the launch of INTELSAT I, better known in the United States as Early Bird. Within one year of that success, 55 nations had joined INTELSAT, and Early Bird provided the first regularly scheduled transmission of live television between Europe and North America.

INTELSAT's role as a facilitator of global telecommunications began slowly but expanded rapidly. Early Bird contained 240 voice/data circuits and

Shown here on its launch pad in French Guiana, this French Ariane-5 rocket took two more telecommunications satellites—Asia Star and Insat 3B—into geostationary orbit in March 2000.

one television channel. By the time the INTELSAT VI generation of satellites was operational in 1989, each had the capacity to handle 120,000 voice/data circuits and three television channels (Stevenson, 1994, p. 322). However, 1982 is considered a benchmark year for global telecommunications because by that time INTELSAT's system was sufficiently large and technologically sophisticated that it could begin offering its television customers full-time leases rather than the customary single-show service. Now, not only did satellites make possible global distribution of telecommunications, they encouraged it. Once a media company had paid for its lease, it incurred no additional cost no matter how much additional content it sent.

In 1973 COMSAT gave up control of INTELSAT, and the latter became an independent consortium of member nations, each paying a yearly fee. However, at the 1999 INTELSAT Assembly of Parties, the official meeting of all 143 member governments, the consortium unanimously decided to privatize its service, largely in response to the growing number of competing systems. Among them are Europe-led EUTELSAT and several regional and national systems. Most significantly, though, are the numerous private, commercial systems that commenced operation beginning in 1984. Today, companies such as Globalstar (partners with Loral and Qualcom), Inmarsat (partners with Lockeed Martin and Matra Marconi Space), Hughes Electronics' Spaceway, and Lockheed Martin's Astrolink manage the majority of the nearly 800 working communication satellites (Lo, 1998). The capacity of these satellites has grown to the point that a single provider such as PanAmSat carries more than 100 television transmissions a day, charging customers as much as $200,000 a month for a dedicated transponder, the part of the satellite that receives and retransmits a specific telecommunications service (Littleton, 1999b). Chapter 5 has additional information on communications satellites.

Global Telecommunications Today

One traditional way to understand the workings of the contemporary global telecommunications scene focuses on the media systems of the different countries around the world. In doing so, we become familiar with how different folks in different places use telecommunications, and we can also better evaluate the workings of our own system. Naturally, not every media system resembles that of the United States, and audience expectations, economic foundations, and the regulation of mass media differ across nations. The study of different countries' mass media systems is called **comparative analysis** or **comparative studies.**

Comparative Analysis: Four Theories and Five Concepts

The mass media systems of a country reflect its level of development and prosperity, values, and the nature of its political systems. Often, a country's geography also influences the type of media system it embraces. The level of diffusion of different communication technologies in different countries offers a clear example. The United States is prosperous; we have much mobility and leisure. We tend to live in bigger homes than do people of most other countries, and we have two radios for every citizen and almost one television per person. Australia, a prosperous country, is very large. Many of its people live in quite remote parts of this nation continent. It, too, has more than one radio per person, but only one television for every two people. Uganda, a developing African nation, has almost no television, and despite the remote living situation of many of its people, its lack of prosperity makes even radio ownership a rarity. In Oman, an oil-wealthy desert nation, even though only one in 11 people has a telephone (the hostile terrain makes stringing and maintaining phone lines difficult), there is one television (served by a domestic satellite system) for every two people (Peterson, 1998; Hilliard & Keith, 1996).

That a country's political system will be reflected in the nature of its media system is only logical. Authoritarian governments need to control the mass media to maintain power. Therefore, they will have a much different media system than will a democratic country with a capitalistic, free economy. The overriding philosophy of how media ideally operate in any given system of social values is called a **normative theory.** The classic work on normative theory is the 1956 *Four Theories of the Press* by Fred Siebert, Theodore Peterson, and Wilbur Schramm. It defined four major normative theories that could be used to classify the world's different media systems.

Authoritarian media systems are characterized by government control of the media. Officials can prevent or allow production and distribution of content. Some emerging democracies, despite claims of a free and open media system, exhibit adherence to the authoritarian model. Since 1955, for example, Argentina has had an elected president, free elections, and a capitalist-based economy. But even today its leaders argue that they must control the media for reasons of national security. As a developing democracy, the leaders feel that the people cannot yet be trusted with too much freedom. Of course, not all its citizens agree.

Communist media systems differ from authoritarian systems only in their stated goals. North Korean government leaders, for example, assert full control over media because their political philosophy, in this case communism, views media as a "tool of the people." As such, media must serve the people. Because the people are the Communist Party, the government controls this tool.

Siebert, Peterson, and Schramm identify the **libertarian media system** as indicative of democratic societies. Because democracies rely on government of the people, their media must be free of official control. This is the libertarian ideal. But as we've seen throughout this book, control can and often does come in other forms—commercial pressure, for example. A truly libertarian media system may not best serve democracy. In the United States the print media operate under the libertarian ideal, but the broadcast media are more closely regulated by the government. Yet, of all the nations in the world, the U.S. media system comes closest to the libertarian ideal.

The fourth of Siebert, Peterson, and Schramm's categories describes media systems as being constructed on a **social responsibility model,** wherein media are deemed a public trust. Media professionals are granted significant freedom in exchange for their commitment to serve the larger societal good. The British media system was consciously developed on this model, and the libertarian-style U.S. system is in fact guided by social responsibility (Baran & Davis, 2000, pp. 102–110). See Chapter 12 for a more detailed discussion.

Siebert, Peterson, and Schramm's categories served the study of telecommunications well for many decades, but there has been profound political and technological change since they were first suggested. For example, communism today exists as a guiding political philosophy in only three countries (North Korea, China, and Cuba). Because of advances in telecommunications technologies such as satellite and the Internet, even the most authoritarian nations in the world now allow (or are forced to allow) greater freedom for both media professionals and audiences. These and other technologies have required virtually every country to reconsider how its media systems operate. The four theories, then, require "a little sprucing up" (Stevenson, 1994, p. 108), which has been provided by William Hachten. Hachten (1992) offers "five concepts" that guide the world's many media systems: Western, development, revolutionary, authoritarianism, and communism. We'll examine each and look at examples that exemplify them.

The Western Concept: Great Britain The Western concept is an amalgamation of the original libertarian and social responsibility models. It recognizes two realities: (1) there is no completely free (libertarian) media system, and (2) even in the most commercially driven systems there exist not only the expectation of public service and responsibility but "significant communication-related activities of government" to ensure that media professionals meet those responsibilities (Stevenson, 1994, p. 109).

Great Britain offers a good example of a media system operating under the Western concept. The British Broadcasting System was originally built on the premise that broadcasting was a public trust (the social responsibility model). Long before television, BBC radio offered several services: one designed to provide news and information, one designed to support high or elite culture such as symphony music and spoken-word plays, and another designed to provide popular music and entertainment. To limit government and advertiser control, the BBC was funded by license fees levied on receivers (currently about $165 a year), and its governance was given over to a nonprofit commission. Many observers point to this goal-oriented, noncommercial structure as the reason the BBC developed and still maintains the most respected news operation in the world.

Eventually, Britain, like all of Western Europe, was forced by public demand to institute a more American style of broadcasting. There are now local commercial radio stations in addition to the BBC's five radio networks: Radio 1 (pop and rock music), Radio 2 (classical music), Radio 3 (light music and entertainment), Radio 4 (drama and talk), and Radio 5 (children's programming). In addition to television networks BBC1 (more serious) and BBC2 (more popular), there are commercial Channels 4 and 5 and several regional commercial networks operating under the auspices of the Independent Television Authority (ITA). There is an ITA channel for Ireland, for example, one for Scotland, one for Wales, one for southern England, and so on. But even these must ac-

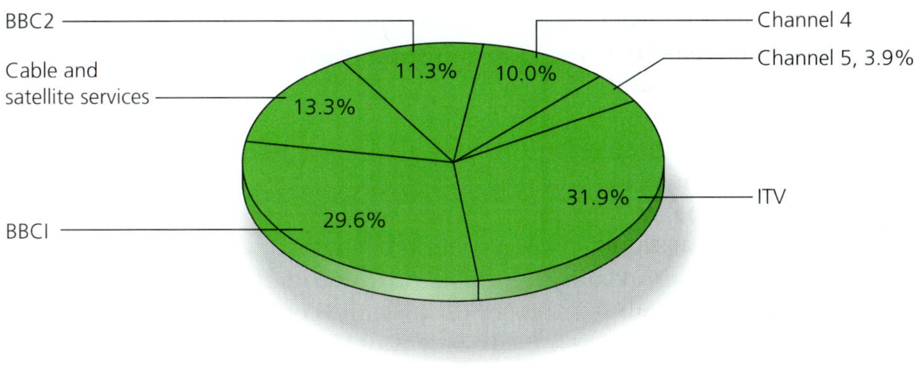

Figure 11.1 **What Services Do Viewers in Great Britain Watch?**
Here are the audience shares of the several television channels and services available to British viewers for the 1997/98 broadcast year. Keeping in mind that BBC1 and BBC2 are noncommercial channels, how would you account for the fact that, taken together, these public channels attract the greatest share of Great Britain's viewing audience? How does this compare to public television's audience share in the United States? Can you account for the difference? *Source:* "United Kingdom," 1998.

cept limits on the amount of advertising they air and agree to specified amounts of public affairs and documentary news programming in exchange for their licenses to broadcast. This is referred to as their **public service remit.**

Cable was slow to develop in Great Britain for two reasons. First, many people simply did not want their beautiful or historic towns sullied by cable's overhead wires, nor did they want their cobbled streets dug up to accommodate underground cables. The second reason is that virtually all of the country has long had access to Rupert Murdoch's DBS system, Sky TV. Murdoch pioneered the minidish receiver, which not only gave viewers multiple channels but satisfied traditional aesthetic sensibilities (no big, obtrusive dish, no unsightly wires, and no excavated streets). Today, satellite provider BSkyB has 3.5 million viewers in Great Britain, and total cable subscribership is 2.6 million (in a country with a population of 60 million adults). Figure 11.1 offers an overview of British television viewership.

The media in Great Britain do not enjoy a First Amendment guarantee of freedom of expression. Prior restraint does occur, but only when a committee of government officials and representatives of the media industry can agree on the issuance of what is called a **D-notice.** British media are also forbidden to report on court trials that are in progress, and Parliament can pass other restrictions on the media. For example, there is a ban on broadcasting the voice of anyone associated with a "terrorist" organization, which is aimed at silencing the Irish independence movement.

The Development Concept: Honduras The media systems of many developing African, Asian, Latin and South American, and now former Eastern Bloc European nations best exemplify the development concept. Here government and media work in partnership to ensure that media assist in the planned, beneficial development of the country. Content is designed to meet specific cultural and societal needs, for example, teaching new farming techniques, disseminating information on methods of disease control, and improving literacy. This isn't the same as authoritarian control. There is less censorship and other official control of content.

Honduras offers one example. This Central American country has government as well as privately owned commercial radio and television stations. A government-authorized and controlled private commission, the Honduran Contractors of Television, coordinates much of the commercial outlets' operations. In addition, the Honduran equivalent of the FCC, HONDUTEL, exerts tight regulatory control over all broadcasting. Together these two bodies ensure that Honduran broadcasting meets the developmental needs of the nation, as determined by the government.

The Honduran government is forbidden by its constitution from taking over or shutting down media outlets because of the content they distribute. Control, however, comes in other forms. One is the requirement that media professionals hold membership in a government-run "official press organization." Naturally, if the government decides who can and cannot function in media organizations, it can easily control those organizations. Another resides in "the close ties of the owners of the most powerful media to the military and to government leaders. The media, therefore, by and large operate with little freedom despite strong constitutional guarantees" (Hilliard & Keith, 1996, p. 161).

The Revolutionary Concept: Poland No country "officially" embraces the revolutionary concept as a normative theory, but this does not mean that a nation's media will never serve the goals of revolution. Stevenson (1994) identified four aims of revolutionary media: ending government monopoly over information, facilitating the organization of opposition to the incumbent powers, destroying the legitimacy of a standing government, and bringing down a standing government. In Yugoslavia, for example, 33 radio and 18 television stations have combined to form the Association of Independent Electronic Media (ANEM) for the purpose of challenging the regime of President Slobodan Milosevic and asserting Serbian independence. ANEM operates with significant public support despite repressive legislation, threats of jail, huge fines, and even military and police confiscation of members' equipment (Aumente, 1999). The experience of the Polish democracy movement Solidarity, however, offers a better-known example of the use of media as a tool of revolution.

By the first years of the 1980s the Polish people had grown dissatisfied with the domination of almost all aspects of their lives by a national Communist Party perceived as a puppet of the Soviet Union. This frustration was fueled by the ability of just about all Poles to receive radio and television signals from neighboring democratic lands. (Poland's location in central Europe made it impossible for authorities to limit what the people saw and heard.) In addition, Radio Free Europe, the Voice of America, and the BBC all targeted Poland

The Polish workers' and democracy movement, Solidarity, was greatly aided by media, both official outlets from beyond Poland's borders and its own extensive network of clandestine new and old technologies.

with their mix of Western news, entertainment, and propaganda. With the people's taste for freedom thus whetted, Solidarity established an extensive network of clandestine revolutionary media. Much of it was composed of technologies traditionally associated with revolution such as pamphlets, newsletters, and audio and video cassettes, but much of it was also sophisticated radio and television technology used to disrupt official broadcasts and disseminate information. Despite efforts of the government to shut the system down, even going as far as to suspend official broadcasting and mail services to deny Solidarity these communication channels, the revolution was a success, marking Poland as the first of the Eastern Bloc nations to defy the party apparatus and install a democratically elected government.

Authoritarianism and Communism Concepts: China Because there are only three remaining communist nations, and because the actual operation of the media in these and other authoritarian systems is quite similar, we can discussed the authoritarianism and communism concepts as one. Both call for the subjugation of media for the purpose of serving the government. China is a good example of a country operating its media under these concepts, and it also offers an example of how difficult it is becoming for authoritarian governments to maintain strict control over media and audiences.

The Chinese media system is based on that of its old ideological partner, the now-dissolved Soviet Union. For a variety of reasons, however, it has developed its own peculiar nature. China has more than a billion people living in more than a million hamlets, villages, and cities. Despite sophisticated life in many big cities, there is near universal illiteracy in the countryside. Because good pulp wood is not native to China and importing it from abroad is too expensive, newspapers are printed on expensive but poor-quality paper made from bamboo. Daily circulation is 116 million copies, around one paper for every 100 citizens. As a result, print is not a major national medium. In fact, face-to-face communication remains a primary means of transmitting news and information among the country's enormous personal communication system. This process is aided by the wide distribution of **wired radio,** which are centrally located loudspeakers, in a town square, for example, that primarily deliver political and educational broadcasts.

The media exist in China to serve the government. Chairman Mao Zedong, founder of the Chinese Communist Party, clarified the role of the media very soon after coming to power in the wake of World War II. The media exist to propagandize the policies of the party and to educate, organize, and mobilize the masses. This is still their primary function.

Radio came to China via U.S. reporter E. C. Osborn, who established an experimental radio station in China in 1923. Official Chinese broadcasting began 3 years later. Television went on the air in 1958, and from its start it was owned and controlled by the party in the form of Central China Television (CCTV) which, in turn, answers to the Ministry of Radio and Television. Radio, now regulated by China People's Broadcasting Station (CPBS), and television stations and networks develop their own content, but content must conform to the requirements of the Propaganda Bureau of the Chinese Communist Party Central Committee.

Chinese broadcasting operates under direct government subsidy. But in 1979 the government approved commercial advertising for broadcasting, and it has now evolved into an important means of financial support. Coupled with the Chinese government's desire to become a more active participant in the international economy, this commercialization has led to increased diversity in broadcast content. Some foreign programming is now permitted, documentaries from the BBC, for example, and cultural programming from Western countries including Germany and Brazil. Since 1994, U.S. programs have been airing alongside the more typical high-level cultural programming—*Little House on the Prairie* has been a particular favorite. CNN and other satellite-delivered services are also available in major metropolitan areas.

Basic government control over major media remains, however, including control of the Internet. In December 1997 the government began enforcing criminal sanctions against those who would use the Net to "split the country" or "injure the reputation of state organs" or for "defaming of government agencies," "promoting feudal superstitions," or otherwise posing a threat to "social stability" ("China Adopts," 1997, pp. C1, C4; Dobson, 1998, p. 20). Internet accounts have to be registered with the police. The state has established a 24-hour Internet task force to find and arrest senders of "counterrevolutionary" commentary. Popular bulletin boards are shut down when their chat becomes a bit too free. Web sites such as Human Rights Watch, the *New York Times,* and publications about China that are independent of government control, such as *China News Digest,* are officially blocked, but not very successfully, as skilled Internet users easily access the Web by routing themselves through distant servers. Many Chinese people are equally disdainful of party restrictions on the reception of satellite television. It is illegal in China for individuals (as opposed to expensive hotels) to directly receive satellite television signals. Despite repeated and severe crackdowns, this rule is openly flouted in most of China's major cities, where DBS providers often advertise their services in local newspapers. In Shanghai alone, "tens of thousands of dishes" dot people's rooftops (Brent, 1999, p. 35).

Programming

Regardless of the particular concept guiding a nation's media systems, those systems produce and distribute content, in other words, programming. In most respects radio and television programming throughout the world looks and sounds much like that found in the United States. This is because the United States is not only among the world's leaders in international distribution of broadcast fare but also because very early in the life of television U.S. producers flooded the world with their programming at very low prices. Foreign operators of emerging television systems were delighted to have access to this low-cost content because they typically could not afford to produce their own high-quality domestic material. For U.S. producers, however, this strategy served the dual purpose of building markets for their programming and ensuring that audience tastes and expectations in those countries would develop along lines not unlike those of viewers in the United States, guaranteeing future sales of programs originally produced for U.S. audiences (Barnouw, 1990). The success of this approach can be seen in comments made by actress/producer Pamela Anderson Lee in discussing her globally syndicated *V.I.P.:* "I took a lower salary on this show because I felt the back end could really be meaningful. I know a lot about syndication and what appeals to the international market. I make sure there are a lot of physical gags and comedy, explosions and beautiful scenery" (as quoted in Stein, 1999, p. 134). Ms. Lee also could have added that action shows such as *V.I.P.* are easily dubbed into local languages.

There are some variations in programming among different countries. The commercial television systems of most South American and European countries are far less sensitive about sex and nudity than are broadcasters in the United States. In Brazil, for example, despite a constitutional requirement that broadcasters "respect society's social and ethical values" (Epstein, 1999, p. A11), television networks such as SBT, TV Record, and TV Globo compete in what critics call the *guerra da baixaria,* the war of the lowest common

V.I.P., a first-run syndication hit in the United States, has been quite successful in foreign distribution, largely because it requires little dubbing of dialogue, and it offers international audiences "a lot of physical gags and comedy, explosions and beautiful scenery," according to its star and producer, Pamela Anderson Lee.

denominator. Guests on variety shows wrestle with buxom models dressed only in bikinis and eat sushi off other women's naked bodies. On game shows contestants who give incorrect answers can be punished by having patches of leg hair ripped out, and those who give correct answers are rewarded with a visit from a nearly naked model who sits in their laps. European commercial operations regularly air shows featuring both male and female nudity, sometimes as integral to the plot, sometimes simply for titillation.

Another difference between U.S. programming and that of its global neighbors is its function in different places. Telecommunications professors Robert Hilliard and Michael Keith (1996) wrote, "In systems relying on commercial advertising for their support, the value of programming is based on how many viewers any given presentation attracts. . . . In systems operated by public corporations that rely on sources of funding other than advertising, the purpose of programming is oriented principally to the educational, social, and/or political purpose of the operating entity" (p. 109). In general this is the case, but many nations, even those with commercially supported systems, use a particular genre, the soap opera, for educational and social purposes. For example, American-style soap operas in countries as diverse as Mexico and Kenya are used to encourage greater use of birth control and family planning. Typically aired in the evening, soap operas are used to this end rather than comedies or dramas primarily because they are inexpensive to produce, often a requirement in developing nations.

The Debate Over Cultural Imperialism

In Chapter 1 we discussed how globalization, concentration of ownership, and hypercommercialism are changing the telecommunications industries and their relationship with their audiences, and we have returned to this theme in subsequent chapters as we've examined the different media. But in the context of global telecommunications, these trends raise the additional issue of **cultural imperialism**—an invasion by media promoting the values of the invading culture. In the past governments wishing to limit the availability of foreign media content within their borders could physically prohibit the introduction and distribution of unwanted newspapers, magazines, and books. They could jam unwanted radio and television broadcasts. Satellites and the Internet have changed matters. Governments cannot disrupt satellite signals. Only the lack of the necessary receiving technology can limit the reach of satellites. In addition, the worldwide expansion of the Internet (see Figure 11.2) provides millions of people in every corner of the world with access to another borderless technology. As a result, difficult questions of national sovereignty and cultural diversity are being raised.

The questions are not new. The debate over national sovereignty and culture reached its height with the 1980 release by the United Nations Educational, Scientific, and Cultural Organization (UNESCO) of the *MacBride Report.* It was named after the chairman of the commission set up by UNESCO to study the question of maintaining national and cultural sovereignty in the face of rapid globalization of mass media. At that time many developing and communist countries were concerned that international news coverage was dominated by the West, especially the United States, and that Western-produced content dominated the media of developing countries that lacked sufficient resources to create their own quality fare. The fear was that Western, especially American, cultural values would overwhelm and displace those of other countries. Critics saw this as a form of colonialism. This cultural imperialism was no less than the invasion of an indigenous people's culture through mass media by a powerful outside empire.

The *MacBride Report,* endorsed by UNESCO, called for establishment of a New World Information Order (NWIO), which is characterized by several elements anathema to Western democracies. Arguing that individual nations should be free to control the news and entertainment entering their lands, it called for monitoring all such content, monitoring and licensing foreign journalists, and requiring prior government permission for direct radio, television, and satellite transmission into foreign countries. Predictably, Western nations rejected these rules as a direct infringement on freedom of the press. Just as predictably, those in favor saw this as additional evidence of the arrogance and conceit of those Western nations.

The Western allies of the United States may have agreed that the restrictions of the NWIO were a threat to the free flow of information, but virtually every ally had rules in place, in the form of quotas, that limited the flow of U.S. media content into their own countries. Canada, for example, required that specific proportions of all content—print and broadcast—either be produced in Canada or reflect Canadian cultural identity. The French made printing certain American words illegal. (France still maintains a government office designed to prosecute those who would "debase" its language.) As recently as 1993, virtually all other European countries supported French resistance to

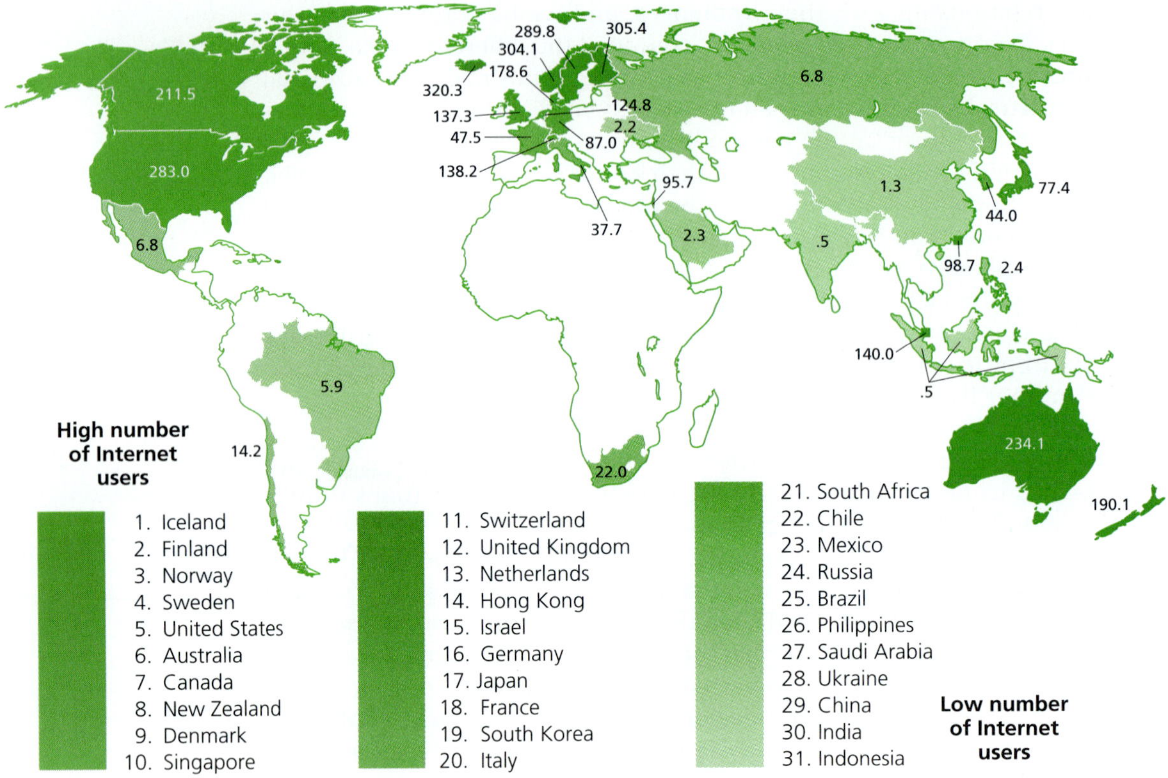

Figure 11.2 Internet Access Around the World

Here is a look at how the Internet has been diffused throughout a number of different countries. Can you explain why some countries have such heavy penetration and others have so little? Do you have any firsthand experience using the Internet in any of these foreign countries? How would you characterize that experience? *Source:* "Leading Countries," 1999.

those parts of the General Agreement on Trade and Tariffs (GATT) that would have eliminated or reduced tariffs on media material entering their continent from the United States. This position was reinforced by a 1998 United Nations recommendation issued from Stockholm that cultural material such as media be exempted from global trade agreements ("Culture Wars," 1998). In that same year the cultural ministers of 20 countries met in Ottawa to develop "ground rules" to protect their nations' cultures from "the Hollywood juggernaut." Included were Brazil, Mexico, Sweden, Canada, and Italy (Turner, 1998, p. 1A). The box titled English as the Global Language has more on this topic.

This resistance to "foreign" media would not exist if countries did not worry about the integrity of their own cultures (Figure 11.3). It makes little sense, then, to argue that non-native media content will have no effect on local culture. The real question is, "How much influence will countries accept in exchange for fuller membership in the global telecommunications community?" In light of instant, inexpensive, and open computer network and satel-

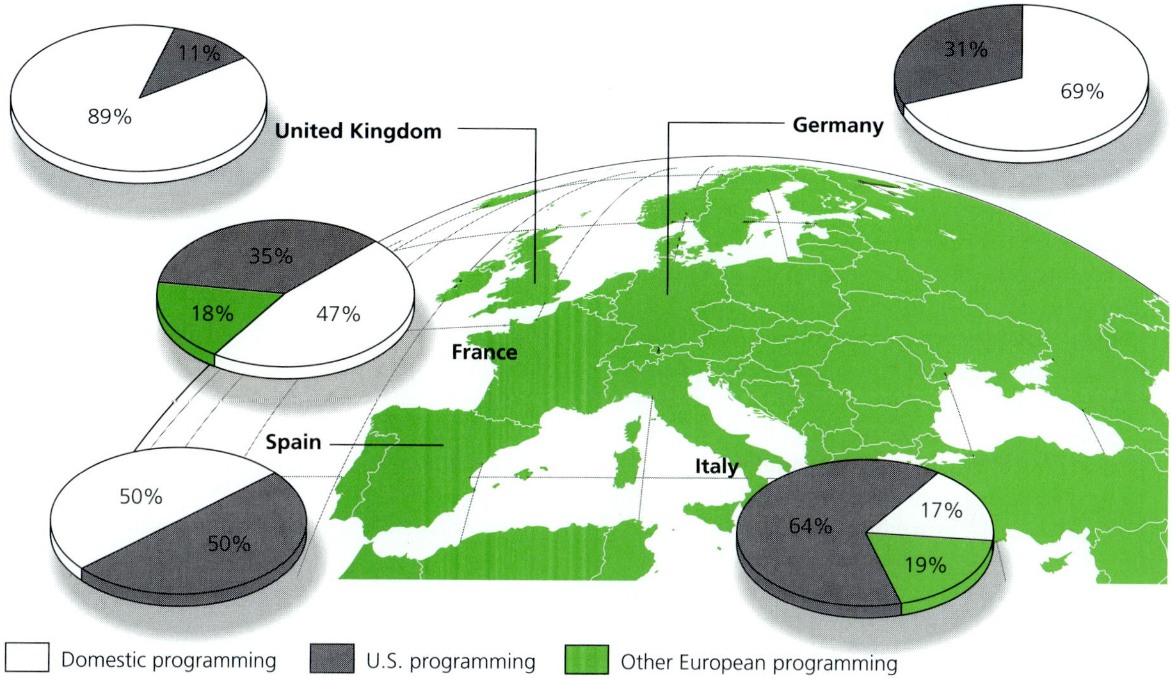

Figure 11.3 Sources of Programming on European Television, 1999
Source: "Loco for Local Fare," 1999.

lite communication, a related question is, "Have notions such as national sovereignty and cultural integrity lost their meaning?" For example, the BBC broadcasts daily to a worldwide audience in 40 languages, as does Radio Beijing from China. CNN uses satellites to transmit to 60 countries. Murdoch's Sky Channel satellites cover the globe. Mexican soap operas dominate the television schedules of much of Latin and South America. Three of the four largest U.S. record companies have international ownership. The Web makes global any radio station that wants to be. In the case of Web-only stations, operators can start and evolve their stations free of their home countries' regulations and traditions and with little concern for their native lands' languages.

There are differing opinions about the benefits of increased globalization, concentration of ownership in the hands of a small number of TNCs, and commercialization of telecommunications. Proponents see the world community coming closer together as a common culture is encouraged and, not incidentally, as we become more economically interconnected. "We are witnessing the revolution of the empowerment of the media consumer," argues Reuters Television Director Enriqué Jara (as quoted in Hilliard & Keith, 1996, p. 1). It is unlikely that individual cultures and national identities will disappear, because the world's great diversity will demand culture-specific, special interest fare. Modern media technology makes the delivery of this varied content not

Critical Issues *English as the Global Language*

The French have established an official government agency, Academie Francaise, charged with keeping their language free of foreign (really, American) influence. In Germany, where English is the most studied second language and the new Volkswagen Beetle is advertised and sold as "the new Volkswagen Beetle" rather than *Der neue Volkswagen Kafer,* the grassroots Society for the Protection of the German Language (*Verein zur Wahrung der Deutschen Sprache*) struggles against English's inroads into its native tongue.

Many people in Germany and other countries fear the rise of English and the parallel decline of their own language because they consider language to be the deepest expression of national identity. A nation's people speak a language, their popular culture reflects it, and their history and literature are preserved in it. But just how far has English encroached on the world's other languages?

English is the first language of 400 million people and the second language of another 400 million (Miller, 1995). The world's air traffic control systems all employ English for their communication. Three quarters of all the world's mail is written in English. English is the primary language for the publication of scientific and scholarly reports and for many international organizations such as the European Union and the Association of South-East Nations.

There are several reasons English has become the global language:

1. English, especially American English, is spread throughout the world through highly attractive and popular U.S. advertisements, movies, television shows, music videos, and other forms of popular culture.
2. Non-English-speaking parents actively encourage their children to learn English. "Parents see minority language speakers experiencing difficulties in breaking into new economic sectors and effectively confined to local economies, often built around (low paying) service and craft industries," reports Britain's The Future of English? Project (1998, p. 2).
3. English has significant "cultural capital." Linguist David Crystal (1997) contends that "the dominant view is certainly that a person is more likely to be in touch with the latest thinking in a subject by learning English than any other language" (p. 102).
4. English has the largest vocabulary of all the world's languages, easily incorporates new and foreign words, and is easy to speak badly (even poor English is understandable). In other words, as *New York Times* columnist William Safire (1996) wrote, English "is the most welcoming, absorbent, easily adaptable language in the world" (p. 61).

But there is a technological reason as well. It resides in the Internet, but it goes beyond the fact that English-speaking countries dominate the Net in terms of the number of users and amount of traffic (Figure 11.4). The Internet was invented in the United States. As a result, the protocols devised to carry data over it were developed for the English alphabet. As the Internet spread, countries wanting to link to it were forced to adopt those protocols. When the World Wide Web was developed, *its* protocol (HTML), was also written in English. HTML employs 258 characters, making it hospitable to most Western European languages, but non-Western languages such as Hindi, Chinese, and Thai are difficult to display. Despite ongoing development of a new character code (Unicode) that will support more than 65,000 characters and permit viewing almost all of the world's languages online, English's domination of the Internet rankles many people. For example, Anatoly Voronov, director of Russian ISP Glasnet, complains, "The product comes from America so we either must adapt to English or stop using it. This is the right of any business. But if you are talking about a technology that is supposed to open the world to hundreds of millions of people, you are joking. This just makes the world into new sorts of haves and have nots" (as quoted in Specter, 1996, p. 1).

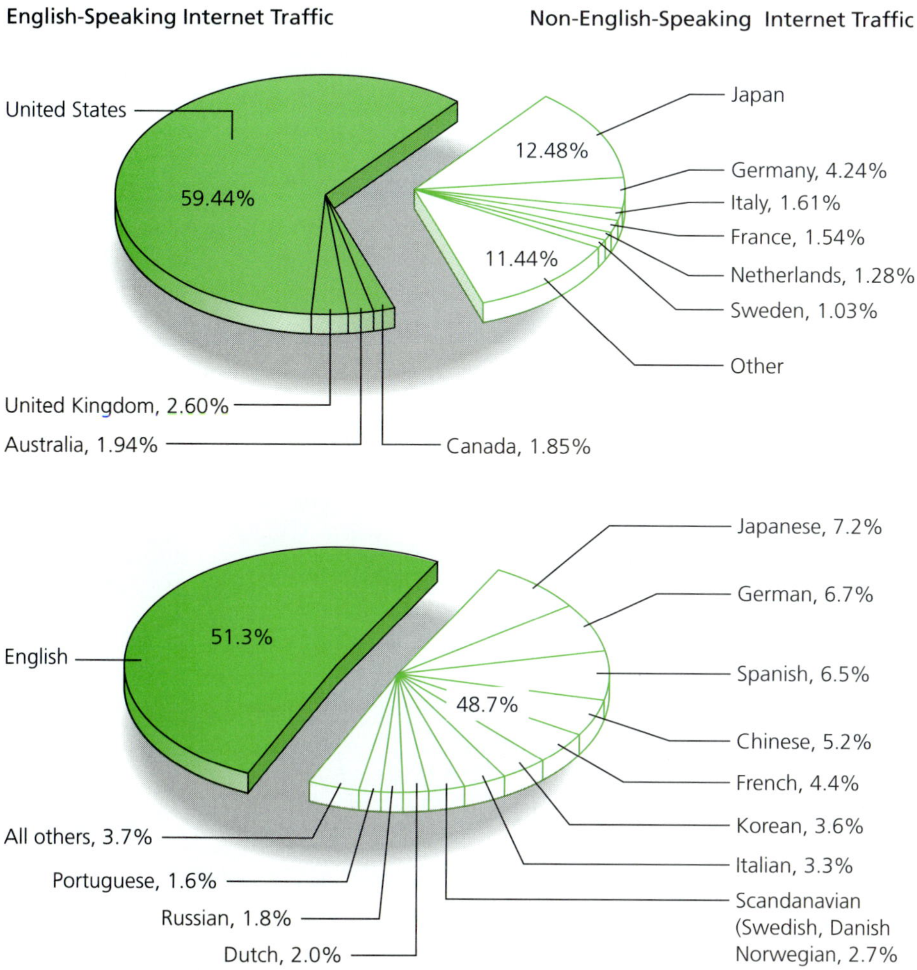

Figure 11.4 **English, the Language of the Internet for How Long?**
Sources: [1]StatMarket, 1999; [2]Global Reach, 2000.

Are the French, Germans, Mr. Voronov, and other critics of global English making too much of the problem? In fact, might it not be best for world peace and understanding if all people spoke a common language? Should we share a common culture and experience? Before you answer, consider rules recently passed by the European Union when it agreed upon **Euro English** as its official language ("Let Them," 1999). Euro English will make the following changes to traditional English, to be phased in over a 5-year span: *soft c* will replace *s; k* will replace *hard c; f* will replace *ph;* double letters will be removed; *z* will replace *th;* the *silent e* will be dropped; *v* will replace *w;* and *ou* will become *o*. In addition, other letter combinations will be halved.

Nov, ask yorclf hov komfortabl yu vuld be and hov vel yu culd komunikat using zis langag. Vuld yu stil fel Amerikan? So, ze qeston remanz, ar zes pepl making to much ovr ze deklin of zer nativ tongz?

Global media does not refer only to the export of American fare to other countries. Many companies, for example Brazil's Globo TV, work hard to have their content distributed in the United States. This ad for a 15-part mini-series appeared in *Broadcasting & Cable*, targeting American broadcasters and cablecasters.

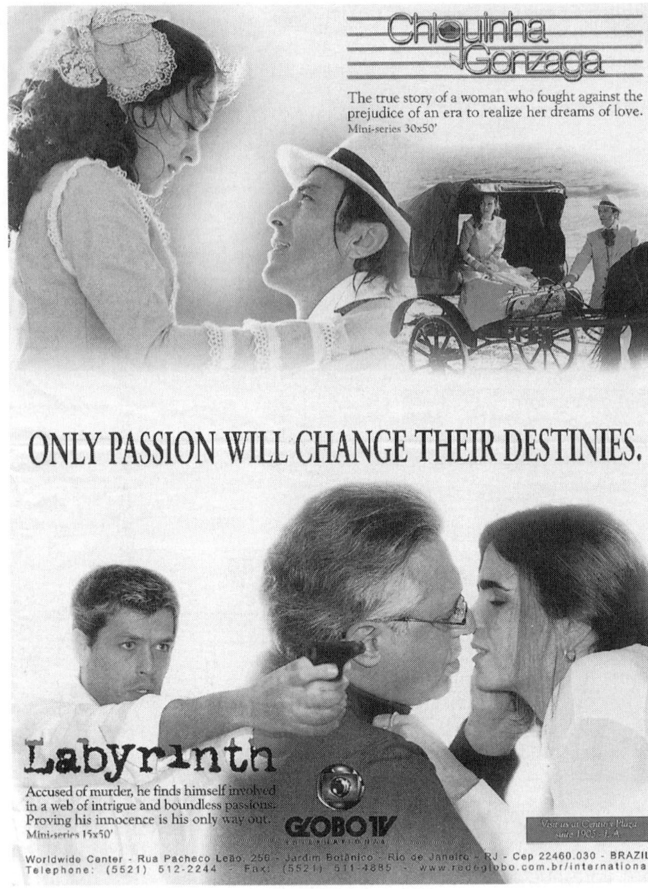

only possible but profitable. For example, native-language versions of *Jeopardy!* and *Wheel of Fortune* exist in most Western European countries.

Global, concentrated, and commercialized media are here, say those with a less optimistic view, but the problem is the nature of their content. Professor Richard Rosenberg (1992) predicts that global telecommunications will erode national sovereignty: "The advanced nations of the world, through their multi-national corporations, will greatly expand their control over the international flow of information. As a result, much of the world may become even more heavily dependent on the Western nations and Japan" (p. 331). Rosenberg also predicts the demise of native cultures: "The ongoing assault on national cultures will continue, fostered by direct satellite broadcasts and worldwide information distribution networks" (p. 332).

Media critic Robert McChesney (1997) likewise fears for worldwide democracy. "The present course," he writes, "is one where much of the world's entertainment and journalism will be provided by a handful of enormous firms, with invariably pro-profit and pro-global market political positions on the central issues of our times. The implications for political democracy, by

Abaya-clad Saudi Arabian women queue up at the single woman's only counter of a McDonalds in a mall in Dhahran. Although critics of cultural imperialism see this as an intrusion of Western culture into the lives of these people, defenders of the globalization of culture see the expansion of opportunity for both the "sending" and "receiving" cultures.

any standard, are troubling" (p. 23). Later McChesney (1999a) elaborated, "The global commercial-media system is radical in that it will respect no tradition or custom, on balance, if it stands in the way of profits. . . . The global media system is better understood as one that advances corporate and commercial interests and values and denigrates or ignores that which cannot be

Careers *Working in Global Telecommunications*

Learn a foreign language. Better yet, learn two. Then travel overseas. English may be the global language, but employers, especially those looking at applicants for positions in international or global telecommunications, favor people who are bilingual or multilingual. Think about it for a moment. What would you do as an employer if an applicant for a job in global media had neither foreign language nor foreign travel experience? What would this say about his or her commitment to the larger world and working in it?

Naturally, a telecommunications company would prefer to hire people who can speak the language of one or more of the countries in which it operates. In addition, skill at foreign languages signals a familiarity with and a commitment to the values and traditions of other people, important characteristics in someone working in global telecommunications. Being multilingual also tells a prospective employer that a person has discipline, as learning a foreign language clearly takes more dedication than not learning one.

As for international travel, it can provide people interested in careers in global telecommunications with experiences that will be useful during their professional lives, and it also tells prospective employers that these travelers are adventurous, know what they are getting into, and that they are not overwhelmed by things that are different.

incorporated into its mission. There is no discernible difference in the firms' content, whether they are owned by shareholders in Japan or Belgium or have corporate headquarters in New York or Sydney" (pp. 13–14). The box titled Working in Global Telecommunications offers advice for those interested in doing so.

Chapter Review

International telecommunications has its roots in the clandestine stations of World War II and the Cold War and in the pirate radio stations operated to challenge the dominance of Europe's noncommercial broadcasters. These efforts, relying largely on shortwave radio, gave rise to external services still with us today, for example, the Voice of America and the BBC World Service. The United States maintains several surrogate services in addition to the VOA, including Radio Marti and Radio Free Europe/Radio Liberty.

Satellites made truly global telecommunications possible. COMSAT, representing the United States, managed the first global system of telecommunications satellites, INTELSAT, which has since been joined by many other commercial and government-run satellite providers.

Comparative analysis is the study of different countries' media systems, and it traditionally examines those systems in terms of the normative theories that guide them. The classic approach identified four theories: authoritarian, communist, libertarian, and social responsibility. But changes in technology and in politics require a more modern approach, one that sees a nation's media as operating in terms of its guiding concepts: Western, for example, as in the United States and Great Britain; development, for example, as found in Honduras; revolutionary, as practiced by Poland's Solidarity Movement; and authoritarianism and communism, still in place in the world's few remaining communist countries.

Regardless of the nature of a country's media system, that system carries programming. Most of the world's radio and television programming looks much like that found in the United States. The United States is not only a world leader in distributing its media fare to other countries, but it was among the first to do so, setting audience expectations around the globe. The global spread of U.S. and other Western media content raises questions about cultural imperialism. Countries are concerned over their ability to maintain national sovereignty and control over their own cultures. Critics of cultural imperialism argue that native cultures are inevitably debased by non-native fare and that global media companies exert undue influence over the flow of information. Others, however, see little to fear in the globalization of media content, arguing that the world might indeed be a better place if its people shared a common culture. They also point out that most nations demand culture-specific content.

Questions for Review

1. What characteristics of shortwave radio make it well suited for international radio transmissions?
2. What are clandestine broadcast stations? Differentiate between indigenous and exogenous stations.
3. What are pirate broadcasters? What differentiates them from traditional clandestine operations?
4. How did World War II and the Cold War shape the efforts of the United States in terms of its external and surrogate services?
5. Describe the goals of the Voice of America, Radio and TV Marti, Armed Forces Radio and Television Services, and Radio Free Europe/Radio Liberty.
6. What are COMSAT and INTELSAT? How are they related?
7. What is comparative analysis?
8. List and define Siebert, Peterson, and Schramm's four theories of the press.
9. Characterize media systems operating under Hachten's Western concept.
10. Characterize media systems operating under Hachten's development concept.
11. Characterize media systems operating under Hachten's revolutionary concept.
12. Characterize media systems operating under Hachten's authoritarian and communism concepts.
13. In British media, what is the public service remit? What is a D-notice?
14. What is meant by wired radio?
15. What is cultural imperialism? What two telecommunications technologies fuel current concern over its operation?

Questions for Critical Thinking

1. If you applaud the efforts of the clandestine and pirate broadcasters to use media to challenge existing political and economic authorities, you must also applaud the efforts of U.S. pirates and microbroadcasters. What similarities do they share? If you do not see them as equivalent, explain why.

2. Do you have experience with another country's media? If so, which one? Can you place that system's operation within one of the concepts listed in this chapter? Describe how that system's content was similar to and different from that with which you are familiar in the United States. Do you favor one system's fare over another's? Why or why not?

3. Do you think countries, especially developing nations, have anything to worry about in terms of cultural imperialism? Would you argue that low-cost Western fare that helps a developing system get "off the ground" is a good way to do so, or do you agree with critics that this unnecessarily predetermines what a system's content will become?

4. Much of the rest of the world allows more sex and nudity in its television programming than is the case in the United States. What does this say about these countries and their people? What does this say about U.S. audiences? Would you favor more openness in sex and nudity on U.S. television? Why or why not?

Engaging Technology

1. Despite doing much good, the Voice of America and its services such as Radio and TV Marti remain controversial. Visit their Web sites at <http://www.voa.gov/> and <http://www.voa.gov/marti/> to investigate what they have to say about themselves and how they respond to criticism. Are you persuaded by their arguments? Why or why not?

2. INTELSAT faces significant competition from government and commercial satellite providers. Still, it is a major force in its industry. Visit its Web site at <http://www.intelsat.com/> to determine how it plans to deal with the competition and its own view of its future. Report on what you learn.

3. The Center for Global Media Studies was formed in the summer of 1999 by a group of international scholars interested in studying and disseminating information on global media. Visit its Web site at <http://www.cgms.org> for a fuller description of its work. Once there, access at least one of its reports on some aspect of global telecommunications that might be of interest to you. Report on what you find.

4. Among the thousands of international distributors of television programming who see the United States as a lucrative market are Brazil's GloboTV at <http://www.redeglobo.com.br/international>, Radio Netherlands Television at <http://www.rnw.nl>, and England's Dorling Kindersley Vision at <http://www.dk.com>. Visit their Web sites to see what kinds of content they distribute. Make predictions on which of their offerings might in fact attract U.S. viewers.

Chapter Twelve

Rights and Regulations

In November 1964 small AM/FM radio station WGCB in Red Lion, Pennsylvania aired its weekly installment of *The Christian Crusade.* The show's host, Reverend Billy James Hargis, offered his review of a book written by a man named Fred J. Cook. Hargis did not enjoy Cook's book, *Goldwater—Extremist of the Right,* an analysis of the career of conservative politician Barry Goldwater. In his comments on the work, Hargis accused Cook of a number of offenses: lying, being fired from his job as a reporter, and being left wing. To Cook this amounted to a personal attack, and under FCC rules he was entitled to time to reply.

The station owner, Red Lion Broadcasting Company, offered to sell time to Cook or, if the writer would plead poverty, to give him free time. Cook refused, claiming the *right* to reply. FCC rules said that reply time must be free in cases of personal attack. The station, and the broadcast industry as a whole that funded its defense, argued that this free time requirement was an infringement of broadcasters' First Amendment rights. As a result, the stakes were high. The question before the courts was nothing less than an affirmation or denial of the FCC's power to promulgate rules regarding the public's access to airwaves that it, in fact, owned.

In 1966 the United States Court of Appeals for the Seventh Circuit in Chicago District ruled in favor of the station. But the FCC persisted, and in 1967 the United States Court of Appeals for the District of Columbia overturned that decision, siding with Cook and the commission. Red Lion Broadcasting and the broadcast industry, through the Radio Television News Directors Association (RTNDA), appealed to the Supreme Court.

On June 9, 1969, the Justices delivered what has become known as the Red Lion Decision. Justice Byron White expressed the Court's support for Mr. Cook and the FCC this way:

> There is nothing in the First Amendment which prevents the Government from requiring a licensee to share his frequency with others and to conduct himself as a proxy or fiduciary with obligations to present those views and voices which are representative of his community and which would otherwise, by necessity, be barred from the airwaves. . . . [T]he people as a whole retain their interest in free speech by radio and their collective right to have the medium function consistently with the ends and purposes of the First Amendment. . . . It is the purpose of the First Amendment to preserve an uninhibited marketplace of ideas in which truth will ultimately prevail, rather than to countenance monopolization of that market, whether it be by the Government itself or a private licensee. (as quoted in Gillmor & Barron, 1974, p. 813)

But Justice White's most memorable and meaningful comment on the clash between broadcaster and audience rights remains, "*It is the right of the viewers and listeners, not the right of the broadcasters, which is paramount*" (p. 813).

In September 1999 the broadcasters, again through RTNDA, went to federal court demanding that the FCC drop its personal attack rules because, in their view, "the requirements violate (our) First Amendment rights" (McConnell, 1999c, p. 13). Red Lion has not been overturned, so why does RTNDA think it can now be successful in killing the personal attack rule?

Much has changed since 1969. New technologies have altered the broadcasting landscape, and deregulation has attracted countless proponents in both the telecommunications industries and government. What RTNDA understands quite well is that it (and we) face a much different regulatory environment than the one that produced Fred Cook's victory, an environment more receptive to its call for regulatory relief.

Preview

In Chapter 2 we saw how government regulation came to broadcasting, and in subsequent chapters we examined how that regulation shaped many aspects of the telecommunications industries we've studied. In this chapter we have a number of additional goals. First, we investigate the First Amendment, the guarantee of media freedom against which all regulation must be evaluated. Then we look at the Federal Communications Commission and other regulatory agencies and how they perform their duties.

The FCC and other regulators' efforts in several contemporary pressing telecommunications issues are then scrutinized. These include federal and local regulation of cable, proposed regulations for digital television, and the complex situation of controlling Internet expression.

The Roots of Telecommunications Regulation

Democracy, government by the people, requires a free and open press. The members of the First Congress of the United States, the authors of the **Bill of Rights,** understood this based on their experience with the European monarchies from which they and their forebears had fled. As a result, the First Amendment to their new Constitution expressly stated that "Congress shall make no law . . . abridging the freedom of speech, or of the press." They based this guarantee of a privileged position for the press on the **libertarian philosophy** that people cannot govern themselves in a democracy unless they have access to the information they need for that governance. The libertarian philosophy is based on the **self-righting principle,** which was forcefully articulated in 1644 by English author and poet John Milton in his book *Areopagitica.* Milton's argument is as follows:

- The free flow or trade of ideas serves to ensure that public discourse will allow the truth to emerge.
- Truth will emerge from public discourse because people are inherently rational and good.

Press freedom demands that government regulation of the media be unobtrusive and sufficiently justified to meet the limits of the First Amendment. As you might imagine, "unobtrusive" and "justified" have been and are open to much interpretation and debate.

Defining and Refining the First Amendment

The First Amendment is constantly undergoing clarification and refinement. One view of media freedom is housed in the **absolutist position,** expressed succinctly by Supreme Court Justice Hugo Black: "No law means no law." Yet the absolutist position is more complex than this would suggest. Absolutists accept that the First Amendment provides a central and fundamental wall of protection for the media, but several questions about its true meaning remained to be answered over time.

What Does "No Law" Mean? The First Amendment said that the United States Congress could "make no law." But could state legislatures? City councils? Mayors? Courts? Who has the power to proscribe the media? This issue was settled in 1925 when the Supreme Court stated in *Gitlow v. New York* that the First Amendment was "among the fundamental personal rights and 'liberties' protected by the due process clause of the Fourteenth Amendment from impairment by the states" (Gillmor & Barron, 1974, p. 1). Given that, "Congress shall make no law" should be interpreted as "government agencies shall make no law." Today, "no law" includes statutes, laws, administrative regulations, executive and court orders, and ordinances from government, regardless of their locale.

What Is "the Press"? Just what "press" enjoys First Amendment protection? In 1952 the Supreme Court in *Burstyn v. Wilson* declared that film was protected expression. Radio and television joined the movies in 1973 with *CBS v. Democratic National Committee.* Advertising, or commercial speech, also enjoys First Amendment protection. This was established by the Supreme Court in 1942 in *Valentine v. Chrestensen.* In 1967 in *Time, Inc. v. Hill,* the Supreme Court ruled that the First Amendment grants the same protection to entertainment content as it does to nonentertainment content.

What Is "Abridgment"? Even absolutists accept the idea that limits can be placed on the time, place, and manner of expression—as long as the restrictions do not interfere with the substance of the expression. Few, for example, would find unreasonable limits on the use of a sound truck to broadcast political messages at 4 o'clock in the morning. But the Supreme Court did find unconstitutional an ordinance that forbade all use of sound amplification except with the permission of the chief of police in its 1948 decision in *Saia v. New York.* The permissibility of other restrictions, however, is less clear cut.

Clear and Present Danger Can freedom of the press be abridged if the likely result is damaging? The Supreme Court answered this question in 1919 in *Schenck v. United States.* This case involved the distribution of a pamphlet urging resistance to the military draft during World War I. Justice Oliver Wendell Holmes wrote that expression could be limited when "the words used are used in such circumstances and are of such a nature as to create a clear and present danger that they will bring about the substantive evils that Congress has a right to prevent" (Gillmor & Barron, 1974, p. 11). Justice Holmes added, "Free speech would not protect a man in falsely shouting fire in a theater and causing panic" (p. 11). This decision is especially important because it firmly established the legal philosophy that there is no absolute freedom of expression; the level of protection is one of degree.

Balancing of Interests This less-than-absolutist approach is called the **ad hoc balancing of interests.** In individual First Amendment cases several factors should be weighed in determining how much freedom media are granted. In his dissent to the Court's 1941 decision in *Bridges v. California,* Justice Felix Frankfurter wrote that free speech and the press are "not so absolute or irrational a conception as to imply paralysis of the means for effective protection of all the freedoms secured by the Bill of Rights. . . . In the cases before us, the claims on behalf of freedom of speech and of the press encounter claims on behalf of liberties no less precious" (Gillmor & Barron, 1974, p. 422).

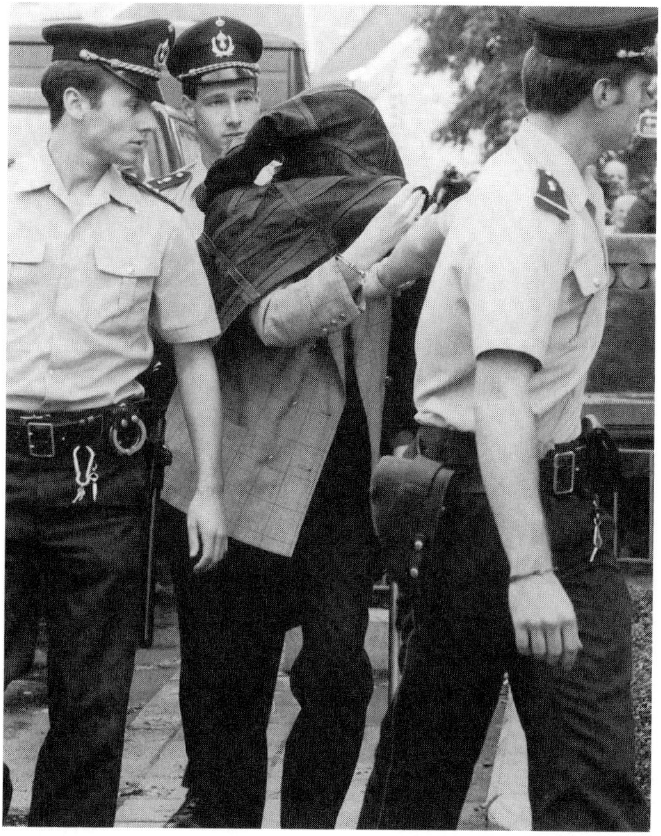

Is the woman here guilty or innocent? If she is guilty, she should want to cover her face to hide her identity. But if she is innocent, wouldn't she be just as likely to want to hide her identity? These so-called perp walks raise the issue of unfair pre-trial publicity.

Free Press Versus Fair Trial One example of the clash of competing liberties is the conflict between a free press (First Amendment) and a fair trial (Sixth Amendment). This debate typically takes two forms: (1) Can pretrial publicity deny citizens judgment by 12 impartial peers, therefore denying them a fair trial? (2) Should cameras be allowed in the courtroom, supporting the public's right to know, or do they so alter the workings of the court that a fair trial is impossible when they are present?

Courts have consistently decided in favor of a fair trial in conflicts between the First and Sixth Amendments. But it was not until 1961, in *Irvin v. Dowd,* that a conviction was overturned because of pretrial publicity.

Print reporters have long enjoyed access to trials, but broadcast journalists have been less fortunate. In 1937, after serious intrusion by newspaper photographers during the 1935 trial of Bruno Hauptmann, accused of kidnapping the baby of trans-Atlantic aviation hero Charles Lindbergh, the American Bar Association (ABA) adopted Canon 35 as part of its Code of Judicial Ethics. This rule forbade cameras and radio broadcasting of trials. In 1963 the ABA amended the Canon to include a prohibition on television cameras. This, however, did not settle the issue of cameras in the courtroom.

Texas was one of three states that did not subscribe to the ABA's Canon 35. Texas financier Billy Sol Estes was convicted of embezzlement, and he took his case to the Supreme Court, where the decision was overturned. The Court ruled that televising his trial had denied Estes a fair trial, and debate flared again. Justice Tom Clark wrote for the majority:

> The free press has been a mighty catalyst in awakening public interest in governmental affairs, exposing corruption among public officers and employees and generally informing the citizenry of public events and occurrences, including court proceedings. While maximum freedom must be allowed the press in carrying on this important function in a democratic society its exercise must necessarily be subject to the maintenance of absolute fairness in the judicial process. (Gillmor & Barron, 1974, p. 455)

Television cameras, then, were out. But Justice Clark continued, "When advances in [broadcast journalism] permit reporting . . . by television without their present hazards to a fair trial we will have another case" (p. 455). Cameras were back in if they posed no hazard to the principle of a fair trial.

In 1972 the ABA replaced Canon 35 with Canon 3A(7), allowing some videotaping of trials for specific purposes but reaffirming its opposition to broadcasting trial proceedings. In 1981, the Supreme Court, in *Chandler v. Florida,* determined that television cameras in the courtroom were not inherently damaging to fairness, and different states have adopted different standards on the issue. So common has televising court proceedings become that Court TV, a cable channel that frequently programs real trials and commentary on them, was launched in 1991.

Libel and Slander **Libel,** the false and malicious publication of material that damages a person's reputation, and **slander,** oral or spoken defamation of a person's character, are not protected by the First Amendment. The distinction between libel and slander, however, is sufficiently narrow that "published defamation, whether it is in a newspaper, on radio or television, in the movies, or whatever, is regarded in the 1990s as libel. And libel rules apply" (Pember, 1999, p. 134). Therefore, if a report (a) defames a person, (b) identifies that person, and (c) is published or broadcast, it loses its First Amendment protection.

A report accused of being libelous or slanderous is protected if it meets any one of three tests. The first is *truth.* Even if a report damages someone's reputation, if it is true, it is protected (of course, those accused of libel must prove the truth of their statements). The second is *privilege.* Coverage of legislative, court, or other public activities may contain information that is not true or that is damaging to someone's reputation. The press cannot be deterred from covering these important news events for fear that a speaker or witness's comments will open it to claims of libel or slander (privilege, however, requires a demonstration that the accused media outlet operated accurately and fairly). The third is *fair comment.* The press has the right to express its opinions or comment on public issues (but the report in question must be clearly labeled as opinion). For example, theater and film reviews, however severe, are protected, as is commentary on other matters in the public eye.

For public figures, however, a different set of rules applies. Because they are in the public eye, public figures are fair game for fair comment. But does that leave them open to reports that are false and damaging to their reputation? The Supreme Court faced this issue in 1964 in *New York Times Co. v. Sullivan.* In 1960 the Committee to Defend Martin Luther King bought a full-page ad in the *New York Times* asking people to contribute to Dr. King's defense fund. The ad detailed abuse of Dr. King and other civil rights workers at the hands of the Montgomery, Alabama, police. L. B. Sullivan, one of three elected commissioners in that city, sued the *Times* for libel. The ad copy was not true in some of its claims, he said, and because he was in charge of the police, he had been "identified."

The Supreme Court ruled in favor of the newspaper. Even though some of the specific facts in the ad were not true, the *Times* had not acted with **actual malice.** The Court defined the standard of actual malice for reporting on public figures as "knowledge of a report's falsity" or "reckless disregard" for whether the report is true or not.

Prior Restraint There is much less confusion about another important aspect of press freedom, **prior restraint.** This is the power of the government to *prevent* publication or broadcast of expression. U.S. law and tradition make the use of prior restraint relatively rare. Yet there have been a number of important efforts by governments to squelch content before dissemination.

In 1931 the Supreme Court ruled in *Near v. Minnesota* that freedom from prior restraint was a general principle. It also determined that it was not an absolute principle. Two of the four exceptions it listed were in times of war when national security was involved and when the public order would be endangered by the incitement to violence and overthrow by force of orderly government. These exceptions were to become the basis of two landmark prior restraint decisions. The first, involving the *New York Times,* dealt with national security in times of war; the second, focusing on protecting the public order, involved publishing instructions for building an atomic bomb.

On June 13, 1971, at the height of the Vietnam War, the *New York Times* began publication of what became commonly known as the "Pentagon Papers." The papers included detailed discussion and analysis of the conduct of that unpopular war during the administrations of Presidents Kennedy and Johnson. President Nixon's National Security Council (NSC) had stamped them Top Secret, but NSC staff member Daniel Ellsberg believed this was an improper restriction of the public's right to know, and he gave copies to the *Times.* After the first three installments had been published, the Justice Department, citing national security, was able to secure a court order stopping further publication.

On June 30 the Supreme Court ordered the government to halt its restraint of the *Times'* and other papers' right to publish the Pentagon Papers.

"All the News That's Fit to Print"

The New York Times

LATE CITY EDITION

Weather: Chance of showers today, tonight. Partly sunny tomorrow. Temp. range: today 74-94; Wed. 72-91. Temp. Hum. Index yesterday 82. Full U.S. report on Page 94.

VOL. CXX . . . No. 41,431 — NEW YORK, THURSDAY, JULY 1, 1971 — 15 CENTS

SUPREME COURT, 6-3, UPHOLDS NEWSPAPERS ON PUBLICATION OF THE PENTAGON REPORT; TIMES RESUMES ITS SERIES, HALTED 15 DAYS

Nixon Says Turks Agree To Ban the Opium Poppy

By JOHN HERBERS
Special to The New York Times

WASHINGTON, June 30—President Nixon announced today that Turkey had agreed to eliminate within a year her production of opium poppies, which account for about two-thirds of the illegal heroin reaching the United States.

Mr. Nixon, in a brief announcement delivered in the White House press room, said that as a result of negotiations between the United States and Turkish Governments, Premier Nihat Erim had agreed to ban altogether the cultivation of opium poppies by June, 1972.

He said the joint announcement, made simultaneously in Washington and Ankara, "represents by far the most significant breakthrough that has been achieved in stopping the source of supply of heroin in our worldwide offensive against dangerous drugs."

Two weeks ago, Mr. Nixon sent a message to William J. Handley, the United States Ambassador in Turkey, saying that the time for talk had passed and the United States must have action by the Turkish Government in ending poppy cultivation.

Today, the President praised Premier Erim for "courageous, statesmanlike action" and said the United States would provide money and technical assistance in helping Turkish farmers shift to other crops.

Officials would not say how much American money would be involved, but the United States has made a $3-million commitment to Turkey on the heroin problem.

Secretary of State William P. Rogers, who helped work

Continued on Page 22, Column 1

Soviet Starts an Inquiry Into 3 Astronauts' Deaths

By BERNARD GWERTZMAN

PRESIDENT CALLS STEEL AND LABOR TO WHITE HOUSE

He Asks Both Sides to Meet With Him Tuesday Before Contract Talks Start

By PHILIP SHABECOFF
Special to The New York Times

WASHINGTON, June 30 — President Nixon has called negotiators of the steel companies and steelworkers union to meet with him next Tuesday before they sit down to begin contract negotiations, a White House spokesman announced today.

It will be the first time that the President will have met with labor and management in any industry prior to nation-wide contract negotiations, according to Ronald L. Ziegler, the White House press secretary.

Discussion Issues Listed

Mr. Ziegler said that the President had called the meeting to discuss general economic developments and trends in the world steel markets.

Earlier today, the chairman

Pentagon Papers: Study Reports Kennedy Made 'Gamble' Into a 'Broad Commitment'

By HEDRICK SMITH

The Pentagon's study of the Vietnam war concludes that President John F. Kennedy transformed the "limited-risk gamble" of the Eisenhower Administration into a "broad commitment" to prevent Communist domination of South Vietnam.

Although Mr. Kennedy resisted pressures for putting American ground-combat units into South Vietnam, the Pentagon analysts say, he took a series of actions that significantly expanded the American military and political involvement in Vietnam but nonetheless left President Lyndon B. Johnson with as bad a situation as Mr. Kennedy inherited.

"The dilemma of the U.S. involvement dating from the Kennedy era," the Pentagon study observes, was to use "only limited means to achieve excessive ends."

Moreover, according to the study, prepared in 1967-68 by Government analysts, the Kennedy tactics deepened the American involvement in Vietnam piecemeal, with each step minimizing public recognition that the American role was growing.

The expansion of that role, over three decades, is traced in the 3,000 pages of the Pentagon's study, which is accompanied by 4,000 pages of documents on the Vietnam era. Previous articles in The Times's presentation of this material have recounted President Johnson's movement to war in 1964 and 1965.

President Kennedy made his first fresh commitments to Vietnam secretly. The Pentagon study discloses that in the spring of 1961 the President ordered 400 Special Forces troops and 100 other American military advisers sent to South Vietnam. No publicity was given to either move.

Small as the numbers seem in retrospect, the Pentagon study comments that even the first such expansion "signaled a willingness to go beyond the 685-man limit on the size of the U.S. [military] mission in Saigon, which, if it were done openly, would be the first formal breach of the Geneva agreement." Under the interpretation of that agreement in effect since 1956, the United States was limited to 685 military advisers in Vietnam. Washington, while it did not sign the accord, pledged not to undermine it.

On May 11, 1961, the day on which President Kennedy decided to send the Special Forces, he also ordered the start of a campaign of clandestine warfare against North Vietnam, to be conducted by South Vietnamese agents directed and trained by the Central Intelligence Agency and some American Special Forces troops. [See text, action memorandum, May 11, 1961, Page 3.]

The President's instructions, as quoted in the documents, were, "In North Vietnam . . . [to] form networks of resistance, covert bases and teams for

Continued on Page 6, Column 1

The Times today resumes its series of articles on the Pentagon's secret study of the Vietnam war. The study was obtained through the investigative reporting of Neil Sheehan, and the articles were researched and written over three months by Mr. Sheehan and other staff members. The fourth and fifth articles, both by Hedrick Smith, are published today and form an account of decisions in the Kennedy Administration.

Three pages of documentary material covering the Kennedy policy begin on Page 3, and documents on the 1963 coup begin on Page 9. A summary of the three earlier articles, covering the Johnson Administration, appears on Page 15.

BURGER DISSENTS

First Amendment Rule Held to Block Most Prior Restraints

Decision, concurring opinions, dissents start on Page 17.

By FRED P. GRAHAM
Special to The New York Times

WASHINGTON, June 30 — The Supreme Court freed The New York Times and The Washington Post today to resume immediate publication of articles based on the secret Pentagon papers on the origins of the Vietnam war.

By a vote of 6 to 3 the Court held that any attempt by the Government to block news articles prior to publication bears "a heavy burden of presumption against its constitutionality."

In a historic test of that principle — the first effort by the Government to enjoin publication on the ground of national security — the Court declared that "the Government has not met that burden."

The brief judgment was read to a hushed courtroom by Chief

The *New York Times* heralds its victory in its First Amendment dispute with the Nixon Administration over the publication of the Pentagon Papers.

Among the stirring attacks on prior restraint written throughout its decision was Justice Hugo Black's opinion:

> In the First Amendment the Founding Fathers gave the free press the protection it must have to fulfill its essential role in our democracy. The press was to serve the governed, not the governors. The Government's power to censor the press was abolished so that the press would remain forever free to censure the Government. The press was protected so that it could bare the secrets of government and inform the people. Only a free and unrestrained press can effectively expose deception in government. (as quoted in Gillmor & Barron, 1974, p. 115)

Then came the case of the magazine *The Progressive.* In 1979 it announced its intention to publish instructions on how to make a hydrogen bomb. President Jimmy Carter's Justice Department successfully obtained a court order halting publication, even though the article was based on information and material freely obtained from public, nonclassified sources. Before the case could come to court, several newspapers published the same or similar material. The Justice Department immediately abandoned its restraint, and six months later *The Progressive* published its original article.

Obscenity and Pornography Another form of press expression that is not protected is **obscenity.** Two landmark Supreme Court cases established obscenity's illegality and definition. The first is the 1957 *Roth v. United States* decision.

The Court determined that sex and obscenity were not synonymous, a significant advance for freedom of expression. It did, however, legally affirm for the first time that obscenity was unprotected expression. The definition or test for obscenity that holds even today was expressed in 1973 in *Miller v. State of California.* Chief Justice Warren Burger wrote these basic guidelines:

> (a) whether the average person, applying contemporary community standards, would find that the work, taken as a whole, appeals to the prurient interest, (b) whether the work depicts or describes, in a patently offensive way, sexual conduct specifically defined by the applicable state law, and (c) whether the work, taken as a whole, lacks serious literary, artistic, political, or scientific value. (as quoted in Gillmor & Barron, 1974, p. 376)

The problem for the courts, the media, and the public, of course, is judging content against this standard. For example, what is patently offensive to one person may be quite acceptable to another. What is serious art to one may be serious exploitation to another. And what of an erotic short story written online by an author in New York City but accessed and read by people in Peoria, Illinois? Whose community standards would apply?

An additional definitional problem resides in **pornography,** expression calculated solely to supply sexual excitement. Pornography is protected expression. The distinction between obscenity and pornography is a legal one. That is, sexually explicit content is pornography (and protected) until a court rules it illegal; then it is obscene (and unprotected). The difficulty of making such distinctions can be seen in Justice Potter Stewart's famous declaration, "I may not be able to come up with a definition of pornography, but I certainly know it when I see it" and his dissent in *Ginzburg v. United States,* "If the First Amendment means anything, it means that a man cannot be sent to prison merely for distributing publications which offend a judge's sensibilities, mine or any others" (as quoted in Gillmor & Barron, 1974, p. 362).

Issues surrounding the definition and protection of obscenity and pornography may never be clarified to everyone's satisfaction. Even though these issues may have limited application to broadcasting, they are of particular significance to cable and Internet programmers and audiences.

Other Issues of Media Freedom

The First Amendment has application for a number of specific issues of media responsibility and freedom.

Indecency Broadcasting's commercial base and wide audience make airing obscene and pornographic material unwise. Yet broadcasters frequently do confront the issue of **indecency,** defined by the FCC as indecent language or material that depicts sexual or excretory activities in a way that is patently offensive to contemporary community standards. The commission censured a Denver radio station in its 1960 Mile High Broadcasting decision because one of the station's disc jockeys persisted in uttering "smutty and suggestive"

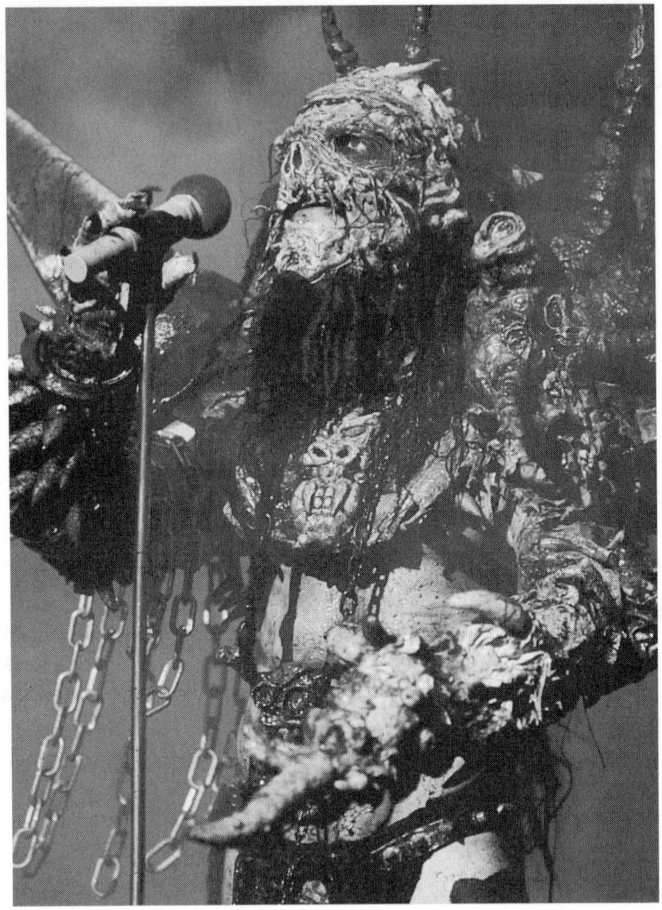

The music of such bands as Gwar is relegated to the safe harbor of late night radio to which children are not usually listening.

language. More recently, in 1978, the FCC censured New York's noncommercial WBAI-FM, owned by Pacifica Broadcasting, for that station's midday airing of George Carlin's recorded comedy monologue, "Seven Dirty Words You Can Never Say on Television." Although the routine was played as part of a serious program about the nature of the English language and only one complaint was made, the commission thought the utterance of those seven words was indecent. The station challenged the FCC's actions in court, appealing all the way to the Supreme Court, which ruled in favor of the commission. In its May 1999 Web release of its planning document, "A New Federal Communications Commission for the 21st Century" (<http://www.fcc.gov>), the FCC reaffirmed its will to "carry out our statutory mandates aimed at protecting the welfare of children, such as the laws governing obscene and indecent programming."

Safe harbor, times of the radio broadcast day when children are not likely to be in the listening audience (typically 10 p.m. to 6 a.m.), is one way the FCC tries to protect young listeners. Because potentially offensive content is in fact protected expression, the FCC recognizes that it should be available to those who wish to hear it during the safe harbor hours. But in its role of trustee (Chapter 2), the commission feels it has not only the right but also the obligation to protect those listeners who do not want such content for themselves or their children.

Copyright The First Amendment protects expression. **Copyright**—identifying and granting ownership of a given piece of expression—is designed to protect the creator's financial interest in that expression. Recognizing that the flow of art, science, and other expression would be enhanced by authors' financial interests in their creations, the framers of the Constitution wrote Article I, Section 8(8), granting authors exclusive rights to their "writings and discoveries." A long and consistent history of Supreme Court decisions has ensured that this protection is extended to the content of the mass media that have emerged since that time.

United States copyright law was extensively rewritten in the late 1970s. Copyright now remains with creators (in all media) for the span of their lives, plus 50 years (extended to 70 years in 1998 for works produced after January 1, 1978, and to 95 years after original copyright for works created before that date). During this time, permission for the use of the material must be obtained from the copyright holder, and if financial compensation (a fee or royalty) is requested, it must be paid. Once the copyright expires, and if the creator does not renew it, the material passes into the **public domain,** meaning it can be used without permission.

The exception to copyright is **fair use,** which identifies instances where material can be used without permission or payment. Fair use includes limited noncommercial use, for example, photocopying a passage from a novel for classroom use; use of limited portions of a work, for example, excerpting a few lines or a paragraph or two from a book for use in a magazine article; use that does not decrease the commercial value of the original, for example, videotaping a daytime football game for private, at-home evening viewing; and use in the public interest, for example, an author's use of line drawings of scenes from an important photograph or piece of film.

Two specific applications of copyright law pertain to recorded music and cable television. Imagine the difficulty cable companies would have in obtaining permission from all the copyright holders of all the material they import and deliver to their subscribers. Yet the cable operators do make money from others' works—they collect material from original sources and sell it to subscribers. The solution to the problem of compensating the creators of the material carried by cable systems was the creation of a Copyright Royalty Tribunal (since replaced by a panel under the auspices of the Library of Congress), to which cable companies pay a fee based primarily on the size of their operations. These moneys are then distributed to the appropriate producers, syndicators, and broadcasters.

Now imagine the difficulty song writers would have in collecting royalties from all who use their music—not only film producers and radio and television stations but bowling alleys, supermarkets, and restaurants. Here the solution is the **music licensing company.** The two biggest are the American Society of Composers, Authors and Publishers (ASCAP, founded in 1914) and Broadcast Music Inc. (BMI started in 1940 to accommodate radio stations unwilling to pay ASCAP's rising fees). Both collect fees based on the users' gross receipts and distribute the money to song writers and artists.

Deregulation The difficulty of balancing the public interest and broadcasters' freedom is at the heart of the debate over deregulation. As we saw in Chapter 2, deregulating radio ownership rules have been controversial, but the relaxation of the regulation of broadcasters' public service obligations and other content controls have provided even more debate.

The courts have consistently supported the FCC's right to evaluate broadcasters' performance in serving the public interest, convenience, and necessity. That evaluation inevitably includes some judgment of the content broadcasters air, and broadcasters have long argued that such "judgment" amounts to unconstitutional infringement of their First Amendment freedoms. Listeners and viewers, however, see this as a reasonable and quite small price to pay for the use of their—the public's—airwaves.

The Supreme Court resolved the issue in 1943 in *National Broadcasting Co. v. U.S.* NBC argued that the commission was no more than a traffic cop, limited to controlling the "flow of traffic." In this view the regulation of broadcasters' frequency, power, times of operation, and other technical matters was all that was constitutionally allowable. Yet the Court turned what is now known as the **traffic cop analogy** against NBC. Yes, the Justices agreed, the commission is a traffic cop. But even traffic cops have the right to control not only the flow of traffic but its composition. For example, drunk drivers can be removed from the road. Potentially dangerous "content," like cars with faulty brakes, can also be restricted. It was precisely this traffic-cop function that required the FCC to judge content. The commission was thus free to promulgate rules such as the **Fairness Doctrine,** which required broadcasters to cover issues of public importance and to be fair in that coverage, and **ascertainment,** which required broadcasters to ascertain or actively and affirmatively determine the nature of their audiences' interest, convenience, and necessity.

The Fairness Doctrine, ascertainment, and numerous other regulations such as rules on children's programming and hypercommercialism disappeared with the coming of deregulation during the Reagan Administration. License renewal, for example, was once a long and difficult process for broadcast stations. Thousands of pages of documents had to be generated to demonstrate that broadcasters not only knew what their audiences wanted and needed but had met those wants and needs. The burden of proof in their efforts to keep their licenses rested with them. Had they been fair? Had they kept commercial time to acceptable levels? What was their commitment to news and public affairs?

Since deregulated, renewal is a much less onerous process. Broadcasters simply file brief quarterly reports with the Mass Media Bureau indicating compliance with technical and other FCC rules (for example, the Children's Television Act of 1990). When their licenses are up for renewal (every 8 years), broadcasters file a short, postcard-like renewal application. Nonbroadcast spectrum users, such as cellular phone and paging licensees, use the **Universal Licensing System (ULS),** which allows them to minimally modify their renewal applications by Internet.

The deregulation drive began in earnest with President Reagan's FCC Chair Mark Fowler in the 1980s. Fowler rejected the trustee model of broadcast regulation. He saw many FCC rules as an unconstitutional infringement of broadcasters' rights and believed "the market" was the audience's best protector. He said that special rules for the control of broadcasting were unnecessary, likening television, for example, to just another home appliance. He called television no more than "a toaster with pictures."

This view was not without its critics. Members of Congress charged with FCC and media oversight, such as Massachusetts' Edward Markey, Michigan's John Dingell, and Colorado's Timothy Wirth, as well as many economists, legal experts, and public interest groups, condemned what they saw as the government's abandonment of viewers and listeners to the bottom-line demands of the financial marketplace. They criticized the 50% drop in the number of radio stations airing news and public affairs that occurred during Fowler's years at the commission as well as increases in the amount of advertising on Saturday morning television (Wilke, Vamos, & Maremont, 1985).

Congress has attempted to "reregulate" content in some areas. The 1996 Telecommunications Act's provisions regarding children's programming and violent television are examples of this (see Chapter 14). Nonetheless, ever since Commissioner Fowler's initial aggressive stance, the trustee model of commission oversight, which he called a "series of legal fictions" (Fowler & Brenner, 1982, p. 205), has been increasingly eroded, prompting Dennis Hastert, upon his elevation to the position of Speaker of the House of Representatives in December 1998, to label the commission a "renegade agency" (McConnell & Albiniak, 1999, p. 15).

Still, discomfort with deregulation persists, even from inside the media industries, and it rests on the belief that holding First Amendment protections brings with it obligations well beyond meeting market demands. As Michael Parks, editor of the *Los Angeles Times* argued, "Some of us [in the media] tend to forget why we have these privileges. It's so we can have informed citizens who know what's going on, and can debate and participate in government. We have these privileges to use in trust" (as quoted in Lambert, 1998, p. 39).

The Federal Communications Commission

The Communications Act of 1934 created the FCC and entrusted it with oversight of what we would now call telecommunications, but at that time this

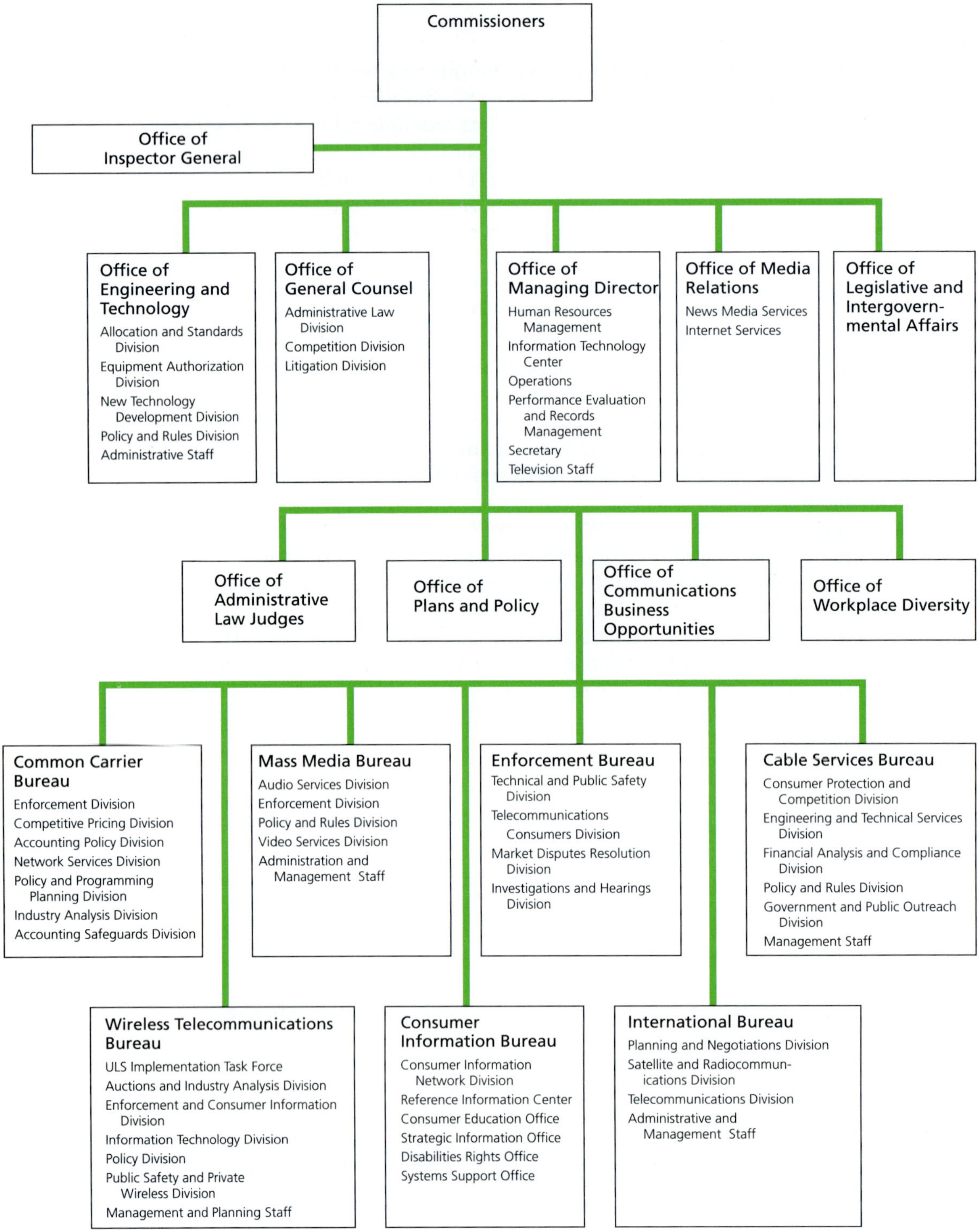

Figure 12.1 The Newly Restructured FCC

Source: Adapted from <http://www.fcc.gov/fccorgchart.html>.

included only AM radio, telephony, and telegraph (Chapter 2). Today's FCC deals with a much more complex and expanded telecommunications environment.

The FCC is headed by five commissioners, no more than three of whom can come from one political party. The president nominates commissioners and the Senate confirms them for staggered 5-year terms. The president also designates one of the five to be Chair. The commissioners supervise all FCC activities, delegating work and responsibilities to its 2,000 employees spread over the seven operating bureaus displayed in Figure 12.1.

The FCC's primary functions have long been licensing new stations, renewing station licenses, and transfering ownership when a station is sold. Granting and renewing licenses affords the FCC two important powers: protection of the spectrum for the benefit of licensees and audiences and enforcement of rules and policies. Theoretically, an applicant must adhere to commission technical and nontechnical rules or be denied access to the spectrum (or in the case of license revocation, forfeit existing access). In reality such denials and revocations are rare and promise to become even more so as deregulation continues to unfold.

Another function of the commission is enforcement of policy. The personal attack rules mentioned at the beginning of this chapter and safe harbor rules are examples of FCC policy. Other examples mentioned in earlier chapters include limits on station ownership, the now defunct bans on importation of distant signals by cable and satellite service providers, and the equally dead concepts of ascertainment and the Fairness Doctrine. As deregulation proceeds apace, the list of commission policies that have been rescinded may well eventually surpass those that still exist. In fact, Section 11 of the 1996 Telecommunications Act requires that the FCC review all of its policies every 2 years for the purpose of determining which rules no longer serve the public interest. Figure 12.2 shows what a broadcast license costs today.

The FCC regulates more than radio and television. We saw in Chapter 6 that it has some involvement in cable, primarily in rate and must-carry issues. The telephone industry's operation and rate structure are also regulated by the commission, as are some technical aspects of wireless telecommunications such as satellite and other mobile communication. Also under FCC jurisdiction are **satellite master antenna television (SMATV),** a type of minicable system serving a single large apartment building or complex, **multichannel multipoint distribution services (MMDS),** a form of "wireless cable" that distributes programming to subscribers by microwave, and DBS.

The commission conducts its business in these areas under the watchful eye of Congress. The Senate confirms potential commissioners, and the Senate and the House also wield power over the FCC through budgeting. All requests for funds, from salaries to operating money, are reviewed and approved (or rejected) by Congress. When Congress wants an activist FCC, it increases funding. When Congress wants the commission to back off its industry oversight, it decreases funding. The courts, too, influence the FCC. All commission rules and policies are subject to review by the U.S. Court of Appeals for the District of Columbia.

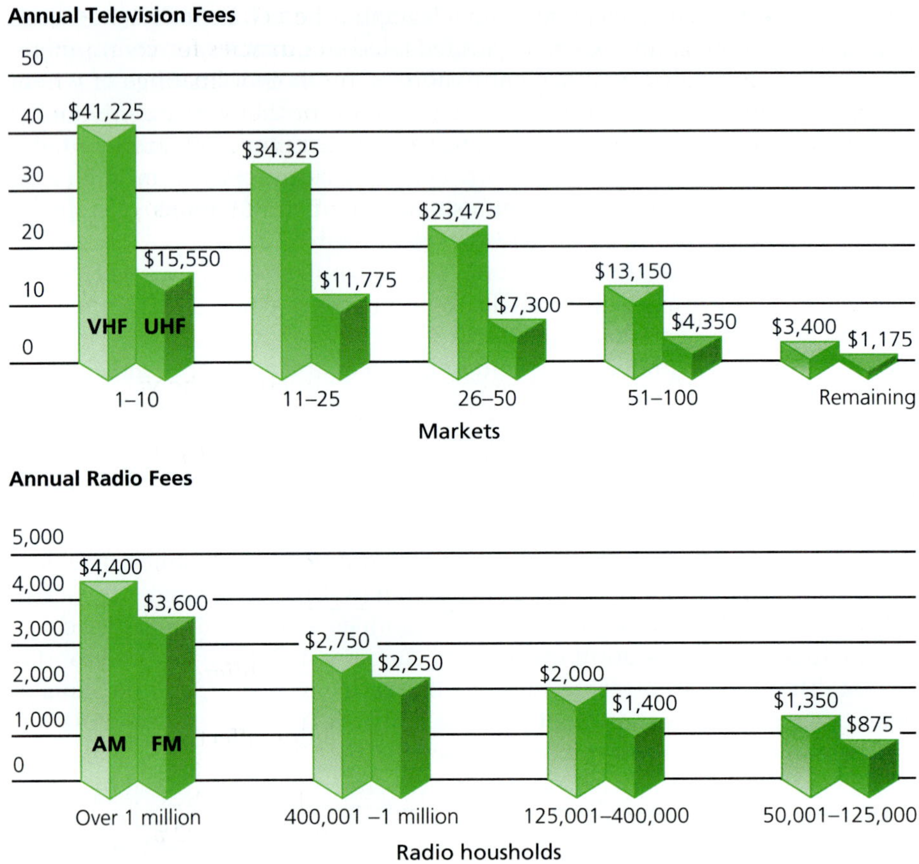

Figure 12.2 What Does a License Cost?
Source: "The Cost," 1999.

Other Regulatory Agencies

Telecommunications businesses deal with other regulators as well. The Federal Trade Commission (FTC) regulates advertising content and practice. Inasmuch as most telecommunications industries are commercially supported, this five-member commission frequently deals with them. The National Telecommunications and Information Administration (NTIA) is a division of the Department of Commerce charged with advising the president on telecommunications policy analysis and development, application (or wider and more efficient use) of telecommunications, use of the spectrum by the federal government, and technical telecommunications research. The report on the Digital Divide discussed in Chapter 1 is an example of the work conducted by the NTIA.

Formed in 1865 and taking its current name in 1932, the International Telecommunication Union (ITU) has moved from oversight of cross-border telegraph to regulating global telecommunications. Based in Geneva, Switzerland, and now an agency of the United Nations, the ITU coordinates worldwide spectrum allocation and orbiting slots and frequencies for communication satellites. It has recently become a major force in coordinating the efforts of engineers to develop global technical standards for the Internet (Molony & Cukier, 1998). The ITU has no official enforcement power. Its influence resides in the realization by its 150 members that global telecommunications can only function if there is clear demarcation of different nations' operations.

Contemporary Issues in Regulation

Despite the movement toward greater regulatory freedom, the various telecommunications industries all operate under some degree of regulation. Sometimes that control is exercised by the FCC and sometimes by other official bodies. Here is an overview of the most important and pressing contemporary regulatory activities.

Contemporary Cable Regulation

The regulation of cable is divided between the FCC, local franchising authorities, and Congress. Each regulator concerns itself with a different aspect of the cable industry.

The two most significant pieces of federal legislation affecting cable and its consumers are the Cable Television Consumer Protection and Competition Act of 1992 and the Telecommunications Act of 1996, both passed by Congress and enforced by the FCC. The most important provisions of the 1992 act are these:

- Gave local broadcasters in a system's service area the right to charge a fee for the system's carriage of their signals (**retransmission consent**), to require the system to carry their signals for free (must-carry), or to deny the system the right to carry their signals at all. These decisions are made every 3 years. In fact, stations do not want to be excluded from local cable systems, and cable systems, who do want to carry those channels because they are important to subscribers, do not want to pay stations for that right. As a result, one common compromise is that systems provide channel space to local broadcasters typically up near Channel 70 or 80. They in turn offer additional programming, for example, sports and talk or, in the case of the Fox-owned stations, the FX Channel.
- Permitted the FCC to regulate cable subscriber rates, which it did, scaling back the industry's egregious rate increases of the previous 5 years.
- Required the FCC to evaluate and certify every local franchising authority in the United States, some 20,000 agencies. The goal was to ensure that local authorities interpret and enforce federal rules in a manner appropriate to both operators and subscribers.

- Required local franchising authorities to permit, within reason, **overbuilds,** that is, establishment of a second cable service in an area. In fact, though, very few businesses ever attempt to overbuild an existing cable system.

The 1996 Act opened cable to competition from the phone companies and encouraged their entry into the business of bundling (Chapter 6). It had other provisions as well. Most significantly for subscribers, it deregulated rates, effectively wiping out federal controls. The FCC continues to monitor rates at the basic tier, challenging the cable industry to keep increases within reason and service reliable, enforcing these "requests" with its power to reregulate.

Cable and Copyright Congress also regulates cable through passage and enforcement of copyright rules. In the early days of cable, copyright was a source of acrimony between broadcasters and cable operators. The operators paid no fees to broadcasters for the use of their signals, nor did they pay the content providers to whom the broadcasters were paying copyright fees. This changed with the passage of the Copyright Act of 1976, which required cable systems to pay a **compulsory license fee** of .07% of their revenue from their basic subscription income to a Copyright Royalty Tribunal. The tribunal would, in turn, distribute those moneys to copyright holders. The tribunal had the right to adjust the size of the fee as necessary (it now sits at 3.75%) and to determine its distribution among the many claimants. Congress abolished the Copyright Royalty Tribunal in 1993, leaving cable copyright issues in the hands of several different arbitration panels under the auspices of the Library of Congress.

Local Franchising Authority Local communities became frustrated with the federal government's muddled and contradictory regulation of the early cable industry, so they entered the fray in the form of **local franchising authorities,** local bodies charged with granting and oversight of cable franchises. After all, potential system operators used cities' and municipalities' rights of way—either on poles or underground—for their wires. Franchising authorities, especially those in highly attractive urban areas, quickly learned that they could demand much in exchange for granting a franchise. Public access channels and well-equipped access studios, channels set aside for government and education, rapid wiring of their communities, improved customer service, and reasonable subscription rates all became part of the give and take.

The Cable Television Consumer Protection and Competition Act of 1992 strengthened the franchising authorities' hand. It permits authorities to charge operators an annual franchising fee of up to 5% of their gross revenues and allows franchising authorities to require equipment and channel capacity upgrades and improved customer service as conditions of franchise renewal or granting. The cable industry challenged these powers as violations of their First Amendment rights, but they were rebuffed by the Supreme Court.

Regulating Digital Television

In Chapter 3 we discussed the promise of digital television. That promise—of riches to broadcasters and service to viewers—is at the base of the dispute over regulation of this technology. In response to controversy following what critics called "the $70 billion giveaway" of the spectrum to broadcasters, President Bill Clinton established the Advisory Committee on Public Interest Obligations of Digital Television Broadcasters. The group, composed of industry and public interest representatives, quickly became known as the Gore Commission, after its head, Vice President Al Gore. At the end of 1998, after a year of deliberation, the Gore Commission referred its recommendations to the FCC, stating, "Marketplace forces do not always deliver important social benefits, such as . . . adequate attention to public affairs. In such circumstances, government can appropriately play a role" (Hickey, 1999, p. 47).

Media columnist Max Frankel (1999) of the *New York Times* hailed this view, "The public's last defense against this robbery is at hand. We must bleed the broadcasters who have run off with $70 billion worth of our spectrum space but still refuse to acknowledge any obligation to render significant public service in exchange" (p. 27). The television industry responded, citing the proposed rules as an infringement of stations' First Amendment rights and arguing that stations weren't getting anything for free because they were required to return their analog spectrum space (which the commission will auction to the highest bidders, typically mobile communications providers). For more on this issue, read the box titled Evaluating the Gore Commission Recommendations.

To date, the only official action taken by the FCC regarding broadcasters' payback for the use of the digital spectrum is the requirement that they pay a fee of 5% of all income earned from digital subscription services. In other words, if a channel is available to all viewers free of charge, income derived from it is not subject to this fee. If it is a subscription-only service, for example, stock market or sports news, then the fee applies. The FCC also promulgated a rule requiring DBS providers offering digital service to set aside 4% of their channel capacity for public interest programming, but it left the determination of just what constituted public interest programming to the providers themselves.

Regulating the Internet

By its very nature the Internet raises a number of important First Amendment issues. Its diffusion—there is no central location, no on and off button—makes it difficult to control for those who want to do so. For free expression advocates, however, this freedom from control is the medium's primary strength. The anonymity of its users provides their expression—even the most radical, profane, and vulgar—great protection, giving voice to those who would otherwise be silenced. Those who want to strengthen Internet control say this

The Real World *Evaluating the Gore Commission Recommendations*

The recommendations of the Gore Commission on how digital broadcasters might repay the public for the use of its spectrum space proved controversial and unsatisfying to almost everyone involved. But the commission's 22 members were free to walk away from the dispute, leaving the eventual disposition of their suggestions to the FCC. The commission itself was not only required to judge the proposals on their merits, trying to find a balance between the interested parties, but it also had to deal with a hostile Congress wanting the FCC to place as few requirements on broadcasters as possible. That pressure was so severe that Vice President Gore was moved to declare, "For its part, Congress ought to help us move forward on this issue on a bipartisan basis and not continue to threaten the FCC with further legislative reprisals" (Albiniak, 1998, p. 7).

What were the Gore group's recommendations? Why did they create so much controversy? Read these 10 proposals made by the Gore Commission and decide for yourself whether you agree with the commission.

1. The FCC should adopt minimum public interest requirements for digital television stations.
2. Stations should be required to disclose their public interest activities every 3 months.
3. Congress should create a permanent trust fund to support public television; public stations should be given additional spectrum space; and commercial broadcasters offering subscription data services should be required to provide local educational information.
4. The National Association of Broadcasters should adopt an updated voluntary Code of Conduct.
5. Broadcasters using their spectrum space to offer multiplexing can fulfill their public interest requirements through payment of a fee or by setting aside a channel for public interest purposes.
6. In the event Congress should undertake comprehensive campaign finance reform, digital broadcasters should voluntarily provide 5 minutes each evening of the 30 days leading up to an election for candidates (only two other democracies in the world, Malaysia and Taiwan, do not make free airtime available to candidates); the FCC should study possible rule making aimed at banning all sale of airtime to state and local candidates.
7. Digital broadcasters should offer emergency warning services.
8. Digital broadcasters should offer closed-captioned services for all programming.
9. Digital broadcasters should consider diversity in programming decisions, political discourse, hiring, promotion, and business opportunities.
10. The commission, Congress, and the president should explore alternative approaches for digital broadcasters to meet their public service obligations.

What do you think? If put in place, would these proposals be an unfair burden on a legal business protected by the First Amendment or a small price to pay for access to $70 billion worth of the public's resources? You can offer your observations on the Gore Commission recommendations and other regulatory matters to the FCC in several ways. Its Web site is <http://www.fcc.gov/>, its e-mail address is <fccinfo@fcc.gov>, and its toll-free phone number is 1-888-callfcc.

anonymity is a breeding ground for abuse. But opponents of control counter that the Net's affordability and ease of use make it our most democratic medium. Proponents of control argue that this freedom brings with it responsibilities that other media and those who create their content understand but that are ignored by many online. Internet freedom of expression issues fall into two broad categories. The first is the Net's potential to make the First Amendment's freedom of press guarantee a reality for greater numbers of people. The second is the problem of setting boundaries of "permissible" control.

Controlling Internet Expression Veteran *New Yorker* columnist the late A. J. Liebling, author of that magazine's "Wayward Press" feature and often called the "conscience of journalism," frequently argued that freedom of the press is guaranteed only to those who own one. Theoretically, anyone can own a broadcast outlet or cable television operation, but their number in any community is limited and they are unavailable to all but the richest people and corporations. Theoretically, anyone can own a newspaper or magazine, but again the expense involved makes this an impossibility for most people. Newsletters, like a soap box speaker on a street corner, are limited in reach, typically are of interest only to those who already agree with the message, and are relatively unsophisticated when compared to the larger commercial media.

The Net, however, turns every user into a potential mass communicator. Equally important, on the Internet every "publisher" is equal. The Web sites of the biggest government agency, the most powerful broadcast network, the newspaper with the highest circulation, the richest ad agencies and public relations firms, the most far-flung religion, and the lone user with an idea or cause all sit figuratively side by side. Each is only as powerful as its ideas. This power, or more specifically its abuse, is behind the argument for greater control of the Internet. This medium that can empower users who wish to challenge those more powerful than themselves can also be used to lie and cheat. The Internet does not distinguish between true and false, biased and objective, trivial and important. Once misinformation has been loosed on the Net, it is almost impossible to catch and correct it.

Lies have been part of human nature for all time; the Internet only gives them greater reach. Therefore, there is little that government can do to control this abuse. Legal remedies already exist in the form of libel laws and prosecution for fraud. Users can help by teaching themselves to be more attentive to return addresses, ignoring messages that are sent anonymously or that have suspicious origins. There is an Internet-based solution as well. Computer security specialists at the United States Department of Energy maintain a Web site designed to track and debunk online misinformation: <http://ciac.llnl.gov/ciac/CIACHoaxes.html>.

Pornography on the World Wide Web Most efforts at controlling the Internet are aimed at indecent or pornographic Web content. But indecent and pornographic expression is protected expression, and the 1996 Communications

Figure 12.3 Enemies of the Internet. The countires shown in dark green use various means to limit their citizens' access to the Internet. *Source:* Reporters Sans Frontiers, 1999.

Decency Act, designed to tightly control access to sexually oriented Web sites, was declared unconstitutional soon after its passage. The 1996 Child Pornography Act does remain in force, and it forbids online transmission of any image that "appears to be of a minor engaging in sexually explicit conduct." Proponents argue that the horrible impact of child porn on the children involved, as well as on society, warrant this legislation. Opponents argue that child pornography per se is already illegal, regardless of the medium. Therefore, they see this law as an unnecessary and overly broad intrusion into freedom of expression on the Net. In August 1997 a federal court upheld the law.

The primary battleground is in the realm of protecting children from otherwise legal content. The Net, by virtue of its openness and accessibility, raises particular concerns. Children's viewing of sexually explicit material on cable television can theoretically be limited by parents (for example, they can control the remote). Moreover, viewers must specifically order this content and typically pay an additional fee for it. The purchase of sexually explicit videos, books, and magazines is controlled by laws regulating vendors. But computers sit in homes, schools, and libraries. Children are encouraged to explore their possibilities.

Proponents of stricter control of the Net liken the availability of smut on the Internet to a bookstore or library that allows porn to sit side by side with books that children *should* be reading. In actual real-world bookstores and libraries, they argue, professionals, whether book retailers or librarians, apply their trained judgment in selecting and locating material, ideally striving for

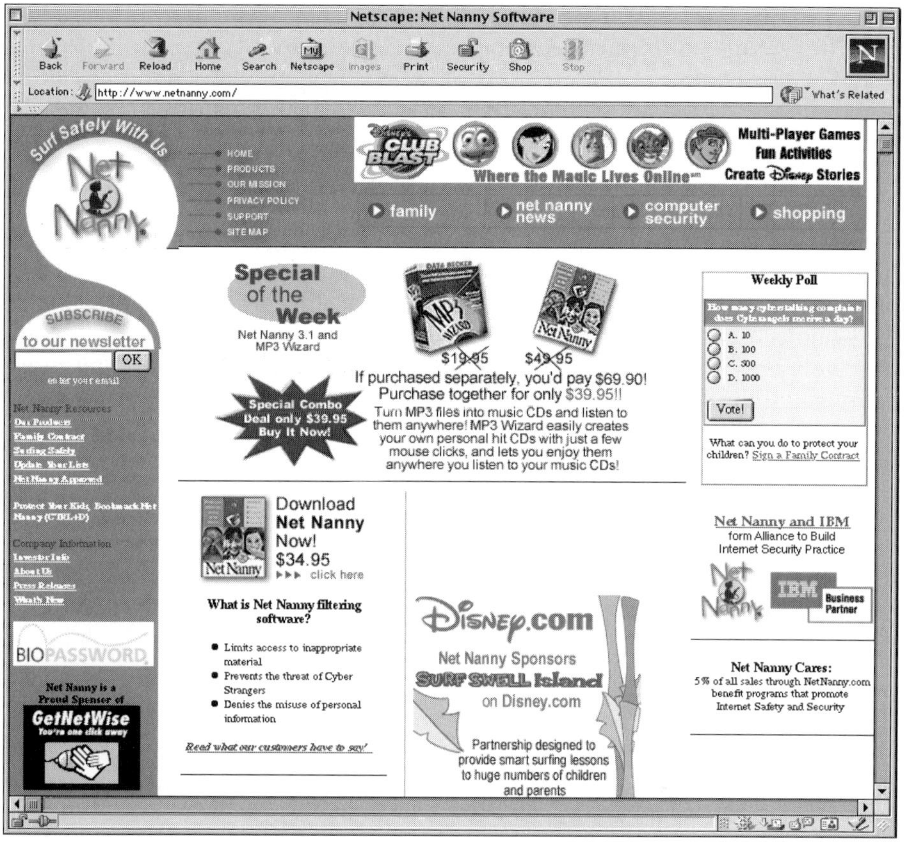

Net Nanny is one of the more popular Web content filters. It can be set to block access to Web sites by title or by the presence of specific words and images.

appropriateness and balance. Children are the beneficiaries of this professional judgment. No such professional selection or evaluation is applied to the Internet. Opponents of control accept the bookstore/library analogy but argue that, as troubling as the online proximity of all types of content may be, it is a true example of the freedom guaranteed by the First Amendment. To see how some other countries are confronting this issue, see Figure 12.3.

One solution to this dilemma is technology. Filtering software, such as Net Nanny and SurfWatch, can be set to block access to Web sites by title and by the presence of specific words and images. Another form of control is **reverse filtering.** Provided by sites Your Own World (for free) and JuniorNet (for a fee), reverse filtering involves using these sites as restrictive portals; that is, they allow access only to preapproved sites, and all links are detailed in on-screen reports, allowing parents to see what sites their children visit.

Few free speech advocates are troubled by either form of filter on home computers, but they do see them as problematic when used on more public machines, for example, in schools and libraries. They argue that software that can filter sexual content can also be set to screen out birth control information, religious sites, and discussions of racism. Virtually any content can be blocked. This, they claim, denies other users—adults and mature teenagers, for example—their freedoms.

The Internet and Copyright Another First Amendment issue that takes on a special nature when applied to the Internet is copyright. The assumption behind copyright is that more "authors" will create more content if assured of monetary compensation from those who use it. When the content is tangible—books, movies, videotapes, magazines, or CDs—authorship and use are relatively easy to identify. But in the cyberworld, things become a bit more complex. John Perry Barlow (1996), a cofounder of the Electronic Frontier Foundation, expressed the situation this way:

> The riddle is this: If our property can be infinitely reproduced and instantaneously distributed all over the planet without cost, without our knowledge, without its even leaving our possession, how can we protect it? How are we going to get paid for the work we do with our minds? And, if we can't get paid, what will assure the continued creation and distribution of such work? (p. 148)

Technically, copyright rules apply to the Internet as they do to other media. Material on the Net, even on electronic bulletin boards, belongs to the author, so its use requires permission and possibly payment. The problem is that material on the Internet is not tangible, and it is easily, freely, and privately copied. This renders it difficult if not impossible to police those who do copy.

Another confounding issue is that new and existing material is often combined with other existing material to create even "newer" content. This makes it difficult to assign authorship. If a user borrows some text from one source, combines it with images from a second, surrounds both with a background graphic from a third, and adds music sampled from many others, where does authorship reside?

To help deal with these issues, President Clinton in 1998 signed the Digital Millennium Copyright Act, bringing the United States into line with the World Intellectual Property Organization's treaty on online copyright protection. Its primary provisions free service providers from liability for illegal use of copyrighted material by their customers and create criminal penalties for those who circumvent antipiracy technologies such as encryption and **digital watermarking,** embedding a digital signature into a piece of digital content.

The problem of applying copyright to the Internet received a very public airing in mid-2000 as record companies battled Napster, a company that provides software enabling registered users to search one another's hard drives for MP3-stored files and transport them to their own computers.

Guaranteeing high visibility for their claims, heavy metal rock band Metallica sued Napster, physically delivering 30 cardboard boxes containing the names of 335,000 Napster users who had, in the band's eyes, stolen its music. Metallica, and later rapper Dr. Dre, framed their arguments in terms of artists' rights—their copyrighted music cost millions of dollars to create and market; those who use Napster to download it for free are taking their rightfully earned profits away from them. Napster, whose software was downloaded by 10 million users in the first 9 months it was available (Surowiecki, 2000, p. 35), countered that it does not itself make any copyrighted content available. It only offers software that permits users to choose and download files from each other. The company even argued that its service helps recording artists sell more music: users can sample an artist's offerings, and then go out and buy it on disc.

Napster, however, was legally required to drop the 335,000 Metallica fans from its service. The Digital Millennium Copyright Act frees service providers from liability for the illegal use of copyrighted material by their customers. But it also requires that once a service provider is informed that copyright infringement is occurring it is itself liable for that violation unless it takes steps to stop it.

Most industry observers believe technology will eventually solve this knotty copyright issue. Recording companies will establish Web sites

> offering unlimited access to all the music they own, old and new, for twenty dollars a month. The music files would be encrypted and of the highest quality, so you'd always know you were getting exactly what you wanted. . . . [There are no such guarantees on Napster.] Once high-speed wireless Internet access became widespread, you'd be able to listen to any song from anywhere. The cost of producing and distributing physical goods would disappear. Revenue would come from both subscriptions and advertising. Think of how profitable MTV is—more profitable than any of the big record labels. Then multiply the number of viewers it has by five or ten times, and you'll get a sense of how lucrative such a service might be. (Surowiecki, 2000, p. 35)

The Internet and Privacy The issue of privacy in telecommunications has traditionally been considered one of individuals' rights to protect their privacy from invasive, intrusive media (Chapter 13). For example, should newspapers publish the names of rape victims and juvenile offenders? When does a person become a public figure and forfeit some degree of privacy? On the Internet, however, the issue takes on a new character. Where privacy was once "the right to be left alone," today it is just as likely to mean "the right to maintain control over our own data." As legal technology writer Simson Garfinkel (2000) wrote:

> Privacy isn't just about hiding things. It's about self-possession, autonomy, and integrity. As we move into the computerized world of the twenty-first century, privacy will be one of our most important civil rights. But this right of privacy isn't the right of people to close their doors and pull down their window shades. . . . It's the right of people to control what details about their lives stay inside their own houses and what leaks to the outside. (p. 11)

Privacy online has two facets. The first is protecting the privacy of communication we wish to keep private. The second is the use (and misuse) of private, personal information willingly given online.

Protecting Privacy in Communication The 1986 Electronic Communication Privacy Act guarantees the privacy of our e-mail. It is a criminal offense to either "intentionally [access] without authorization a facility through which an electronic communication service is provided; or intentionally [exceed] an authorization to access that facility." In addition, the law "prohibits an electronic communications service provider from knowingly divulging the contents of any stored electronic communication." The goal of this legislation is to protect private citizens from official abuse; in other words, giving e-mail "conversations" the same protection that phone conversations enjoy. If a government agency wants to listen in, it must secure permission, just as it must get a court order for a telephone wiretap.

If a person or company feels that more direct protection of communication is necessary, encryption is a possibility, but it is controversial. According to the FBI and the Clinton Administration, encryption-provided privacy is an invitation for terrorists, drug lords, and mobsters to use the Net to threaten national security. As such, in January 2000 they proposed "relaxed" rules—relaxed from initial plans to allow the government to hold the key to *all* encryption technologies. The new rules required makers of encryption software to turn over a copy of their code to a designated third party. The government may access it only with a court order. Software makers were also prohibited from exporting sophisticated encryption software. The Internet industry objected on the ground that the rules amounted to unnecessary government intrusion into the market. Civil libertarians objected on the ground that the rules placed too much power in the hands of the government. The administration quickly reversed itself, agreeing to drop even its "relaxed" controls, freeing manufacturers to distribute encryption software as they saw fit.

Protecting Privacy of Personal Information Every online act leaves a "digital trail," making possible easy "**dataveillance**—the massive collection and distillation of consumer data. . . . Information gathering has become so convenient and cost-effective that personal privacy has replaced censorship as our primary civil liberties concern" (Shenk, 1997, p. 146). So troublesome has the problem become that the Federal Trade Commission "censured" the Internet industry in a June 1998 report to Congress "for doing little to insure adequate consumer privacy protection online" ("FTC Faults," 1998, p. 1C). Dissatisfied with the industry response, in May 2000 the FTC took the rare step of asking Congress to grant it the power to officially regulate online privacy. The commission's own study demonstrated that only 20% of Web sites with more than 39,000 visitors and 42% of the 100 most popular U.S. Web sites had adopted its voluntary, four-part privacy standard ("Tighter Online," 2000). The FTC wants sites to notify users that information is being collected and how it will be used, grant users the option to choose whether that information can be shared with others, give users the right to review the information on them that a site has collected, and guarantee users that their personal information is secure during storage and transmission.

One form of dataveillance is distributing and sharing personal, private information among organizations other than the one for whom it was originally intended. Information from every credit card transaction (online or at a store), credit application, phone call, supermarket or other purchase made without cash (for example, with a check, debit card, or "club" card), newspaper and magazine subscription, and cable television subscribership is digitally recorded, stored, and most likely sold to others. The increased computerization of medical files, banking information, job applications, and school records produces even more salable data. Eventually, anyone who wants to know something about a person will be able simply to buy the necessary information—without that person's permission or even awareness. These data can then be used to further invade people's privacy. Evening meals and family conversations at home may be interrupted by targeted phone solicitations. Employers may withhold jobs for reasons unknown to applicants. Insurance companies may selectively deny coverage to people based on supermarket-provided data on grocery choices.

Recognizing the scope of data collection and the potential problems that it raised, in 1974 Congress passed the Federal Privacy Act, which restricts governments' ability to collect and distribute information on citizens. The act, however, expressly exempted businesses and other nongovernmental organizations from control. This stands in stark contrast to European Union countries, where it is illegal for an organization of any kind to sell the name and other personal information of a customer or client for any reason without that person's permission. The box titled Whose Life Is It Anyway? explores some of the cross-national differences in how personal data are treated.

A second form of dataveillance is electronically "tracking" the choices people make when they are on the Web; this is called the click stream. Despite the anonymity online users think they enjoy, every click of a key can be, and often is, recorded and stored. This happens whether or not the user actually enters information, for example, a credit card number to make a purchase or a Social Security number to verify identity. Software records what sites users visit, what they look at in those sites, and how long they stay at a given place in that site as they click from Web page to Web page. In and of itself, this is not a problem, because all the Web site reads about the user is a computer address. But once a user enters personal information, that "anonymous" address has a name.

Since 1996, Congress, the Federal Trade Commission, and public interest groups have attempted to draft legislation that will offer greater protection for online privacy. These efforts have largely been unsuccessful because of resistance on two fronts. Web site creators and others who want access to information will not back laws that limit their profit potential, and many Internet users are fiercely opposed even to well-intentioned government intrusion in the workings of the Internet. Still, in October 1999 the FTC, with White House

Critical Issues *Whose Life Is It Anyway?*

How to protect the privacy of personal information is an issue that other nations as well as our own face. In the United States the FTC threatens federal regulation if Web sites do not clearly and prominently publish their policies on how personal data are used (Financial Times, 1999). Some companies sign on to TRUSTe, and others, such as IBM, announce that they will not advertise on any site that does not post its privacy policies or allow users to "opt out" of having their data sold or given to a third party without their permission ("Big Blue," 1999). Still, there is no nationwide standard for protecting personal information. Countless online companies collect personal data on visitors to their Web sites and sell that information to countless others.

The situation in Europe is completely different. Countries that belong to the European Union (EU) require all companies, not just those doing online business, to guarantee their customers these rights:

- The right to see any and all information a company has about them.
- The right to be informed about how their personal information will be used.
- The right to access their personal information and make changes.
- The right to be notified before personal data are shared or sold.
- The right to veto any sharing or selling the user does not want.
- The right to sue a company that violates these rules.

Why do you think this difference between the United States and the EU exists? Would you prefer that the United States adopt controls similar to those of the EU nations? Do you accept industry arguments that it is income from the sale of customers' personal data that allows vendors to keep costs to consumers down or to provide interesting Web sites at no cost? Explain your answer. Do you believe your personal data belong to you? If you do, what do you consider legitimate use of that data?

---

and Congressional backing, put the Children's Online Privacy Protection Act in place, requiring that Web sites that collect personal data from children under 13 years of age first obtain parental permission.

Two forms of industry self-regulation are in place. One, offered by the Electronic Frontier Foundation and CommerceNet, an association of Internet businesses, is TRUSTe. Its guiding philosophy has two parts: users have a right to informed consent, and no single privacy principle is adequate for all situations. To gain the TRUSTe "seal of approval," Web sites must agree to certain restrictions on their collection and use of user information, presumably making them more attractive to privacy-conscious users.

A second effort at self-regulation was sanctioned by the Clinton Administration. Beginning in January 1998, 14 companies—including Lexis-Nexis and the biggest credit reporting companies in the United States—agreed to block dissemination of all personal information to anyone but "qualified customers." The 14 companies account for more than 90% of all online personal information selling. The voluntary plan has two serious handicaps, according to privacy advocates. First is the definition of "qualified." Government agencies and private investigators would seem to be qualified—but under *all* conditions and situations? The second problem is that individual users must request to

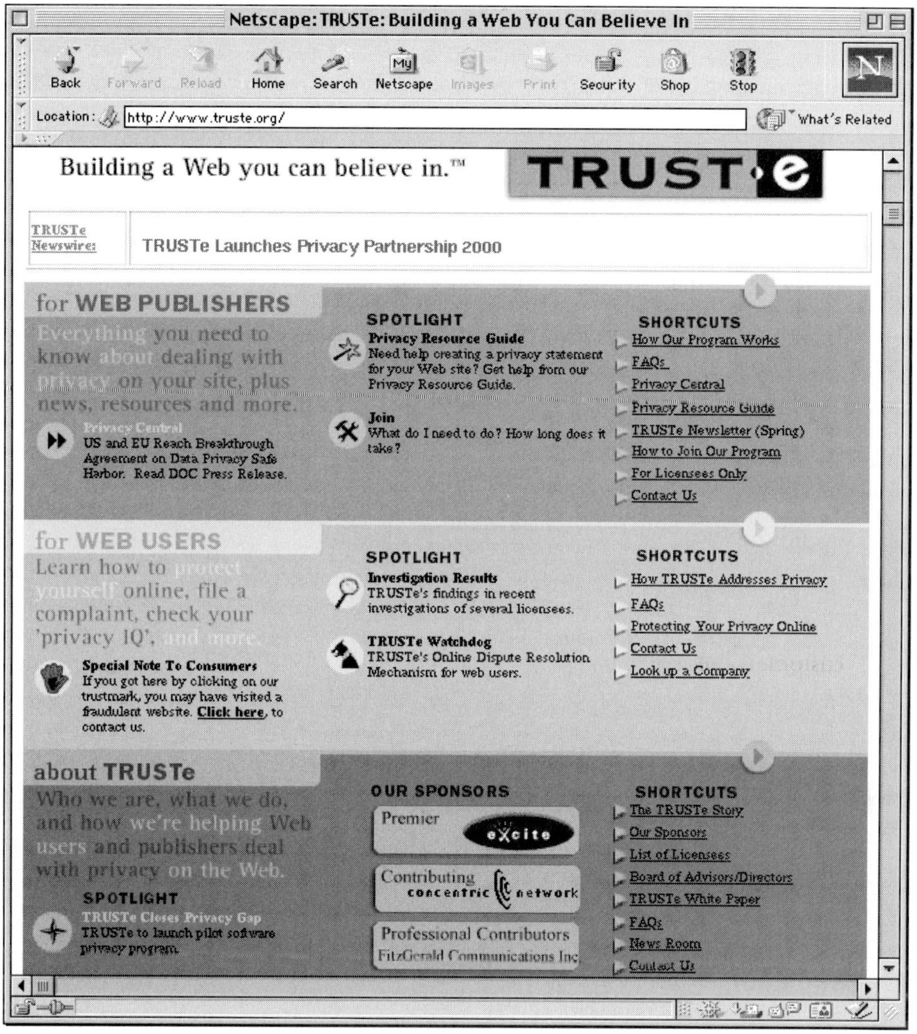

Many Web sites boast the TRUSTe logo in an effort to assure users of the security of their personal data.

opt out; that is, they must contact each of the participating companies and ask that their information be guarded. Privacy proponents argue that the standard, not only in this but in all instances, should be *opt in;* that is, companies can only collect and distribute personal data if users authorize them to do so in advance of that collection and use.

For those uneasy with government regulation as well as industry self-regulation, there is a technological solution. Companies such as Anonymizer.com, Zero-Knowledge, and Privada offer software that blocks information from the recording eyes of Web sites.

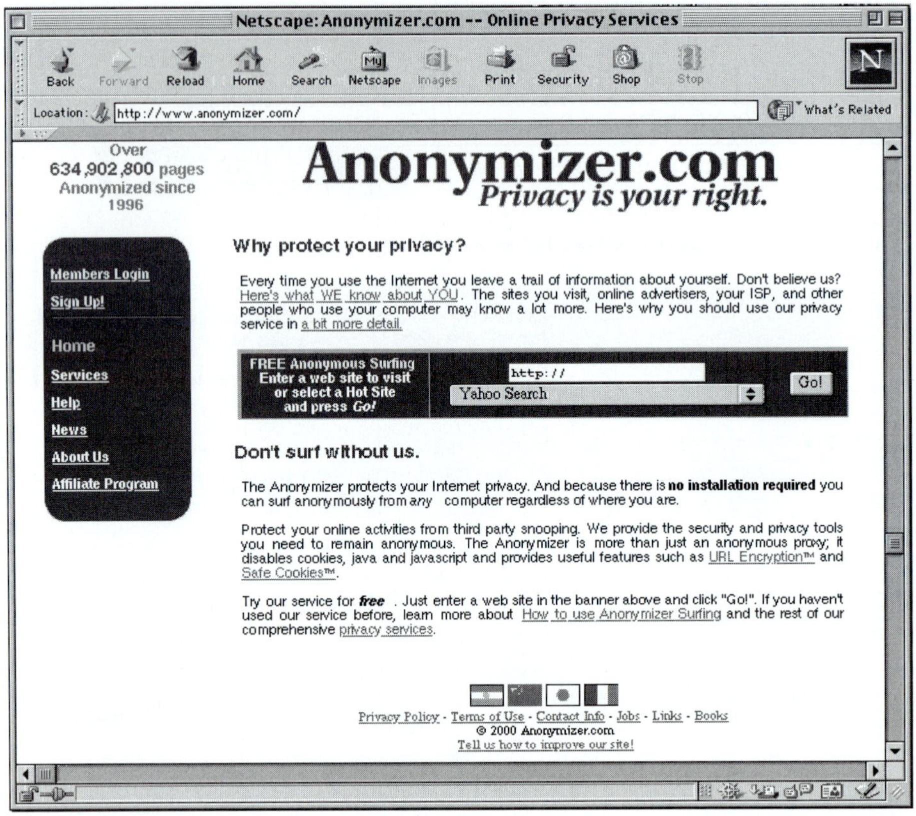

Anonymizer.com is among the many companies/Web sites offering a technological means of protecting personal privacy online.

Chapter Review

The First Amendment, based on libertarian philosophy, guarantees freedom of press and speech. Supported by the 14th Amendment, this protects the press from official intrusion by all levels of government. That protection carries to all media but can be suspended in cases of clear and present danger and when competing interests must be balanced, for example, in free press versus fair trial conflicts. Some expression, primarily libel, slander, and obscenity, is not protected. Pornography, however, is protected expression. In almost no case is prior restraint acceptable. Indecent expression is protected but limited in broadcasting and typically relegated to late night hours.

Copyright does not protect the expression of ideas; rather, it protects the creator's financial interest in that expression, considered an important incentive in maintaining the free and ongoing flow of art, science, and other public speech. The 1978 copyright law set new rules and formalized exceptions such as public domain, fair use, and use in the public interest.

Contemporary regulation of telecommunications is characterized by widespread deregulation, resulting in, among other things, the demise of the Fairness Doctrine and ascertainment. The FCC has established mechanisms to relieve licensees of many former controls, for example, by streamlining the licensing process. The FCC itself is composed of five commissioners overseeing 2,000 employees spread over seven operating bureaus. It is joined by other regulatory bodies involved in telecommunications, including the Federal Trade Commission, the National Telecommunications and Information Administration, and the International Telecommunication Union.

Interesting contemporary regulatory issues involve cable regulation, digital television, and controlling the Internet. Regulation of cable is divided between the FCC, Congress, and local franchising authorities. The Cable Television Consumer Protection and Competition Act of 1992 and the Telecommunications Act of 1996 are the most important recent efforts at regulating cable. Regulation of digital television centers on how broadcasters will compensate the public for the use of the digital spectrum. Controlling the Internet has multiple facets. The power and reach of the Internet, plus the anonymity it gives users, demonstrate its potential—a potential that excites proponents and frightens others. Pornography is protected expression, so protecting children from porn is one Internet regulatory battleground. Copyright is another, and it involves solutions such as encryption and digital watermarking. Much attention, too, is paid to the Internet and privacy, to protect both online communication and users' personal information.

Questions for Review

1. What are the basic tenants of libertarianism? How do they support the First Amendment?
2. What is the relationship between the First and Fourteenth Amendments?
3. Name important court cases involving the definition of "no law," "the press," "abridgment," "clear and present danger," "balancing of interests," and "prior restraint."
4. What are libel and slander? What are the tests of libel and slander? How do the rules change for public officials?
5. Define obscenity, pornography, and indecency.
6. What is safe harbor?
7. What is copyright? What is its intended function? What are the exceptions to copyright protection?
8. What is the traffic-cop analogy, and why is it important to broadcast regulation?
9. What are the arguments for and against deregulation of telecommunications?
10. How many FCC commissioners are there? How are they chosen and confirmed?
11. What are SMATV and MMDS?
12. What are the Federal Trade Commission, the National Telecommunications and Information Administration, and the International Telecommunication Union?
13. With what aspects of cable regulation does the FCC involve itself? With what aspects of cable regulation do local franchising authorities involve themselves?
14. What are the arguments of those for and against increased public service requirements for digital television broadcasters?

15. Why is the protection of children a primary battleground in the ongoing debate over controlling the Internet?
16. What are encryption and digital watermarking?
17. What is dataveillance? How does TRUSTe relate to it?

Questions for Critical Thinking

1. Are you a libertarian? That is, do you believe people are inherently rational and good and that they are best served by a fully free press? Defend your position.

2. What is your position on pornography? It is legally protected expression. Would you limit that protection? When?

3. How much regulation (or deregulation) do you think broadcasters should accept?

4. Imagine yourself in Fred Cook's shoes. That is, imagine that you have been attacked, by name, as a racist and a liar on a morning drive time radio show. If the RTNDA and the broadcast industry have their way, you will have no right of reply. Broadcasters will not even be required to sell you time to present your side. Do you think this is fair? What about your First Amendment rights? How did you answer Question 3? Does this situation shake your confidence in increased deregulation? Why or why not?

5. Do you ever make personal information available online? If so, how confident are you in its security? Do you take steps to protect your privacy? If so, what are they?

Engaging Technology

1. The First Amendment is considered our "first freedom." Numerous Web sites are devoted to its study and protection. Among the better are Vanderbilt University's Freedom Forum First Amendment Center at <http://www.fac.org/default.htm>, the Thomas Jefferson Center for the Protection of Free Expression at <http://www.tjcenter.org/whatsnew.html>, and the First Amendment Cyber-Tribune at <http://w3.trib.com/FACT/>. Visit these sites and determine how "absolute" each is in its defense of our first freedom.

2. Some Web sites are devoted specifically to ensuring that the Internet, like other media, enjoys First Amendment protection. Among the more notable are the Center for Democracy and Technology at <http://www.cdt.org>, the Electronic Frontier Foundation at <http://www.eff.org>, and the Electronic Privacy Information Center at <http://www.epic.org>. Access these sites to investigate their arguments for freedom of Internet expression. How well does each make its case for free Internet expression?

3. Three official regulatory bodies that we've discussed in this chapter are the FCC (<http://www.fcc.gov>), the FTC (<http://www.ftc.gov>), and the ITU (<http://www.itu.int>). Visit these sites to see what they have to say about themselves. Pay particular attention to how they explain their regulation of media that are otherwise protected by the First Amendment.

4. Numerous public interest groups work to protect freedom of the press and of expression. The American Civil Liberties Union at <http://www.aclu.org> and Fairness and Accuracy in Reporting at <http://www.igc.org/fair/> each present their guiding philosophies online. Go to these sites to see if you accept their rationales for their activities toward that end.

Chapter Thirteen

Ethics and Self-Regulation

In July 1998 a gunman entered the Capitol Building in Washington, D.C., and opened fire, killing two policemen, Detective John Gibson and Officer Jacob Chestnut. Television reporters were immediately on the scene and soon had their stories ready for broadcast.

Anchor Bernard Shaw opened CNN's report with an explanation: "Apparently one of the wounded officers' family has not been told . . . and, because of that, it would be obscene for us to be reporting the man's name with his wife and children and relatives not knowing that he has been injured." Later in that same story Shaw explained to viewers, "We had to back off reporting one name because we . . . had learned that the officer's family had not been notified. So obviously, out of decency, we would not report that name" (Voukydis, 1998, p. 36).

Obviously? Maybe not. At the same time that the CNN report was airing, Tim Russert, head of NBC's Washington bureau, appeared on that network's cable channel MSNBC with as much information as he could gather, including the names of both victims. Then MSNBC cut away from Russert's on-camera report to present live coverage of a patient being unloaded from a helicopter, following him and the attending paramedics into the emergency room. NBC anchor Tom Brokaw provided commentary: "This is the medevac helicopter arriving at the hospital. We believe that one of those gurneys contains one of the officers that was shot, either Officer Gibson or Officer Chestnut" (Voukydis, 1998, p. 36).

Both news operations had to make a judgment, and they had to make it fast—on the spot. What guided the two respected news teams to completely different conclusions? There are no laws against reporting victims' names before their families have been notified, so what NBC did was not illegal. But was it right? What must Officer Chestnut's children have felt if they first learned from television that their Dad had been murdered? How might they have reacted, not only to seeing his body off-loaded from a helicopter but to seeing television cameras following the corpse into the hospital? What must their reaction have been when they later learned that MSNBC had misidentified the person on the stretcher—that it was not their father at all?

Preview

This type of judgment and the ethical decisions telecommunications professionals make every day form the basis of this chapter. We study social responsibility theory, the normative theory that requires moral decision making in our media system. We identify different levels of ethics and theories of moral reasoning and examine the conflicting interests that often complicate ethical decision making. We then discuss different processes for arriving at ethical decisions and apply these models to a number of vexing situations often faced by people working in telecommunications. The issues explored include truth and honesty, privacy, confidentiality, conflict of interest, profit versus social responsibility, and the use of potentially offensive content.

Finally, we look at some aids to ethical decision making, such as codes of ethics, and study their advantages and disadvantages. Then we examine other forms of self-regulation and their strengths and limitations.

Social Responsibility

The First Amendment is based on the libertarian philosophy assuming a fully free media and a rational, good, and informed public (Chapter 12). But we have also seen throughout this text that the media are not necessarily fully free. Government control is sometimes allowed, and corporate control is assumed and accepted. During the 1930s and 1940s, serious doubts were also raised concerning the public's rationality and goodness, an important component of libertarian belief. As World War II spread across Europe at the end of the 1930s, libertarians were hard pressed to explain how Nazi propaganda could succeed if people could in fact tell right from wrong. As the United States was drawn closer to the European conflict, calls for greater government control of press and speech at home were justified by less-than-optimistic views of the "average American's" ability to handle difficult information. As a result, libertarianism came under attack for being too idealistic.

Time magazine owner and publisher Henry Luce provided money to establish an independent commission of scholars, politicians, legal experts, and social activists that would study the role of the media in U.S. society and make recommendations on how they should best operate in support of democracy. The Hutchins Commission on Freedom of the Press, named after its chairperson, University of Chicago Chancellor Robert Maynard Hutchins, began its work in 1942 and, in 1947, produced its report, *The Social Responsibility Theory of the Press* (see Davis, 1990).

Social responsibility theory is a normative theory—an idea that explains how media should *ideally* operate in a given system of social values—and is now the standard for U.S. media. As we saw in Chapter 11, other social and political systems adhere to different normative theories.

Social responsibility theory asserts that media must remain free of government control, but in exchange they must serve the public. The theory's core assumptions are a cross between the libertarian principles of freedom and practical admissions of the need for some form of external control on media (McQuail, 1987):

1. Media should accept and fulfill certain obligations to society.
2. Media can meet these obligations by setting high standards of professionalism, truth, accuracy, and objectivity.
3. Media should be self-regulating within the framework of the law.
4. Media should avoid disseminating material that might lead to crime, violence, or civil disorder or that might offend minority groups.
5. The media as a whole should be pluralistic, reflect the diversity of the culture in which they operate, and give access to various points of view and rights of reply.
6. The public has a right to expect high standards of performance, and official intervention can be justified to ensure the public good.
7. Media professionals should be accountable to society as well as to their employers and the market.

In rejecting government control of media, social responsibility theory calls for responsible, ethical industry operation, but it does not free audiences from their responsibility. People must be sufficiently aware of the workings and effects of the various telecommunications industries to develop firm yet reasonable expectations and judgments of media performance. But ultimately it is practitioners, through the conduct of their duties, who are charged with operating in a manner that obviates the need for official intrusion.

Media Industry Ethics

A number of formal and informal controls, both external and internal to the various media industries, are aimed at ensuring that telecommunications professionals operate in a manner consistent with social responsibility theory. Among the formal controls are laws and regulations, codified statements of what can and can't be done and what content is permissible and not permissible (see Chapter 12), and industry codes of practice. Among the informal controls are pressure groups, consumers, and advertisers. Throughout this text we have seen how these informal controls operate. Our interest here is in the media's internal controls, or ethics.

Defining Ethics

Ethics are questions of behavior or moral principles that guide our actions in given situations. The word comes from the Greek *ethos,* which means the customs, traditions, or character that guide a particular group or culture. In our discussion ethics specifically refers to the application of rational thought by media professionals when they are deciding between two or more competing moral choices.

For example, it is not against the law to broadcast the name of a rape victim. But is it ethical? It is not illegal to stick a microphone in a crying father's face as he cradles the broken body of his child at an accident scene. But is it ethical? It is not illegal to report the name of a deceased person before her family has been notified. But is it ethical?

The application of media ethics almost always involves finding the *most morally defensible* answer to a problem for which there is no single correct or even best answer. Return to the grieving father. The reporter's job is to get the story; the public has a right to know. The man's sorrow is part of that story, but the man has a right to privacy. As a human being, he deserves to be treated with respect and to be allowed to maintain his dignity. The reporter has to decide whether to get the interview or leave the grief-stricken man in peace. That decision is guided by the reporter's ethics.

Three Levels of Ethics

Because ethics reflect a culture's ideas about right and wrong, they exist at all levels of that culture's operation. **Metaethics** are an expression of a culture's fundamental values. What do we mean by justice? What does it mean to be good? What is fairness? We need to examine these questions to know ourselves and the context within which we must make ethical judgments. But as valuable as metaethics are for self-knowledge, they provide only the broadest foundation for the sorts of ethical decisions people make daily. They define the basic starting points for moral reasoning.

Normative ethics are more or less generalized theories, rules, and principles of ethical or moral behavior. The various media industry codes of ethics or standards of good practice are examples of individual industries' expressions of their normative ethics. They serve as real-world frameworks within which people can begin to weigh competing alternatives of behavior. Fairness is a metaethic, but journalists' codes of practice, for example, help define what is meant by fairness in the world of reporting. How far must a reporter go to ensure fairness, and how must fairness be applied when being fair to one person means being unfair to another?

Ultimately, media practitioners must apply both the big rules and the general guidelines to very specific situations. This is the use of **applied ethics,** which is our focus for the remainder of this section.

Balancing Conflicting Interests

In applying ethics the person making the decision is called the **moral agent.** For moral agents, sticky ethical issues invariably bring together conflicting interests. In the case of the slain Capitol guards that opened this chapter, these interests include the privacy of the victims' families, protecting the media organization's integrity and reputation, and maintaining the public's trust in the profession of journalism itself.

Media ethicist Louis Day (1997) identified six individual or group interests that often conflict:

- The interests of the moral agent's *individual conscience;* media professionals must live with their decisions.
- The interests of the *object of the act;* a particular person or group is likely to be affected by media practitioners' actions.
- The interests of *financial supporters;* someone pays the bills that allow the station to broadcast or the Web site to operate.
- The interests of *the institution;* media professionals have company loyalty and pride in the organization for which they work.
- The interests of *the profession;* media practitioners work to meet the expectations of their colleagues and have respect for the profession that sustains them.
- The interests of *society;* media professionals, like all of us, have a social responsibility. Because of the influence their work can have, they may have even greater responsibilities than do many other professionals.

In telecommunications these conflicting interests play themselves out in a variety of ways. Some of the most common yet thorniest require us to examine such basic issues as truth and honesty, privacy, confidentiality, conflict of interest, profit, and the use of offensive content. But before we look at these issues, we must investigate how ethical decisions can best be made.

Theories of Moral Reasoning

Moral reasoning is the use of systematic thought in making ethical decisions. Although there are numerous bases for moral reasoning, most ethicists identify three theories or categories of moral reasoning: deontological, teleological, and virtue theories.

Deontological moral reasoning emphasizes principles or moral duties, regardless of their good or bad outcomes. As a result, deontological moral reasoning is duty-based; that is, moral agents' decisions emphasize adherence or commitment to principles that should be universally applied. Based in the 18th-century philosophy of Immanuel Kant, duty-based ethical thought assumes a high standard of conduct, regardless of the situation.

Deontological moral reasoning has some important advantages for moral agents:

- Hard and fast standards free them from having to consider the outcomes of their behavior.
- These standards offer predictability and consistency to moral agents' behavior.
- Ambiguity in vexing ethical dilemmas is reduced.

Deontological moral reasoning has some disadvantages as well:

- In very difficult ethical quandaries where two equally valid standards or duties conflict (for example, the public's right to know and a grieving father's right to privacy), strict standards of moral thought may be of little help.
- It is often difficult to apply hard and fast, universal standards to specific or unusual situations.
- How do you answer this question: When is it moral to ignore the outcome of your behavior, even if it is based on duty and principle? Many ethicists reject duty-based moral reasoning because, in their view, there is no moral answer to this question.

Teleological moral reasoning is consequence-based; that is, the most moral action is that which results in the best outcome. If moral agents seek the best outcome for the greatest number of people, they are called *utilitarians*. Based in the 19th-century work of British philosopher John Stuart Mill, utilitarianism's focus on outcomes offers moral agents some benefits:

- It stresses minimizing harm.
- It is far more flexible than duty-based moral reasoning.
- It can be applied relatively by listing alternative courses of behavior and judging the likely impact of each.

Teleological moral reasoning's drawbacks are these:

- Its application depends on unknown or speculative outcomes.
- It ignores negative outcomes to those not among "the greatest number"; what is the greatest good for the greatest number may unfairly hurt others.

Unlike duty- and consequence-based moral reasoning, which require moral agents to judge the ethics of their behavior, **virtue theories** require moral agents to evaluate the kind of people they are; that is, they emphasize character. Virtue theories of moral reasoning are often criticized for their lack of specificity in helping moral agents sort through especially tricky dilemmas. For example, how can the kind of person you are (or want to be) help you decide a course of action when covering that car crash and its grieving father?

Greek philosopher Aristotle's (384–322 B.C.) *golden mean* is one example of a virtue theory. It argues that "virtuous conduct involves learning to avoid the extreme in any given situation" (Day, 1997, p. 59). In Aristotle's words, this means finding the balance between excess and deficiency. To return to our grieving father, ethical behavior would reside somewhere between jamming a microphone in the crying man's face and ignoring the pain and sadness of the situation altogether.

Applying Media Ethics

Most media professionals routinely make ethical decisions almost every day. Air the video of the burning school or don't? Interview the grieving parent or don't? Accept the speaking fee from a business your organization covers or don't? Run that sugared cereal commercial or don't? *Los Angeles Times* television critic and occasional media ethics instructor Howard Rosenberg says that most of the time "it's so obvious what's right and wrong" (as quoted in Heyboer, 1999, p. 15). But sometimes a particular situation rises above those routinely faced in the day-to-day operation of a media enterprise.

One such situation arose in January 1999. The Internet news site the *Drudge Report* announced, "He's been told all of his life that Bill Clinton is his father, and late last month, 13-year-old Danny Williams of Arkansas underwent a DNA test to find out the truth. And now, just 24 hours after the *Drudge Report* first revealed that the White House was being rocked by a new DNA terror, the story was out of control and landing on front pages of newspapers all over the world!" (Shepard, 1999, p. 28). A few newspapers did pick up the story, and television comedians and radio talk show hosts made their own use of it.

In fact, there had been a DNA test, paid for by supermarket tabloid newspaper *The Star*, but *The Star* refused to run its story until the results were in. This created an ethical question for many media organizations. Should they report an unverified story from a media outlet openly antagonistic toward the president that even a supermarket scandal sheet was unwilling to release? This was an easy one. Few responsible media outlets would report the unsubstantiated rumor that the president of the United States had fathered an illegitimate son with a prostitute.

But this situation was not as ethically straightforward as it first appeared. Some news sources were reporting the story: the *New York Post* ran a front page headline CLINTON PATERNITY BOMBSHELL. Radio and television personalities were also talking about it. How could a media organization legitimately ignore this story now that it was "out there"? Wasn't it news, whether true or

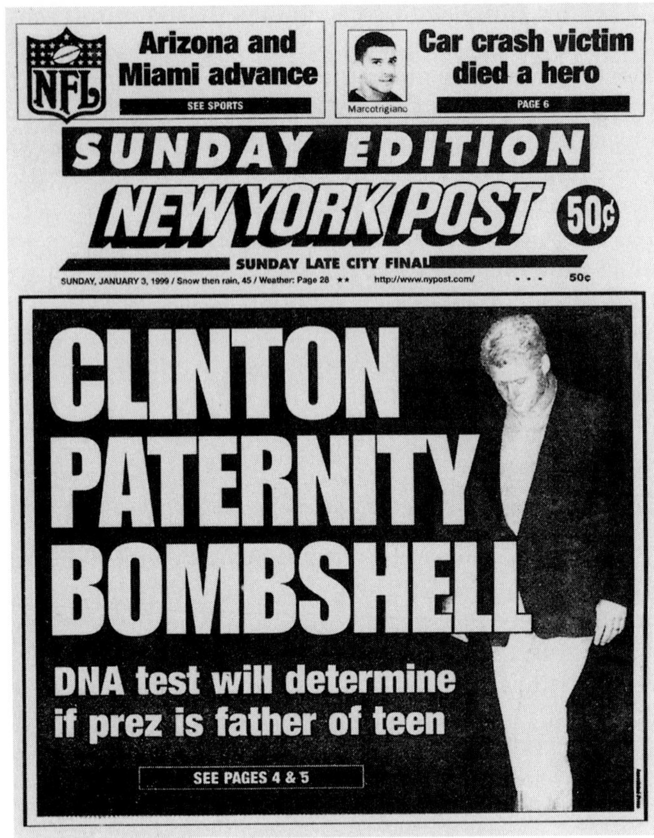
Arizona and Miami advance
SEE SPORTS
Marcotrigiano
Car crash victim died a hero
PAGE 6
SUNDAY EDITION
NEW YORK POST 50¢
SUNDAY LATE CITY FINAL
SUNDAY, JANUARY 3, 1999 / Snow then rain, 45 / Weather: Page 28 ★★ http://www.nypost.com/ • • • 50¢
CLINTON PATERNITY BOMBSHELL
DNA test will determine if prez is father of teen
SEE PAGES 4 & 5

To the credit of the U.S. media, very few outlets followed the lead of the *New York Post* in publishing this unconfirmed story.

not? If a media outlet did run the story—even to say that although the charge was unconfirmed it was appearing in some media—and it turned out to be false (as the DNA test later demonstrated), might it not do harm to the president, his family, and its own reputation? If it did not run the story—and it later turned out to be true—wouldn't the outlet be open to the charge of selective reporting, of favoring the president, and thereby damaging its reputation? If it did not run the story—and it later turned out to be false—wasn't it still open to the charge that it had ignored what other news organizations considered important?

As difficult as these choices were, virtually every major media outlet in the country, save a few avowedly anti-Clinton newspapers, restrained themselves. "Even under tremendous pressure, the story never popped into the mainstream news. You have Jay Leno doing monologues three or four nights in a row and Rush Limbaugh putting this stuff out undigested and as fact, yet it never saw the light of day. . . . That's pretty good," said Tom Fielder, editorial page editor of the *Miami Herald* (as quoted in Shepard, 1999, p. 29).

Making Ethical Decisions

Professionals can take many routes and apply various theories of moral reasoning in arriving at an ethical decision. Different people confront difficult situations in different ways. But as we've already seen, moral reasoning is the use of *systematic* thought in making ethical decisions, especially when confronted by particularly troublesome dilemmas. There are, therefore, a number of different systems for analyzing ethical problems.

Louis W. Hodges (1999), Knight Professor of Ethics in Journalism at Washington and Lee University, proposes seven questions media professionals should ask themselves when confronted by an ethical dilemma:

1. What is the moral issue at stake?

2. Who will be helped or hurt by my action?

3. Do I have all the morally relevant facts?

4. What are the possible courses of action open to me?

5. What are the consequences of each of those courses of action; what are the predictable effects?

6. Which set of possible outcomes is relatively better?

7. Will my decision pass the test of publicity? That is, if "the decision and how I made it were plastered on the front page of the *Wall Street Journal,* would reasonable people likely approve of it?" (p. 4)

Alex S. Jones (1999), Pulitzer prize–winning journalist and media ethics professor at Duke University, proposes what he calls the "Integrity/Judgment Grid." In explaining the grid, Jones argues that media practitioners "function at all times on three moral planes: as professionals, as citizens, and as human beings. Sticky ethical problems almost always involve making a judgment about which of these standards to apply when they are in conflict" (p. 63).

Return to the account of the false story regarding President Clinton's illegitimate child. The moral agent as professional, seeing the original *Drudge Report* claim, might want to beat the opposition to the story or might fear that to ignore the story would suggest a bias. The moral agent as citizen, recognizing the people's need to know about the possible misdeeds of their president, might want to report the story as quickly as possible. The moral agent as human being might want to spare the president, his family, and the whole country the agony of a sex scandal based on an unproved charge.

But a second standard must be considered. Ethical media performance "has two elements that are always present: integrity versus judgment. But unless they are both present in the proper proportion, the outcome can be a mess" (Jones, 1999, p. 63). Back to the love child story. The story was "out there," so the moral agent might show integrity by reporting it. But would this show good judgment? This is a damaging story, from an avowed opponent of the president, based on unsubstantiated rumor. The moral agent

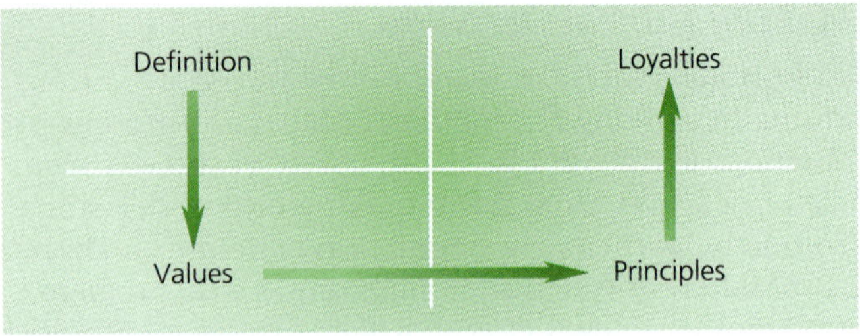

Figure 13.1 The Potter Box
The Potter Box offers one means of systematizing moral reasoning. With permission of Ralph B. Potter.

might decide that ignoring the story showed good judgment—in fact, as we saw, almost all the media did.

Ralph Potter (1972), an ethicist at the Harvard Divinity School, suggests a systematic analysis of *loyalties,* what we identified previously as competing interests. His argument is that it is the competition between these loyalties that most frequently bedevils ethical decision making. Potter's process for making ethical decisions and acting on them involves four dimensions and, when graphically represented, has come to be called the Potter Box (Figure 13.1).

Again using the dilemma of the unsubstantiated report of President Clinton's illegitimate child, we can use the Potter Box to arrive at an ethical decision. The first quadrant, *definition,* requires moral agents to define the situation as accurately and deeply as possible. This raises questions such as these: What is the *Drudge Report*? Who is its author? What are his biases? What are other news outlets doing with the story?

Once a solid definition of the situation is in place, the second quadrant, *values,* calls for moral agents to assess their personal values as they relate to the situation. How highly do they regard privacy, the public's right to know, and professional advancement? What are the moral agents' personal politics?

The third quadrant, *principles,* calls for a consideration of moral agents' principles, or what Christians, Rotzoll, and Fackler (1991) call the *"ethical imperative."* For example, do they subscribe to the principle of the greatest good for the greatest number? Truth above all, regardless of outcome? The Golden Rule—do unto others as you would have them do unto you? Aristotle's golden mean—virtue resides somewhere between two extremes? An advocate of truth above all might come to a much different decision that a follower of the Golden Rule.

The fourth quadrant, *loyalties,* requires moral agents to consider whose interests are at stake and to whom they are most loyal. How will the story, if run on a moral agent's television network, for example, affect the president (the object of the act)? The moral agent him- or herself (individual conscience)? The public's respect for the media (the profession)? Of these, which is the most important to the moral agent? In other words, to which does he or she owe the greatest loyalty?

But systemization is only one strength of the Potter Box. Perhaps its greater value rests in its requirement that the moral agent process the situation again . . . and again . . . and again, each time substituting a different definition,

Heads rolled at NBC News soon after the broadcast of this artificially induced explosion on *Dateline.* But what was the ethical violation? Wasn't it true that GMC trucks like this one sometimes exploded on impact?

value, principle, and loyalty. This forces moral agents to see the situation from a number of different perspectives, possibly some they otherwise might have ignored. Potter (1999) himself suggested an additional benefit of his box. Its use, he wrote, makes it clear "that there can be no fully 'non-moral' view of any . . . conduct (and likewise) that there can be no 'purely moral' view. . . . Those who wish to evade the hard work of coming to grips with facts, with loyalties, with deeper sets of quasi-theological perceptions, are lazy" (p. 6).

Some ethical problem areas frequently faced by telecommunications professionals are presented in the next sections. Regardless of your own preference for systematizing your ethical reasoning, try to use each of the three processes to form your own judgment. Do you come to different conclusions using different means?

Truth and Honesty Can the media ever be completely honest? As soon as a camera is pointed at one thing, it ignores another. As soon as a video editor combines two different images, she has imposed her definition of the truth. Truth and honesty are overriding concerns for media professionals. But what is truth? In 1992 *Dateline NBC* reported on the safety problems of GMC pickups.

Many people believed the trucks were prone to explode in a crash because of deficiencies in design. In fact, several of the trucks had done so. For the story program producers hired a company to stage what would happen in a wreck; in other words, they re-created what they believed to be the truth. To ensure the proper footage, the firm used spark igniters to guarantee the blast. GMC threatened to sue NBC, the network apologized, the president of the news division was forced to resign, and several people were fired. The broadcast report was not false; the videotaped truck exploded, as had other similar vehicles. But was it honest? Fortunately, there is some guidance. The default convention in contemporary telecommunications is to never fake a news event.

How is the *Dateline* story different from the not uncommon practice of **news staging,** or re-creating some event that is believed to have happened or could have happened? Sometimes news staging takes simple forms. For example, a reporter may narrate an account of an event he or she did not witness while video of that event is played. The intended impression is that the reporter is on the scene. What harm is there in this? It's common practice on virtually all U.S. television news. But how much of a leap is it from that to ABC's broadcast of reporter Cokie Roberts, wrapped tightly in winter clothes, seemingly reporting from Capitol Hill on a blustery January 1994 night when in fact she was standing in a nearby Washington studio, her "presence" at the scene staged by digital technology?

The broadcasters' defense is that this is not staging in the sense that the *event* was staged. What does it matter if the reporter was not actually on the spot? What was reported actually *did happen.* Some viewers might accept this argument, but others might not. What differentiates this routine news staging from the news staging practiced by *Dateline*? The producers of the *Dateline* truck explosion justified their actions with the argument that similar explosions *had happened.* A Denver news show in the mid-1990s ran footage of a pit bull fight it had arranged and defended its action on the ground that these things *do happen. ABC Evening News* simulated surveillance camera recordings of U.S. diplomat Felix Bloch handing over a briefcase to a shady character on the street to accompany a 1989 report on Bloch's arrest on espionage charges. The simulation was accurate down to the detail of a surveillance camera's dark, grainy look and the racing time code numbers in the corner of the picture. This staging was justified with the claim that it *could have happened.*

Where do media professionals draw the line? What happens to the public's trust in its news sources as the distinctions between fact and fiction, reality and illusion, and reporting and re-creating disappear? What becomes of their trust in the media in general? Does labeling staged events as "re-creations" or "simulations" solve the problem? If viewers see a televised news story labeled as re-creation or simulation, what leads them to trust the re-creator's or simulator's version? News producers must balance service to the public against ratings and profit, but viewers must balance their desire for interesting, stimulating visuals against their confidence that the news is reported rather than manufactured.

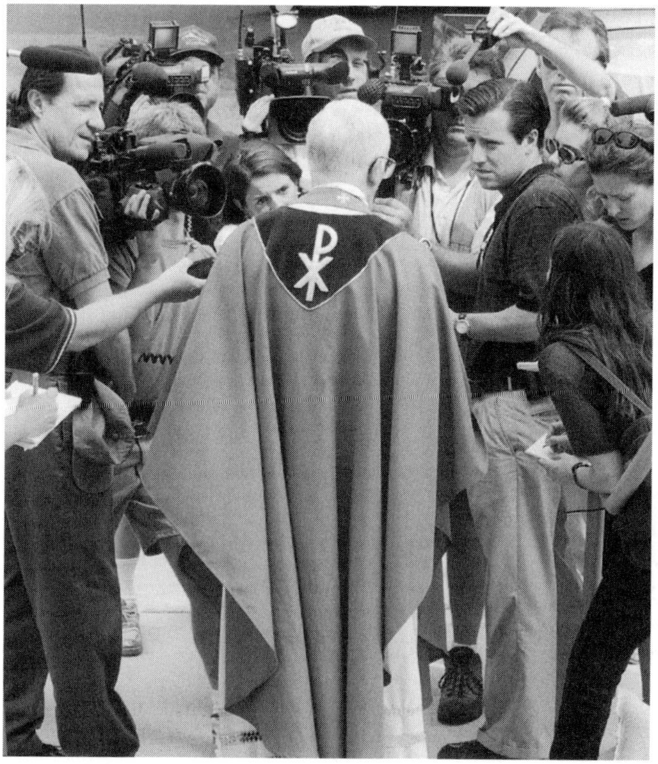

Ethical issues in media, especially the issue of personal privacy, were thrust into open debate with the media frenzy that followed the plane accident and death of John F. Kennedy, Jr. Here, a pack of reporters swarms around the Reverend Edward Byington, the priest of the church where the Kennedys worship when summering on Cape Cod.

Privacy Do public figures forfeit their right to privacy? If so, in what circumstances? Are the president's marital problems newsworthy if they do not get in the way of the job? Who is a public figure? When are people's sexual orientations newsworthy? Should the names of juvenile offenders or women who have been raped be reported? What about sex offenders? When is secret taping permissible?

Our culture values privacy. We use privacy to control the extent and nature of interaction we have with others. Privacy protects us from unwanted government intrusion. By their very nature, however, the media are intrusive. Privacy proves to be particularly sensitive because our culture sees it as a fundamental value. Yet the applied ethics of the various telecommunications industries support, and in fact sometimes demand, that privacy be denied.

The media have faced a number of very important tests regarding privacy in the last few years. Media pursuit of celebrities and the propriety of ride-alongs have generated much comment inside and outside the telecommunications industries.

The death of John F. Kennedy, Jr. in July 1999 put one issue of media ethics—balancing press freedom against personal privacy—squarely in the public spotlight. Kennedy, his wife, and his wife's sister had died in a plane crash, and hundreds of still and video photographers descended on the grieving

Reality television shows like *Cops* may be popular with viewers, but the ride-alongs that provide their video footage raise important ethical questions concerning rights of the accused.

family. The problematic ethical issue became the right to privacy of the Kennedy family children. Reputable media organizations such as CBS's *48 Hours* aired aerial video of the family home as loved ones gathered to console one another. *People* magazine ran cover and inside pictures of the distraught children. Even the venerable *New York Times* put a photo of one of the children leaving the memorial church service on its front page.

Another privacy issue became a hotly debated ethical issue because of a 1999 Supreme Court decision. **Ride-alongs,** the practice of allowing television reporters to accompany police in the conduct of their duty, are the staple of low budget reality television shows like *Cops.* But lawsuits against two well-respected media companies, CNN and the *Washington Post,* forced the Supreme Court to decide if the practice posed a legal rather than an ethical problem. Media professionals need access to police activities, argued the media; journalists are a powerful check on police abuse. Furthermore, they argued, the decision of when to ride along with police and when to enter an alleged crime scene with them should depend on the media professional's moral judgment.

In the case of CNN, reporters accompanied federal Fish and Wildlife agents on a raid against Montana ranchers suspected of poisoning eagles. CNN ran the footage of the raid numerous times. The ranchers were eventually acquitted of killing eagles but were found guilty of misusing pesticides. The *Washington Post* case involved a reporter and a photographer accompanying Maryland police to the home of a couple whose son they were hunting. The *Post* never used photos of the couple—he in his underwear, she in her nightgown—forced to the floor at gun point. The suits of both the ranchers and the couple (whose son did not live with them) produced a unanimous Supreme Court decision taking ride-alongs out of the realm of ethics and making them illegal unless subjects grant permission for the use of the collected pictures. In balancing the public's right to know against the privacy interests of those suspected of wrongdoing, privacy prevailed.

Confidentiality An important tool in contemporary news gathering and reporting is **confidentiality,** the ability of media professionals to keep secret the names of people who provide them with information. Without confidentiality employees could not report the misdeeds of their employers for fear of being fired; people would not tell what they knew of a crime for fear of retribution from the offenders or unwanted police attention. The anonymous informant nicknamed "Deep Throat" would never have been free to divulge the Nixon White House's involvement in the Republican break-in of the Democratic Party's Watergate campaign offices were it not for the promise of confidentiality from the *Washington Post*'s Carl Bernstein and Bob Woodward.

But how far should reporters go in protecting a source's confidentiality? Should they go to jail rather than divulge a name? Thirty-one states and the District of Columbia have established **shield laws,** which is legislation that expressly protects reporters' rights to maintain sources' confidentiality in courts of law. There is no shield law in federal courts, and most journalists want it that way. Their fear is that once Congress makes one "media law" it may want to make another. For example, media professionals do not want the government to legislate the definition of "reporter" or "journalist."

The ethics of confidentiality are also tested by reporters' frequent use of quotes and information from "unnamed sources," "sources who wish to remain anonymous," and "inside sources." The guarantee of anonymity is often required to get the necessary information. But is this fair to those who are commented on by these nameless, faceless news makers? Don't these people—even if they are highly placed and powerful themselves—have a right to know their accusers? You can read more about the use and abuse of confidentiality in the box titled Media Ethics in the Age of the Internet.

Conflict of Interest Ethical decision making requires a balancing of interests. But what of a media professional's own personal conflicts of interest? Should media personalities accept speaking fees, free travel, and other gifts from groups and corporations that they may later have to examine? Is it proper for them to fail to disclose the sources and amounts of such gifts? For example, in mid-1999 CNN senior analyst Jeff Greenfield moderated a conference on the interaction between journalists and public officials during a biotechnical disaster. He later learned that the meeting had been funded by the United States Department of Defense (DoD). He immediately returned his $10,000 fee, prompting the cable network's vice president of public affairs to comment,

The Real World *Media Ethics in the Age of the Internet*

The 1998 sex scandal that embroiled the Clinton White House offers a good case study of the ethical dilemma inherent in the use of unidentified sources. It also provides some insight into how media ethics are being changed by the Internet.

Although sometimes quite valuable, most major news operations try to avoid the use of anonymous sources. Sources who request anonymity may do so based more on their desire to disguise inaccuracy and falsehood than on their need for protection. When the use of an anonymous source is necessary to an investigation, most media organizations will not use the information the source provides unless there is corroboration from at least one other source. In addition, major journalistic operations have a series of internal checks and balances—reporters, a hierarchy of editors, and ombudspeople—to ensure that the material they publish is factual, even when the source cannot be identified. Audience members can then make their own decision on whether to accept the account by evaluating the ethical standards being applied by media professionals. Some readers and viewers may not accept information from anonymous sources; others may. But all understand that the outlet that produced a given story did its best to assure veracity and accuracy. The Internet, however, confounds this situation.

Much of the reporting on the alleged moral and ethical lapses of the Clinton White House was based on information provided by unnamed or anonymous sources and was not subjected to the news industry's normal internal ethical checks and balances. Several of the most dramatic stories originated with the *Drudge Report*, an online newspaper published by Matt Drudge and made available for free by America Online to its subscribers.

Drudge, whose only previous media experience was working in a CBS gift shop, had no editors to check his work. No advertisers were dependent on his credibility for their own. Bound by none of the conventional institutional controls faced by most journalists, Drudge operated free of traditional notions of media ethics. There were "no editors casting a skeptical eye over one's copy, no in-house counsel sweating over the possibility of lawsuits, no publisher worrying about confrontations with aggrieved sources at a dinner party" (Levy, 1998, p. 78). For example, the *Drudge Report* released a

"We don't accept money from people we report on (the DoD). We try to avoid even the appearance of a conflict of interest. Jeff did the right thing by giving the money back" (Wolper, 1999, p. 9). So concerned is CNN over the ethical implications of speaking fees that it has an internal ethics committee charged with approving all network personnel's paid and unpaid appearances. Many media organizations have adopted similar safeguards, but others have not.

Close cooperation between advertisers and media organizations raises the ethical issue of institutional rather than personal conflict of interest. For example, the November 1998 return to space of Senator John Glenn received much media attention. Cable network CNN gave the shuttle flight significant coverage. The sponsor of that coverage was aerospace company Lockheed Martin, not only maker of the space shuttle's external fuel tank but manager, under contract from NASA, of the launch itself. In essence, argued many media ethicists, CNN permitted a company to sponsor news coverage of its own business activities. Would CNN, for example, have allowed a pesticide manufacturer to sponsor newscasts of the Montana ride-along discussed previously?

story that Clinton aide Sidney Blumenthal was a wife beater. Sources were "unnamed top GOP operatives." There was no corroboration, and Drudge printed his story without talking to Blumenthal. The story was false and Drudge later retracted it. Yet before the retraction the traditional media were forced to report, if not the charge, at least the fact that an online newspaper had made the charge (the story was "out there"). Meanwhile, Mr. Blumenthal and his wife were the focus of false and embarrassing scrutiny, and the White House suffered another hit to its reputation.

The old ethical standards may no longer apply. As *Newsweek* media writer Steven Levy (1998) reported, with online "journalists" like Drudge, "unchecked allegations are news simply because they are alleged" (p. 78). Especially in the age of the Internet's instant 24-hour-a-day, seven-days-a-week reporting, media audiences must develop their own standards for accepting news based on anonymous sources and apply them with diligence across all media, making sure that they understand how different media bring with them varying degrees of commitment to ethical performance.

But media professionals have a difficult task as well. Imagine yourself as the news director of a major market, top-rated All News radio station. You are alerted to the *Drudge Report* accusations against Blumenthal by a wire service story. You read that it is based on anonymous sources and that there has been no other verification offered by any other news organization anywhere. Do you air the story? What would you emphasize about this story? Would you focus on Blumenthal as a wife beater or on the fact that an Internet news site was making this accusation? Remember, just about every other news organization around you is running some version of the story. If you run the story, and it turns out to be false, your station's credibility is jeopardized. If you don't run it, and it turns out to be true, your station's credibility is jeopardized. Make your decision now. Yes, now. Then revisit this dilemma using one of the ethical decision-making schemes discussed earlier in this chapter to see if you might come to a different conclusion.

Another institutional conflict of interest issue involves **checkbook journalism,** in which sources are paid for information or interviews. Critics contend that these payments undermine sources' reliability. The interests in conflict here are truth and money. Might sources embellish their accounts to command higher fees? One problem media organizations face is determining just what constitutes checkbook journalism. Paying for information and interviews is checkbook journalism. But what about paying for news leads and tips or background information? Does paying for photos and video constitute checkbook journalism? Is it checkbook journalism when ABC, for example, paid $16,000 to 18-year-old Nathan Dykeman, best friend of Columbine High School gunman Dylan Klebold? The network claimed it paid only for a homemade music video Dykeman made of Klebold in which the murderer acted out a violent fantasy at Columbine. The tape was necessary, said ABC, "to bring context to the Dykeman interview" for which Dykeman was not paid ("Put Away," 1999, p. 12). In other words, ABC paid the subject of a news interview for his videotape, it didn't pay him for the interview.

Profit and Social Responsibility Media industries are just that, industries. They exist not only to entertain and inform their audiences but also to make a profit for their owners and shareholders. What happens when serving profit conflicts with serving the public?

In 1994 ABC's *Day One* ran an investigation called "Smoke Screen," detailing tobacco industry efforts to manipulate the levels of nicotine in cigarettes. Its reward was a 16-month, multimillion-dollar court fight with cigarette maker Philip Morris, which sued the network for $10 billion. The network defended the accuracy of the report and its value to the public. Outside experts agreed that the network would prevail.

In August 1995, however, ABC stunned observers by settling out of court and issuing an apology for the story. Many people believed ABC needed to cut its financial losses because it was about to become part of the Disney communications empire. With the case settled, ABC became more highly valued by its new corporate heads. ABC's decision came under even more scrutiny in the following months as the court released tobacco industry documents demonstrating that the story, as the network had maintained, was true and correct.

A similar set of circumstances became the focus of the Academy Award–nominated 1999 movie *The Insider.* In late 1995 executives of CBS killed an exclusive *60 Minutes* interview with Jeffrey Wigand, a former Brown & Williamson tobacco company executive, who told anchor Mike Wallace that cigarette manufacturers manipulated nicotine levels and lied under oath before Congress. Network officials claimed they only wished to save CBS from a multibillion-dollar lawsuit brought by Brown & Williamson with whom Wigand had signed a nondisclosure agreement. Many observers at the time—and many movie-goers 4 years later—believed the real fear was that a potential lawsuit would reduce the value of CBS executives' stock.

Concerned that this self-censorship went beyond the well-publicized *Day One* and *60 Minutes* incidents, *Columbia Journalism Review* commissioned a survey of 300 journalists. The survey revealed that nearly 25% of "local and national journalists say they have purposely avoided newsworthy stories, while nearly as many acknowledged that they have softened the tone of stories to benefit the interests of their news organizations" (Alterman, 2000, p. 12). Lowell Bergman (2000), executive producer of the now-famous *60 Minutes* smoking story, places the blame of this state of affairs in part on conglomeration and concentration:

> No assignment editor . . . or executive producer . . . would hesitate to pursue, or spend money pursuing, enterprise stories about the Monica Lewinsky scandal, O. J. Simpson, the Princess Di car crash, or the death of JFK Jr. Unfortunately, that commitment is not matched when it comes to stories about who wields unaccountable power over our lives; or who is behind the corporate mergers that now dominate the media. (p. 50)

Balancing profit and social responsibility is not just a concern for journalists. Practitioners in other aspects of telecommunications also face this dilemma. Should a television station advertising manager accept commercials for sug-

The Academy Award–nominated *The Insider* tells the story of the ethical dilemmas (and failings) of CBS's venerable news show *60 Minutes* in its coverage of Big Tobacco.

ared children's cereals, even though doctors and dentists consider these products unhealthy? Does a syndication company distribute the 1950s television show *Amos 'n Andy,* knowing that it embodies many offensive racist stereotypes?

Balancing profit and the public interest does not always involve big companies and millions of dollars. Media practitioners often face this ethical dilemma at a very personal level. What would you do in this situation? The news director at the station where you work has ordered you to produce a story about the 14-year-old daughter of your city's mayor. The girl's addiction to amphetamines is a closely guarded family secret, but it has been leaked to your station. You believe this child is not a public figure. Your boss disagrees, and the boss *is* the boss. By the way, you've just put a down payment on a lovely condo, and you need to make only two more installments to pay off your new car. What would you do?

Offensive Content Media professionals must often make decisions about the potentially offensive nature of content. Other than the particular situation of broadcasters discussed in Chapter 12, this is an ethical rather than a legal issue. Offensive content is protected. Logically, we do not need the First Amendment to protect sweet and pretty expression. Freedom of speech and freedom of the press exist expressly to allow the dissemination of material that *will* offend. But what is offensive?

Howard Stern's radio and television shows are offensive to many people. Critics call them rude, sexist, and racist. Yet the shows are successful in terms of ratings—people watch and listen, advertisers buy time, and broadcasters make money. But is that enough? Programming decisions often entail ethical dilemmas.

Clearly, what is offensive to one person may be quite satisfactory to another. Religious leaders have attacked Mighty Mouse cartoons for their glorification of drug usage (Mighty's flower sniffing was akin to snorting cocaine, they reasoned) and the Teletubbies for their espousal of homosexuality (the purple Teletubby carries a purse and wears a triangular antenna, the triangle being a symbol for homosexuality). Television stations and networks regularly bleep cuss words that are common on cable television and in the school yard. Our culture sanctions the death penalty but is unwilling to view it on television. Where do professionals draw the line? *How* do they draw the line? Do they consider the tastes of the audience? If so, which members of the audience—the most easily offended? What of the rights of the others? These are ethical rather than legal determinations. Revisit the programming dilemma presented in Chapter 10. Would you air "Goodbye Earl"? You can further test your own ethics as a programmer using the box titled Doing Bad to Do Good.

Codes of Ethics and Self-Regulation

To aid practitioners in their ethical reasoning, all major groups of media professionals have established formal codes or standards of ethical behavior. Among others, there is the Society of Professional Journalists' (SPJ) *Code of Ethics,* the Radio-Television News Directors Association's (RTNDA) *Code of Broadcast News Ethics,* and the National Association of Broadcasters' (NAB) *Statement of Principles of Radio and Television Broadcasters.* These are prescriptive codes that tell media practitioners what they should and should not do. For example, the SPJ code states that "[j]ournalists must be free of obligation to any interest other than the public's right to know the truth." The SPJ code then specifies these rules:

- Gifts, favors, free travel, special treatment or privileges can compromise the integrity of journalists and their employers. Nothing of value should be accepted.
- Secondary employment, political involvement, holding public office, and service in community organizations should be avoided if it compromises the integrity of journalists and their employers. Journalists and their employers should conduct their personal lives in a manner that protects them from conflict of interest, real or apparent. Their responsibilities to the public are paramount. That is the nature of their profession.
- So-called news communications from private sources should not be published or broadcast without substantiation of their claims to news value.
- Journalists will seek news that serves the public interest, despite the obstacles. They will make constant efforts to assure that the public's business is conducted in public and that public records are open to public inspection.
- Journalists acknowledge the newsman's ethic of protecting confidential sources of information.
- Plagiarism is dishonest and unacceptable. (Baran, 1999, pp. 477–478)

The RTNDA code requires that members follow these guidelines:

- Strive to present the source or nature of broadcast news material in a way that is balanced accurate and fair.
- Strive to conduct themselves in a manner that protects them from conflicts of interest, real or perceived. They will decline gifts or favors which would influence or appear to influence their judgments.
- Respect the dignity, privacy and well-being of people with whom they deal.
- Recognize the need to protect confidential sources. They will promise confidentiality only with the intention of keeping that promise.
- Respect everyone's right to a fair trial.

Critical Issues *Doing Bad to Do Good*

A major debate over media ethics broke out in 1998 when it was discovered that a *Cincinnati Enquirer* reporter had illegally accessed voice-mail messages at Chiquita Brands International. His story was a good one and an important one; it detailed serious wrongdoing by the giant banana company. But the reporter's reliance on illegally obtained material tainted the story. The reporter was fired for violating an important ethical standard—journalists should not break the law to get a story. This seems logical. People shouldn't do bad to do good. But sometimes the bad isn't that bad, and the good is very good. Where should media professionals draw the line?

Media professionals face such a dilemma every time they run a public service announcement for African famine relief. There is significant debate among international relief organizations and media outlets about the ethics of using highly emotional media appeals to raise money for starving people. Which images should be shown, and what is the "truth" they convey?

Rock 'n' roller Bob Geldof recruited his friends in the music business to start the relief organization Band Aid. Together they produced the record and video *We Are the World,* raising money and awareness of the plight of starving people around the world.

There is no question that using pictures of emaciated, starving children can generate interest, sympathy, and dollars. During the 1980s, for example, horrific news pictures of dying Ethiopian children and their helpless skin-and-bones parents generated millions of dollars of aid for that famine-struck region. After seeing these televised images, Irish rock 'n' roller Bob Geldof leapt into action, calling on his friends in the music business to lend a hand. They quickly created the relief organization Band Aid and its megahit record and video *We Are the World.* Writing in *The Nation,* Africa observer and author James North (1999) commented, "Geldof and his friends ended up changing history. . . . [T]he singers raised millions themselves, but even more

- Broadcast the private transmissions of other broadcasters only with permission.
- Actively encourage observance of this Code by all journalists, whether members of the Radio-Television News Directors Association or not. (Baran, 1999, pp. 482–483)

The NAB statement of principles offers several programming guidelines, including these on violence:

important, they helped create a wave of publicity that shamed the Reagan Administration and other Western governments into greatly increasing aid. In the end, hundreds of thousands of Africans did die. But hundreds of thousands more survived" (p. 28).

So where's the problem? The ethical dilemma resides in the fact that although there are starving people in most developing countries, these countries consist of much more than children dying in the dust. Images of helpless people in need can generate millions of dollars to help them, but just as certainly these images reinforce a stereotype—a deeply entrenched "truth" held by outsiders that these are hopeless people struggling in hopeless places. North calls this "a big lie." "Over the past couple of decades, the most significant development in the Third World has been the rise of independent grassroots organizations—labor, education, ecology, feminist and human rights groups, which are changing the political scene" in these regions (p. 29). Relief organizations and media professionals know that images of industrious people using aid money to improve their situation—farming, harvesting, building homes and barns, constructing wells and bridges—can paint a more complete and accurate picture of those people and their condition, but they also know that these images produce little giving.

What do you do? Imagine yourself as the advertising director for a new ad-based cable channel. You haven't sold all your avails, so you have quite a few spots open for public service announcements. Among those you are considering for use is one that presents startling pictures of grief and hopelessness, designed to raise money for drought and famine relief in central Africa. You're a skilled advertising professional, and you know this is a powerful piece, one that will do the job, bring in a lot of money, and help these folks out. This emotional spot ends with the words, "Your few pennies can save a life." But you are troubled because you believe reinforcing harmful stereotypes, especially those with racial and ethnic overtones, is wrong. It is a bad thing to do. But you also know that this spot will do good; it will produce money that can alleviate much suffering.

You consider a second PSA; this one shows a sub-Saharan village making flat breads in a bakery built with and funded by relief money. The PSA concludes, "Your contributions have put these people on the road to self-sufficiency. As Jesus said, 'Give a man a fish, he eats for one day. Teach him to fish, and he eats forever.'" Running this spot frees you from doing bad (no negative stereotypes here), but you question the amount of good it will do (it will not generate as much money).

You can run both. Each is from a reputable and long-established relief organization. Doesn't each show a different "truth" about the region? But your general manager has expressed concern that you're too fixated on "those people." There are just as many needy people and groups here in the United States, she warns you. You're the moral agent; what do you do?

Violence, physical or psychological, should only be portrayed in a responsible manner and should not be used exploitatively. Where consistent with the creative intent, programs involving violence should present the consequences of violence to its victims and perpetrators. Presentation of the details of violence should avoid the excessive, the gratuitous and the instructional. The use of violence for its own sake and the detailed dwelling upon brutality or physical agony, by sight or by sound, should be avoided. Particular care should be exercised where children are involved in the depiction of violent behavior. (<http://www.nab.org>)

To some, these codes are a necessary part of a true profession. They offer at least two important benefits to thoughtful media practitioners: they are an additional source of information to be considered when making moral judgments, and they represent a particular media industry's best expression of its shared wisdom. To others, however, they are little more than unenforceable collections of clichés that restrict constitutional rights. Ethicists Jay Black and Ralph Barney (1985–86) argue, for example, "The fact should be evident that the First Amendment has a primary purpose of protecting the distribution of ideas . . . from restriction efforts by legions of 'regulators.' Ethics codes should be considered among those 'regulators'" (p. 28). They continue, "It is indeed not difficult to find examples of codified professional ethics that ultimately become self-serving. That is, they tend to protect the industry, or elements of the industry, at the expense of individuals and other institutions, even of the full society" (p. 29).

Another criticism of ethics codes is that they are an open invitation for lawsuits from outsiders. Media industry lawyers "worry that plaintiffs' lawyers will focus on deviations from the guidelines as evidence of actual malice, negligence, or some other fault standard. Depending on how the guidelines are worded, you might even cobble together a breach-of-contract theory based on 'I promised my (audience) I would do X, I didn't, and therefore, I breached my contract with my (audience),'" explained Jane Kirtley, executive director of the Reporters Committee for Freedom of the Press. This argument, she continued, "would probably be thrown out and could be averted by careful wording. . . . But in these days of crazy lawsuits, I don't think I would rule it out" (as quoted in Noack, 1999, p. 9). There is fear that published codes or standards could be used against the organization as evidence that it failed to follow its own rules.

Other Forms of Self-Regulation

In addition to industry professional codes many individual media organizations have formulated their own institutional policies for conduct. In the case of the broadcast networks these are enforced by **Standards and Practices departments.** Local broadcasters have what are called **policy books,** which set out the standards for that station. Some media organizations also utilize **ombudspeople,** practitioners internal to the company who serve as "judges" in disputes between the public and the organization. Some subscribe to the small number of existing **media councils,** which are panels of people from both the media and the public who investigate complaints against the media and publish their findings. There are few media councils because media organizations fear that their information will be used in lawsuits (similar to the fear of using published codes). American Society of Newspaper Editors' president Edward Seaton expressed the view that councils "soon are subverted into quasi-extensions of our legal system. Judges and lawyers have used ethical guidelines (and) council statements . . . as evidence and the basis for court decisions against the press. What is intended as voluntary becomes coercive" (as quoted in Childs, 1998, p. 11).

Limitations of Self-Regulation

These various mechanisms of normative ethics are a form of self-regulation, designed in part to forestall more rigorous or intrusive government regulation. In a democracy dependent on mass communication, they serve an important function. We are suspicious of excessive government involvement in media. Self-regulation, however, has certain limitations:

- *Media professionals are reluctant to identify and censure colleagues who transgress.* To do so might appear to be admitting that problems exist; whistle-blowers are often met with hostility from their peers.
- *The standards for conduct and codes of behavior are abstract and ambiguous.* Many media professionals see this flexibility as a necessary evil; freedom and autonomy are essential. Others believe their lack of rigorous standards renders the codes useless.
- *Media practitioners are not subject to standards of professional training and licensing as stringent as those of law or medicine.* Some practitioners view standards of training and licensing as limiting media freedom and inviting government control. Others argue that licensing has not had these effects on doctors and lawyers.
- *Media practitioners often have limited independent control over their work.* Media professionals are not autonomous, individuals but are part of large, hierarchically structured organizations. It is often difficult to punish violations of standards because of the difficulty in fixing responsibility.

Critics of self-regulation argue that these limitations are accepted willingly by many media practitioners because the "true" function of self-regulation is "to cause the least commotion" for those working in the media industries (Black & Whitney, 1983, p. 432). Nonetheless, even for telecommunications professionals who accept this negative or cynical view of self-regulation, the decision to perform their duties in an ethical manner ultimately rests with themselves. As Black and Barney (1985–86) explained, an ethical media professional "must rationally overcome the status quo tendencies (in order) to become the social catalyst who identifies the topics and expedites the negotiations societies need in order to remain dynamic" (p. 36).

Chapter Review

Because the press is not totally free, libertarianism gave way in the 1940s to social responsibility theory, the idea that to earn their freedom the media must perform responsibly, that is, ethically. Ethics operate at three levels: metaethics, normative ethics, and applied ethics. Applied ethics demand that media professionals, as moral agents, balance the conflicting interests of their consciences, people affected by their actions, their financial supporters, the institutions for which they work, their profession, and society as a whole. Moral decision making can be guided by a number of theories of ethical behavior, including deontological, teleological, and virtue theories.

Several systematic techniques can be used in making ethical decisions. Hodges suggests asking seven questions: (1) What is the moral issue at stake? (2) Who will be helped or hurt by my action? (3) Do I have all the morally relevant facts? (4) What are the possible courses of action open to me? (5) What are the consequences of each of those courses of action? (6) Which set of possible outcomes is relatively better? (7) Will my decision pass the test of publicity?

Jones offers the Integrity/Judgment Grid in which media practitioners approach ethical dilemmas on three moral planes—as professionals, as citizens, and as human beings—and then attempt to find a balance between integrity and judgment for the answers they find there. The Potter Box provides a system for passing an ethical dilemma through four phases of analysis—definition, values, principles, and loyalties—and then passing it through again, altering elements of each phase.

These and other means of ethical reasoning can be applied to a number of situations typically faced by media professionals. The issue of truth and honesty can encompass outright fakery and news staging. Privacy is a thorny issue when covering celebrities as well as when media intrude on people who are not public figures. Confidentiality is necessary for much good journalism, but it is often controversial. Conflict of interest can operate at both personal and institutional levels. The use of potentially offensive content raises many ethical questions, and media practitioners are sometimes called on to choose between profit and public service.

Media practitioners are aided in their moral decision making by formal codes of conduct and their own institutional policies. These forms of self-regulation are controversial, however, in that they are seen by some as necessary and by others as limiting, abstract, and unenforceable. Even self-regulation itself is problematic in that many observers see it, at best, as unenforceable and, at worst, as a self-serving cover for questionable media activities.

Questions for Review

1. What is normative theory?
2. What are the basic tenets of social responsibility theory?
3. What is the relationship between social responsibility theory and ethics?
4. What are metaethics, normative ethics, and applied ethics?
5. What are duty-based, outcomes-based, and virtue theories of moral reasoning?
6. What does it mean that a story is "out there"?
7. What are some of the individual and group interests that often conflict in the application of media ethics?
8. What are Hodges seven questions?
9. What are the three moral planes of Jones' Integrity/Judgment Grid?
10. What is the relationship between integrity and judgment in Jones' Integrity/Judgment Grid?
11. What are the four elements of the Potter Box?
12. What is news staging? What ethical issues does it raise?
13. What are video news releases? What ethical issues do they raise?
14. What are ride-alongs? What ethical issues do they raise?

15. What is confidentiality? Why is it important to media professionals and to democracy?
16. What are shield laws? Why do some media professionals oppose them?
17. What are some examples of personal and professional conflicts of interest faced by media practitioners?
18. What is checkbook journalism? What ethical issues does it raise?
19. What are the strengths and weaknesses of published codes of ethics?
20. What are Standards and Practices departments, policy books, ombudspeople, and media councils?
21. What are the strengths and limitations of self-regulation?

Questions for Critical Thinking

1. Consider your reaction to the Clinton love child example discussed in this chapter. How do you feel about those few media outlets that did run this unsubstantiated and later disproved story? What do you think of those, the vast majority, that did nothing with the information from the *Drudge Report*? Now consider your answer to these questions in terms of your personal political beliefs. How easily can you put aside your own politics when making these judgments? How well do you think most media professionals do at setting aside their own personal politics? Should they? Defend your position.

2. Can you identify two or three metaethics that you use to guide your moral reasoning? Which is most important to you? Why?

3. Social responsibility theory is the normative theory that theoretically guides the operation of U.S. media industries. How well do you think most media organizations fulfill its spirit? Do some media organizations do a better job than others? Which ones? Why?

4. Of all the groups whose interests must be balanced by media professionals, which ones do you think would have most influence over you, all else being equal?

5. In general, how ethical do you believe media professionals to be? Specifically, online journalists? Television journalists? Advertising professionals? Public relations professionals? Television and film writers?

Engaging Technology

1. A number of public groups are dedicated to monitoring the ethical performance of the various media industries. Among them are the Committee of Concerned Journalists (<http://journalism.org/concern>), the Center for Media & Democracy (<http://www.prwatch.org>), and Fairness and Accuracy in Reporting (<http://www.fair.org>). Visit their Web sites to determine the issues of greatest interest to each. Do you share their concerns? Do you agree with their ethical judgments?

2. The National Association of Broadcasters and the Radio-Television News Directors Association both have published ethical standards. Visit their Web sites at <http://nab.org> and <http://rtnda.org>, respectively, to access the full text of the NAB statement of principles and the RTNDA code. Having read them in their entirety, do you agree with the criticisms of such standards discussed in this chapter? Explain why or why not.

3. A number of Web sites provide links to dozens of other sites devoted to media ethics. Two of the more thorough are maintained by the Canadian Center for Applied Ethics (<http://www.ethics.ubc.ca/resources/media/topics.html>), and the

communication faculty at the California State University at Fullerton (<http:commfaculty.fullerton.edu/lester/ethics/media.html>). Visit these sites and travel through the scores of links they provide. Identify for yourself those sites that prove most valuable to your own understanding of media ethics. Explain your choices.

4. The Poynter Institute, in its own words, "is a school for journalists, future journalists, and teachers of journalism. It promotes excellence and integrity in the practice of craft and in the practical leadership of successful businesses. It stands for a journalism that informs citizens and enlightens public discourse." Founded by Nelson Poynter, former publisher of the *St. Petersburg Times*, its Web site provides many pages of interest to people studying telecommunications. One in particular, <http:www.poyner.org/research/biblio/bib-me.htm>, offers a wealth of information on media ethics, including links to other ethics-related sites and a fine bibliography of written work on media ethics. Visit this site to locate those resources that can enrich your understanding of media ethics. Which did you find most valuable? Explain.

Chapter Fourteen

The Effects of Telecommunications

The horrific 1999 events at Littleton, Colorado's, Columbine High School once again forced the issue of media effects into public debate. Senator Joseph Lieberman argued, "The public is so fed up with the killing of innocent children and the violence generally in our society, and they're coming to a hard conclusion that the entertainment industry is contributing to those killings" (as quoted in Albiniak, 1999, p. 7). Senator John McCain added, "I think TV is worse than it used to be. I just think the Internet is an added layer. [And] video games are incredibly graphic and violent and there's a competition to see who can be the most graphic and violent" (p. 7). First Lady Hillary Rodham Clinton was equally hard on the media, "We can no longer shut our eyes to the impact that the media are having on all of our children and the potentially violent impact they're having on some" (p. 7), as were Linda Ellerbee, host of Nickelodeon's *Nick News,* "You can't have millions of kids every day watching people get killed on TV, and then see the same people alive the next week without [desensitizing] kids to violence" (p. 7), and Elise O'Shaughnessy of the *New York Daily News,* "All we can say is that violence on the screen—movie, TV, computer—is probably an accelerant to violence in the mind. And that Eric Harris, 18, and Dylan Klebold, 17, had seen more than their share" (p. 8).

The April 1999 shootings at Littleton, Colorado's, Columbine High School sparked outrage and debate over media portrayals of violence.

But not everyone blamed the media for the mayhem. *Los Angeles Times* television writer Howard Rosenberg offered, "Some myopic Americans see media as the root of all evil. The mantra: If it's bad, television—or the movies—caused it. This becomes their knee-jerk reaction to every violent tragedy" (p. 8). USA Network copresident Steve Brenner defended his industry, "Every American has seen hundreds of films, hundreds of news stories, hundreds of depictions, thousands of cartoons. Millions don't go out and shoot people" (p. 8).

Preview

Media's effects are the focus of this chapter. We examine the arguments of those who believe mass media can and do have important personal and societal effects and those of people who see media's impact as limited. To help evaluate these claims and counter claims about media influence, we discuss the difference between *effects* and *cause*. Then we look in depth at the effects of issues that receive the greatest amount of attention: violence, with special attention to social cognitive theory; drugs and alcohol; gender and racial stereotyping; political campaigns and voting; and media's prosocial effects. Finally, we turn our attention to the Internet and one of the most hotly debated effects arguments, the Net's impact on democracy and self-governance.

The Effects Debate

Whether the issue is Columbine and televised violence, online hate groups, teen drinking, the absence of minorities in prime time network programming, or the decline in the quality of political discourse, the impact of the mass media is hotly debated today. Legislation to address some of these effects was passed in 1996, 1997, and 1998: the Telecommunications Act, which in part addressed the issue of televised violence; the introduction of the television program content rating system; passage of the Child Pornography Act, designed to protect kids from online images of sex; Congressional plans to control cigarette advertising and to require filtering software on school and library computers; and the Supreme Court's dismissal of the Communications Decency Act, intended to control Internet content. In 1999 there were Senate hearings on violent popular culture and youth; President Clinton's "White House Summit on Children and Violence"; stepped up efforts by the FCC to educate parents about television content program ratings and to force producers to employ them more conscientiously; and a $130 million lawsuit against several media companies by the parents of three dead teens in Paducah, Kentucky, killed in yet another school shooting. The assumption behind these efforts at control of media use and content is that media have effects. But the presence, strength, and operation of those effects have long been controversial.

The Pros and Cons of Media Effects

The debate over media's impact has a number dimensions. Here are several of them, with the argument that media have little or no effects stated first, followed by the counterargument.

Argument 1 *Media content has limited impact on audiences because it's only make-believe; people know it isn't real.*

The counterarguments: (a) News is not make-believe (at least it's not supposed to be), and we are supposed to take it seriously. (b) Most film and television dramas, for example, *ER* and *NYPD Blue*, are intentionally produced to seem real to viewers, with documentary-like production techniques such as handheld cameras and uneven lighting. (c) Much contemporary television is expressly *real*—reality shows such as *Cops* and *America's Most Wanted* and talk shows such as *The Jerry Springer Show*, which purport to present real people. (d) Advertising is supposed to tell the truth. (e) Before they develop the intellectual and critical capacity to know what is not real, children confront the world in all its splendor and vulgarity through television, what media effects researchers call the **early window.** To kids, what they see is real. (f) To enjoy what we consume, we **willingly suspend disbelief;** that is, we willingly accept as real what is put before us.

Argument 2 *Media content has little impact on audiences because it is only play or just entertainment.*

The counterarguments: (a) News is not play or entertainment (at least it's not supposed to be). (b) Even if media content is only play, play is very important to the way we develop our knowledge of ourselves and our world. When we play organized sports, we learn teamwork, cooperation, the value of hard work, obedience to authority, and respect for the rules. Why should play be any less influential if we do it on the Internet or with television?

Argument 3 *If media have any effects at all, they are not the media's fault; media simply hold a mirror to society. They reflect the status quo, showing us and our world as they already are.*

The counterargument: Media hold a very selective mirror. The whole world, in all its vastness and complexity, cannot possibly be represented, so media practitioners must make choices (they function as gatekeepers). For example, according to the Center for Media and Public Affairs, coverage of murder on the three evening network news shows *rose* 721% from 1993 to 1996, even though the actual U.S. homicide rate *fell* 20% in the same period (Overholser, 1999). Even though violent crime in the United States dropped to a 23-year low in 1997, the number of violent crime stories broadcast on the Big Three evening news shows rose from 542 in 1990 to 1,392 in 1998. Media, at best, hold a fun-house mirror to society and distort what they reflect. Some things are overrepresented, others underrepresented, and still others disappear altogether.

Argument 4 *If media have any effect at all, it is only to reinforce preexisting values and beliefs. Family, church, school, and other socializing agents have much more influence.*

The counterarguments: (a) The traditional socializing agents have lost much of their power in our complicated and fast-paced world. (b) Reinforcing effects are not the same as having no effects. Media that can reinforce the good in our culture can just as easily reinforce the bad. Is racism eradicated yet? Sexism? Disrespect for others? If our media are doing no more than reinforcing the values and beliefs that already exist, then they are as empty as many critics contend. Former Federal Communications Commission member Nicholas Johnson has long argued of television in particular that the real crime of that medium is not what it's doing *to* us, but what it could be doing *for* us but isn't.

Argument 5 *If media have any effects at all, they are only on the unimportant things in our lives, like fads and fashions.*

The mirror that media hold up to culture is like a fun-house mirror—some things appear bigger than they truly are, some things appear smaller, and some disappear altogether.

The counterarguments: (a) Fads and fashions are not unimportant to us. The cars we drive, the clothes we wear, and the way we look help define us; they characterize us to others. In fact, it is media that have helped make fads and fashions so central to our self-definition and happiness. Kids don't kill other kids for their $150 basketball shoes because their mothers told them Air Jordans were cool. (b) If media influence only the unimportant things in our lives, why are billions of dollars spent on media in efforts to sway our opinions about social issues such as universal health care and nuclear power?

The grist for these debates over media influence comes from scientists who test for the presence or absence of effects using **research,** the objective search for knowledge. Research is conducted by both media industry researchers and academic researchers. A variety of research methods are available for this inquiry. Some are better suited to individual effects questions than others, and some are more objective than others. This is one reason so much controversy exists about our willingness to accept the presence and extent of media effects. People research media's impact from different perspectives, and they use different tools and apply different degrees of objectivity. These various methods often produce contradictory results. Another reason people disagree about whether or not media cause effects is that there is disagreement about what exactly is meant by the words *cause* and *effects.*

Micro-Level Effects Versus Macro-Level Effects

Does television cause violence? Do beer ads cause increased alcohol consumption? Does pornography cause rape? What is meant by the word *cause* in each of these questions? If it means a direct effect—that is, a person sees television violence or even a lot of television violence and then behaves aggressively toward others—then there is much research that demonstrates this causal link and much that does not.

Media researchers attempt to resolve this problem by thinking about effects as operating on either the **micro level** (that is, on individuals) or the **macro level** (that is, on the culture or society in general). At the micro level we know that effects can occur, but they don't necessarily occur for everyone. Return to Steve Brenner's argument from the chapter opening. He is correct: millions of people watch television violence, and the vast majority do not commit violent acts. Dick Wolf, executive producer of the popular but controversial television hit *Law & Order,* made the same point in an interview with cable channel A&E's *Investigative Reports.* He argued persuasively that he "grew up on Tom and Jerry and I haven't climbed into a tower and pulled a gun on people."

Who can argue with this? For most people media have relatively few *direct* effects at the personal or micro level; that is, they don't *cause* people to behave violently. But we live in a culture in which there are people who have shot people or who are willing to use violence to settle disputes, at least in part because of the messages embedded in our media fare. The hidden but much more important impact of media operates at the cultural or macro level. Violence on television contributes to the cultural climate in which real world violence becomes more acceptable. Sure, none of us have shot people from atop a tower. But do we have bars on the windows of our homes? Are there parts of town where we would rather not walk alone? Do we vote for the "tough on crime" candidate over the "education" candidate? The micro-level view is that televised violence does not *cause* real world violence because most people are not directly affected. The macro-level view is that televised violence has a great impact because it influences the cultural climate, making it more likely that violence will occur.

Transmissional Perspective Versus Ritual Perspective

Effects researchers confront the meaning of *effects* by differentiating between two distinct views of how media function in a given society. The **transmissional perspective** sees media as senders of messages, discrete bits of information, across space. The **ritual perspective,** wrote influential media theorist James Carey (1975), views media not as a means of transmitting "messages in space" but as central to "the maintenance of society in time." Mass communication is "not the act of imparting information but the representation of shared beliefs" (p. 6). In other words, the ritual perspective is necessary to understand the *cultural* (or macro-level) importance of mass communication.

Consider a commercial for Michelob beer. What message is being transmitted? "Buy Michelob," and people either do or don't buy Michelob. The message either controls or does not control people's beer-buying behavior; it either has an effect or it doesn't. So, if two people are debating media effects while relying on the transmissional view of media, it is likely that they would come to the conclusion that media have little effect (or else everyone who sees a Michelob spot would run out to buy Michelob). But if these same debaters were to adopt the ritual perspective, they would be less interested in whether or not people bought Michelob and more concerned with the cultural "truths" reflected in that commercial. What reality about alcohol and socializing is shared? Can young people really have fun in social settings without alcohol? What constitutes a good-looking guy or good-looking woman?

Media Effects

Most contemporary effects researchers and theorists combine aspects of the ritual perspective on media and the macro-level view of effects to accept the potential for powerful media impact. They point to a number of weaknesses in the no-effects argument: The same media professionals who promise advertisers significant impact for their money claim little or no impact when it comes to issues such as violence. The same media professionals who boast of the good effects of programming such as their public service announcements deny that their other content can have any harmful effects. Most of the country's companies, governments, and nonprofit groups employ public relations to ensure favorable media coverage. Why would they bother if media had little or no impact? Why would we have a First Amendment protecting media if those media had little impact? Why do we grant media special protection if their contributions to society are so insignificant?

Despite the persistence of disagreement, most serious observers of media and their influence—and this does include many thoughtful media practitioners—have come to agreement on a number of generally accepted explanations of how, when, and to what extent media effects occur. We will look at those that have received the greatest amount of public attention: violence, drugs and alcohol, gender and racial stereotyping, political campaigns, and prosocial effects.

Violence

No media effects issue has captured public, legislative, and industry attention like the relationship between media portrayals of violence and subsequent aggressive behavior. Among the reasons for this focus are the facts that violence is a staple of all our media and that the United States experienced an upsurge in real violence in the 1960s at just about the time television entrenched itself as the country's dominant mass medium.

The prevailing view during the 1960s was that *some* media violence affected *some* people in *some* ways *some* of the time. Given the dominance at the time of the transmissional perspective, researchers believed that for normal people (that is, those who were not predisposed to violence) *little* media violence affected *few* people in *few* ways *little* of the time. However, increases in youth violence, the assassinations of Robert F. Kennedy and the Reverend Martin Luther King Jr., and the violent eruption of cities during the civil rights, women's rights, and anti–Vietnam War movements led in 1969 to creation of the Surgeon General's Scientific Advisory Committee on Television and Social Behavior. Members of this committee had to be approved by the television networks. After 2 years and one million dollars worth of research, the committee produced findings that led Surgeon General Jesse L. Steinfield to report to the U.S. Senate:

> While the . . . report is carefully phrased and qualified in language acceptable to social scientists, it is clear to me that the causal relationship between televised violence and antisocial behavior is sufficient to warrant appropriate and immediate remedial action. The data on social phenomena such as television and violence and/or aggressive behavior will never be clear enough for all social scientists to agree on the formulation of a succinct statement of causality. But there comes a time when the data are sufficient to justify action. That time has come. (Ninety-Second Congress, 1972, p. 26)

Thirty-three years later, in 1992, psychologist Leonard Eron told another Senate hearing, "There can no longer be any doubt that heavy exposure to televised violence is one of the causes of aggressive behavior, crime, and violence in society. The evidence comes from both laboratory and real-life studies. Television violence affects youngsters of all ages, of both genders, at all socio-economic levels and at all levels of intelligence. The effect is not limited to children who are already disposed to being aggressive" (as quoted in Levine, 1996, pp. 3–4). After reviewing the available research, in 1996 child development expert Madeline Levine wrote:

> The debate is over. . . . Every major group concerned with children has studied and issued position papers on the effects of media violence on children. The Surgeon General's Scientific Advisory Committee on Television and Social Be-

No media effects issue has generated as much debate, disagreement, and study as the impact of TV violence. Here, wrestler Hulk Hogan feels the impact of a metal folding chair during a broadcast of *WCW Nitro Monday.*

> havior, the National Institute of Mental Health, the U.S. Attorney General's Task Force on Family Violence, the American Psychological Association, the American Academy of Pediatrics, and the National Parent Teachers Association have all called for curbing television and movie violence. Their findings represent the inescapable conclusions of decades of social science research. (pp. 3–4)

Despite the apparent certainty of these statements, disagreement persists over the existence and extent of the media's contribution to aggressive behavior. Few would argue that media violence *never* leads to aggressive behavior.

The disagreement is about what circumstances are needed for such effects to occur, and to whom.

Under What Circumstances A direct causal relationship between violent content and aggressive behavior, the **stimulation model,** has been demonstrated in laboratory experiments. So has the **aggressive cues model,** the idea that media portrayals can suggest that certain classes of people, for example, women or foreigners, are acceptable targets for real world aggression, thereby increasing the likelihood that some people will act violently toward individuals in these groups.

Both the stimulation and aggressive cues models are based on **social cognitive theory** (originally called social learning theory), which is the theory that people learn through observation. Social cognitive theory argues that people model (copy) the behaviors they see and that **modeling** happens in two ways. The first is **imitation,** the direct replication of an observed behavior. For example, a child might see cartoon cat Tom hit cartoon mouse Jerry with a stick and then hit his sister with a stick. The second form of modeling is **identification,** which is a special form of imitation in which observers do not exactly copy what they have seen but make a more generalized but related response. For example, the child might still be aggressive to his sister but dump a pail of water on her head rather than hit her with a stick.

The idea of identification is of particular value to media researchers who study television's impact on individual, micro-level behavior. Everyone admits that people can imitate what they see on television, but not all do. When this imitation does occur in dramatic instances—for example, when someone hijacks a plane after seeing it done on a made-for-TV movie—it is so outrageous that it is considered an aberration. Identification, although harder to see and to study, is the more likely way that television influences individual behavior.

Social cognitive theorists demonstrated that imitation and identification are the products of three processes:

- **Observational learning:** observers can learn new behaviors simply by seeing those behaviors performed. Many of us who have never fired a handgun can do so because we've seen it done in the media.
- **Inhibitory effects:** seeing a television character punished for a behavior reduces the likelihood that the observer will perform that behavior. In the media we see good Samaritans sued for trying to help others, and it reduces our willingness to help in similar situations. That behavior is inhibited by what we've seen.
- **Disinhibitory effects:** seeing a model rewarded for prohibited or threatening behavior increases the likelihood that the observer will perform that behavior. This is the basis for complaints against the glorification of crime on television, for example. Behaviors people might not otherwise make, because they are inhibited, now become more likely to occur. They are disinhibited.

Social cognitive theory has made three additional contributions to the violence debate. First, it deflated the notion of **catharsis,** the idea that watching mediated violence reduces people's innate aggressive drive. Viewing people eating, for example, does not reduce hunger; viewing people making love does not reduce the drive to reproduce. Scientists were already skeptical, but social cognitive theory provided a scientific explanation for the research that did show a reduction in aggression after viewing violence. This phenomenon was better explained by inhibitory effects rather than by some cathartic power of the media. That is, mediated aggression, if it is portrayed as punished or prohibited, can indeed lead to the reduced likelihood that that behavior will be modeled.

Some people, typically media industry practitioners, to this day defend catharsis theory. But nearly 30 years ago respected media researcher and theorist Joseph Klapper, who at the time was the head of social research for CBS television, told the U.S. Senate, "I myself am unaware of any, shall we say, hard evidence that seeing violence on television or any other medium acts in a cathartic . . . manner. There have been some studies to that effect; they are grossly, greatly outweighed by studies as to the opposite effect" (Ninety-Second Congress, 1972, p. 60).

Second, social cognitive theory introduced the concept of **vicarious reinforcement,** the idea that observed reinforcement operates in the same manner as actual reinforcement. This helped direct researchers' attention to the context in which mediated violence is presented. Theoretically, inhibitory and disinhibitory effects operate because of, respectively, negative and positive vicarious reinforcement. That is, seeing the bad guy punished is sufficient to inhibit subsequent aggression on the part of the viewer. Unfortunately, researchers discovered that when the bad guys are punished in contemporary media presentations they are usually punished by good guys who are more aggressive. Even when media portray punishment for aggressive behavior, they may in fact be positively reinforcing that very same behavior.

Finally, social cognitive theory introduced the concept of **environmental incentives,** the notion that real world incentives can lead observers to ignore the negative vicarious reinforcement they have learned to associate with a given behavior. In 1965 psychologist Albert Bandura conducted a now classic experiment in which nursery school children saw a video aggressor, a character named Rocky, punished for his behavior. These children subsequently showed lower levels of aggressive play than did those who had seen Rocky rewarded. This is what social cognitive theory would have predicted. Yet Bandura later offered "sticker-pictures" to the children who had seen Rocky punished if they could make the behaviors they had watched him make. They all could. Vicarious negative reinforcement may reduce the likelihood that the punished behavior will be performed, but that behavior is still observationally learned. It's just that, at the same time it is observed and learned, observers also learn not to make it. But when the real world offers sufficient reward, the originally learned behavior can be demonstrated.

The Real World *The V-Chip and Program Ratings*

The Telecommunications Act of 1996 contained a controversial provision designed to deal with televised violence. It required that the V-chip, a content filtering technology, be included in all new television sets with screens larger than 13 inches sold in the United States beginning in January 2000, and it tied the operation of that device to a program content rating system.

The V-chip was required, but it was important to government regulators and the television industry that the rating system be industry-generated. Broadcasters have First Amendment rights that the government respects and the industry wishes to protect. "I feel very strongly that the government should not be involved in this," said President Clinton (as quoted in Fleming, 1996, p. 6). So major broadcasting, cablecasting, and production professionals established a TV ratings implementation group to devise a labeling scheme. They put Jack Valenti, head of the Motion Picture Association of America and author of the MPAA movie code, in charge. The implementation group met for a year with television effects researchers, parent groups, and child advocates.

These interested parties unanimously wanted the rating scheme to be content-based. That is, they wanted the labels to inform parents about what was actually in the shows. A national study by the National Parent Teacher Association found that 80% of surveyed parents wanted content-based ratings, as did a poll conducted by the Roper Center for Public Opinion Research (Kunkel, 1997).

When the industry group presented its ratings scheme in December 1996, it was age-based rather than content-based. A content-based code would have been descriptive, labeling programs for their degree and type of violence, sex, adult language, and nudity. The age-based code presented the program producer's overall judgment about the suitability of a show for a specific age group. That's when the controversy started. Mr. Valenti told the public, "I can tell you right now, we're not going to use any other television rating system except the one we're going to announce" (Fleming, 1996, p. 6).

Why would the television industry be opposed to providing as much information to parents as possible to help them make wise decisions about their children's viewing? The industry offered two explanations. The first was its desire to keep the rating scheme as simple as possible, and age provided recognizable, easily understood categories. The second was the sheer volume of content that had to be rated—all programming other than news and sports.

Critics rejected these arguments. The video game industry labels material by content, as does the Internet. Are parents of television-watching kids somehow less capable than those whose children play video games or surf the Net? Volume of content should not be a problem, they added, because all content has to be rated anyway. If a producer previews his or her show to give it an age-based label, the same amount of time could be used to give it a content-based code.

Advocates of content-based codes point to profits as the primary reason the industry favored its age-based scheme:

> By clouding the reasons why a program is rated in a certain way, as well as by providing only the vaguest information about what types of violence or sex are shown, the controversies surrounding actual program ratings can be minimized. Imagine the economic implications for broadcasters if people who didn't want their children to see violence on TV actually had an effective way of blocking all violent programs. Ratings would go down and advertising dollars would decline. That's an unacceptable outcome for the broadcasters. So the strategy is to deliver a rating system, but to limit it to categories so imprecise that they never really reduce viewership of a program. (Kunkel, 1997, p. B5)

Most broadcasters and cablecasters, however, did listen to the public. In October 1997 almost all of the nation's largest broadcast and cable networks voluntarily introduced content-based codes in addition to the age-based labels. The age-based codes appear on the upper-left corner of the screen for 15

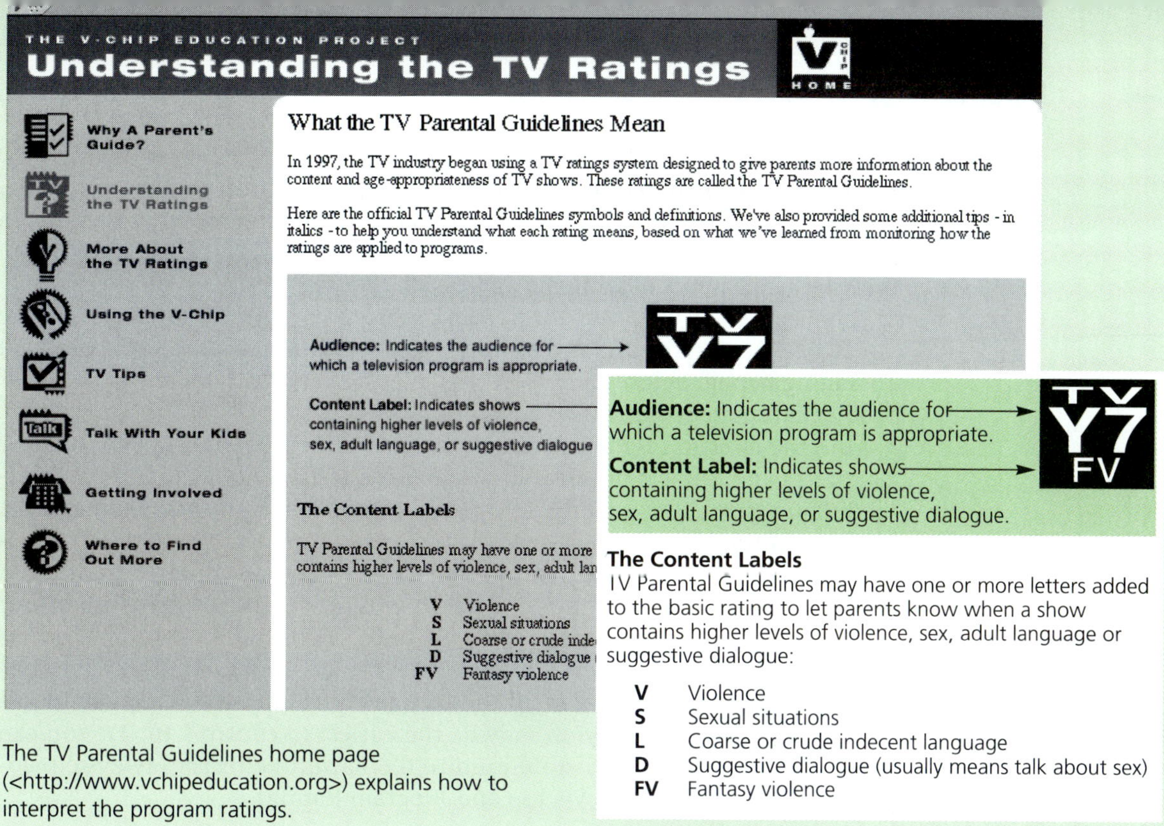

The TV Parental Guidelines home page (<http://www.vchipeducation.org>) explains how to interpret the program ratings.

seconds at the beginning of each program. There are six age-based categories:

TV-Y *All Children:* Content is designed for a very young audience, children 2 to 6 years old.

TV-Y7 *Older Children:* Content is designed for children age 7 and older and can include some mild comedy or fantasy violence.

TV-G *General Audience:* Most parents would find this content acceptable for all ages; it contains little or no violence, strong language, or sexual material.

TV-PG *Parental Guidance Suggested:* Some parents might find this content unsuitable for their younger children; it contains moderate violence, some sexual situations, and infrequent coarse language.

TV-14 *Parents Strongly Cautioned:* Parents might find this content unsuitable for children under age 14; it contains intense violence, sexual situations, and suggestive dialogue.

TV-MA *Mature Audiences Only:*—Content is unsuitable for children under age 17; it contains graphic violence, explicit sex, or crude and indecent language.

On channels that add the content-based codes, the age-based code is accompanied by the letters **FV** (fantasy violence), **S** (sexual situations), **L** (coarse language), **V** (violence), or **D** (suggestive dialogue).

What would you do if you were head of a fledgling network such as the WB, which voluntarily adopted the expanded codes in advance of the government's 2000 V-chip deadline? Would you use the age-based codes alone, or would you employ the additional content-based ratings? Explain your decision.

Keep in mind that many industry and public interest observers believe the codes have little impact on people's exposure to potentially troubling content. They charge that the codes are meaningless because program producers and distributors rate them themselves and do so unevenly and inconsistently. They also point out that parents concerned enough about their children's viewing to go to the trouble to program their V-chips are the very parents least in need of such help. You may want to visit the Web site <http://www.vchipeducation.org> to get more information about the V-chip and codes before you make your decision.

For Whom The compelling evidence of social cognition researchers aside, it's clear that most people do not exhibit aggression after viewing media violence. There is also little doubt that those predisposed to violence are more likely to be influenced by media aggression. Yet viewers need not necessarily be "predisposed to violence" for this link to occur, because at anytime anyone can *become* predisposed. For example, experimental research indicates that frustrating people before they view mediated violence can increase the likelihood of subsequent aggressive behavior.

But the question remains, Who, exactly, is affected by mediated violence? If a direct causal link is necessary to establish effects, then it can indeed be argued that some media violence affects some people in some ways some of the time. But if the larger, macro-level, ritual view is applied, then we are all affected because we live in a world where there is more violence than there would be otherwise. We live in a world where we are less trusting of our neighbors and more **desensitized** to violence in our midst. This need not be the case. As researcher Ellen Wartella (1997) said, "Today, we find wide consensus among experts that, of all the factors contributing to violence in our society, violence on television may be the easiest to control" (p. 4). You can read more on official efforts to minimize the harmful effects of televised violence in the box titled The V-Chip and Program Ratings.

Drugs and Alcohol

Concern about media effects reaches beyond the issue of violence. The claims and counterclaims surrounding media portrayals of drugs and alcohol parallel those of the violence debate.

The wealth of scientific evidence linking media portrayals of alcohol consumption, especially in ads, to increases in youthful drinking and alcohol abuse is sufficient to have moved another Surgeon General, Antonia C. Novello, to declare at a 1991 press conference:

> Of the 20.7 million 7th through 12th grade students nationwide, 10.6 million students drink . . . 8 million drink weekly and . . . 454,000 binge once a week. . . . For the longest time, sex, glamour, and action have been used as time-honored techniques to sell everything from toothpaste to laundry detergent. But let's remember, your laundry detergent doesn't affect your alertness and your toothpaste doesn't break down your barriers against risky behavior. Alcohol does. And alcohol, when mixed with the exhilaration and turmoil of youth, is a brew that can lead to disaster. . . . Current research has documented that youth are especially attracted to ads which *make lifestyle appeals* with attractive role models, imaginary peers, and attractive lifestyles to emulate; *make sexual appeals* about attracting, watching, and even conquering the opposite sex; *use sports figures* who usually are heroes to youth; and, *show risky activities,* leading many people, and in particular the young, to think that it is not only acceptable for people to drink while participating in that activity, but that it is also safe to do so. (emphasis in original, p. 4)

A good deal of scientific research, typically from alcohol industry scientists, discounts the causal link between media portrayals and real world drinking. Again, researchers who insist on demonstration of this direct, micro-level effect will rarely agree on media's influence on behavior. The larger macro-level perspective, however, suggests that media portrayals of alcohol, both in ads and in entertainment fare, tell stories of alcohol consumption that predominantly present it as safe, healthy, youthful, sexy, necessary for a good time, effective for dealing with stress, and essential to ceremonies and other rites of passage.

The same scenario exists in the debate over the relationship between media portrayals of nonalcohol drug use and behavior. Relatively little contemporary media content presents the use of illegal drugs in a glorifying manner. In fact, the destructive power of illegal drugs is often the focus of television shows such as *NYPD Blue* and *ER* and a central theme in movies such as *Trainspotting* and *Boyz in the Hood.* Scientific concern has centered on the impact of commercials and other media portrayals of legal over-the-counter drugs. Many studies show a causal link between this content and subsequent abuse of both legal and illegal drugs, but other research has failed to uncover a causal link. Proponents of the ritual perspective of media hasten to point out that our media often present legal drugs as a cure-all for dealing with that pesky mother-in-law, those screaming kids, that abusive boss, and other daily annoyances.

Gender and Racial Stereotyping

Stereotyping is the application of a standardized image or concept to members of certain groups, usually based on limited information. Media cannot show all realities of all things, and the choices media practitioners make when presenting specific people and groups may well facilitate or encourage stereotyping.

Numerous **content analyses,** which are systematic and quantitative descriptions of media content, have been conducted over the last 50 years. They demonstrate that women and people of color are regularly underrepresented in all media; that when they are portrayed, they are more often than not presented in traditional, inferior roles; that, in the case of women, they are more likely to be presented as the victims of aggression than is the case in the real world; and in the case of people of color, they are more likely to be presented as the perpetrators of crime and aggression than is the case in the real world. When women and minorities are presented favorably, they are often stereotyped as perfect—the "perfect Asian student" or the "perfect African American social worker." These portrayals are as narrow and limiting as the "dumb blonde" or "dangerous dark outsider."

Concern over stereotyping resides in the probable outcome of repeated and frequent exposure to these limited and limiting representations. They influence people's perceptions, and people's perceptions influence their behaviors. Examine your own conceptions not only of women and people of color

The heavy use of African American athletes in commercials is controversial because, say critics, it reinforces the stereotypic belief among African American (and many other) children that sports (rather than education) is the surest route to success in life. In this soft drink commercial, basketball players Grant Hill and Tim Duncan use the debate to sell soda . . . and make a point. During the NBA lock out of its players in 1998, these all-star athletes were reduced to doing odd jobs to make ends meet.

but of the elderly, lawyers, college athletes, and people sophisticated in the use of computers. What images or stereotypes come immediately to mind?

Sure, you're skeptical. You're a smart, progressive, college educated individual. Test yourself on your stereotype of people from the American Heartland, or conversely, of people from big urban centers: Should a female feel more secure walking among residents of Oklahoma City or New York City? Among people from Rapid City, South Dakota, or Washington, D.C.? The FBI tells us that the incidence of forcible rape in Oklahoma City is 73.9 per 100,000 females, three times the levels of New York City and Washington, D.C. The

The televised 1960 debates between presidential candidates John F. Kennedy and Richard M. Nixon changed thinking about the influence of media—especially television—on politics.

U.S. city with the highest rate of forcible rape is Rapid City (Zoellner, 1995, p. 8). How did you develop your stereotypes of the people who live in these places? Where did you find the building blocks to construct your realities of their lives? You can read more about stereotyping in the box titled Invisibility.

Political Campaigns and Voting

Media impact on political campaigns and voting was at the center of much media research in the 1930s and 1940s. Given that television had yet to develop into a true mass medium, the overall conclusion drawn from this early work was that media had little *direct* impact on campaigns and voting and, when and if they did, that impact was in the form of reinforcement.

But with the fuller diffusion of television and the interest generated in that medium by the 1960 debates between presidential candidates John F. Kennedy and Richard M. Nixon, thinking about media and campaigns began to change. For example, research after those inaugural televised debates began to focus on candidates' use of media for image building and the subsequent "cheapening" of campaigns as personalities became more important

Critical Issues *Invisibility*

A particularly insidious form of media stereotyping—invisibility—became front page news when the four major television networks (ABC, CBS, NBC, and Fox) unveiled their schedules for the 1999–2000 season. Of the 26 new series that would enter U.S. homes when the fall season began, not a single one featured a minority character in a lead role. Reaction was swift and furious. The NAACP and other public interest groups threatened a boycott of those networks and their sponsors, leading to a meeting of programmers and representatives of the offended groups in December 1999. The networks agreed to include, immediately and in future programming, greater numbers of people of color in prominent roles. But questions remained. Among them were "What does it mean to be invisible on the screen?" and "Why are minorities so underrepresented in prime time television?"

It is one thing to be stereotyped in narrow, unflattering ways, said the networks' critics, but it is equally wrong to be rendered invisible, stereotyped as unimportant and not worthy of attention. The likely outcome of this invisibility is that members of ignored groups feel they are socially invisible, and to other people they *are* invisible. This makes minority people's assimilation into the larger culture all the more difficult. As Sonny Skyhawk, a Sicangua Lakota and founder of American Indians in Film, explained, "We are concerned, just like everyone else, with the fact that when a young Indian child watches TV and does not see his people, it manifests itself into being a non-person, or non-contributing member of society" (as quoted in Gay, 1999, p. 23A).

The four networks disputed the charge of "whitewashing" prime time television. CBS and ABC said that despite problems with their new shows 11 of their 19 prime time shows featured minority actors in either a lead or costarring role. NBC and Fox each had 8 series with people of color in lead roles (McClellan, 1999a). Nonetheless, Fox Entertainment President Doug Herzog admitted, "I think the diversity problem is something that is widespread throughout the media, both behind the scenes and clearly on the camera. Now, the responsibility we have is that we come directly into people's homes every night. We're piped right in there, and so in that regard we have a big responsibility. So, should we be doing a better job of presenting a more diversified front in terms of programming? Absolutely" (as quoted in Schlosser, 1999b, p. 12). But the issue of why there were relatively few minorities in central prime time roles remained.

Economics, more specifically, advertisers' desire for the attractive upscale, professional demographic was cited as one reason. But, critics argued, the

than issues. Another important development in contemporary thinking about media and campaigns was **agenda setting theory**—the idea that media don't tell us what to think but that they certainly do tell us what to think about. This concept was introduced after research on the 1968 presidential elections. Agenda setting was used to explain how media can influence campaigns and voting. Media help set the issue agenda for the campaign, and these issues become the issues on which many people ultimately base their voting decisions.

The growing influence of television news throughout the 1960s and 1970s turned attention to its role in the electoral process. Early thinking saw news reports as more important than political ads in shaping preferences because of their perceived impartiality. But recent research has demonstrated dramatic declines in the amount of time given to candidates' *positions* in typical news programs. Vote seekers now construct more visually interesting, content-free

belief that people of color are absent from this demographic is, in itself, proof of the harmful effects of their invisibility. Not only do African Americans comprise just over 12% of the total U.S. population, Hispanics 11%, and Asian Americans 6%, but a September 1997 Census Bureau report indicated that middle and upper income African Americans and Hispanics were indistinguishable from Whites in terms of such economic indicators as home ownership and consumer purchasing. That same report also concluded that the average household income for African Americans exceeded all previous levels and that of Hispanic households was growing at a rate 5 times faster than that of all other citizens. Asian Americans, another group invisible in prime time, had a median household income of $44,460 a year, 19% above the national average (Taylor & Stern, 1997).

A second explanation for the invisibility problem is that the overwhelming majority of network executives and those who write network television are White. Said Doug Alligood, senior vice president of special markets for advertising giant BBDO, "A lot of people are dealing with people like themselves and programming to people like themselves" (McClellan, 1999a, p. 18). *Village Voice* writer Ed Morales (1999) concurred, "White writers tend to write what they know. . . . Of the twenty-one shows produced last fall in New York, where people of color are a majority, 81.5 percent of the 130 characters were white, 15 percent African-American, and 3 percent Latino" (p. 9). The numbers would seem to support this view. Only 7% of all members of the Writers Guild of America are non-White, and at the start of the 1999–2000 season "CBS had two African-American writers out of 144; ABC seven of 172, NBC one of 186, and Fox three of 163" (Schlosser, 1999a, p. 53).

What do you think? Is this just a case of a few disgruntled minorities making a mountain out of a mole hill? At least the portrayals, when they are there, are not negative. How would you feel if you belonged to a group of people who were invisible on prime time television? How do you explain the startling absence of people of color from the ranks of those who write for network television? Do you agree with Fox's Doug Herzog that the networks have a responsibility to program a more ethnically diverse schedule? Return to the discussion in Chapter 13 of interests that must be balanced when making ethical decisions. What are the competing interests involved in producing a more balanced image of race? If you were programming chief of one of the Big Four networks, what would you do to rectify the current situation (if you believe it should be rectified)?

pseudo-events—events that have no real informational or issue meaning but exist merely to attract media attention. Candidates recognize that the public tends to trust news more than commercials. The goal of pseudo-events, therefore, is to earn candidates and their campaigns media coverage while allowing the campaigns to control the content of that coverage.

Contemporary research and theory concerning media and campaigns show that media do indeed have important effects. On the micro level media coverage of candidates and campaigns tends to reconfirm support among already committed voters and to solidify support among those leaning toward particular candidates. The macro-level view sees media, especially television with its attack ads and ever briefer sound bites, debasing and trivializing the electoral process, turning off voters, and generating disinterest (at best) or disgust (at worst) with the political process.

This latter situation has given rise to a number of proposals for improvement. Primary among them is campaign finance reform. The media, especially television, have become essential to modern campaigning, and raising money to buy media time and space has become a full-time obsession for many politicians. This state of affairs has led to widespread suspicion among voters that the money raised to buy media access for candidates ultimately buys access to the candidate for the sources of the money. A second suggestion, one embraced by the Fox television network among others and touted by President Clinton in his 1997 State of the Union address, is that broadcasters should make free time available to candidates and campaigns, reducing the influence of money and increasing the length of time candidates have to present full, cogent explanations of their positions. (As we saw in Chapter 12, this was also at the heart of the Gore Commission proposals for broadcaster payment for digital spectrum space.) A third suggestion is that the United States should follow the British election practice of not only granting free airtime to candidates but also limiting all campaigning—personal and mediated—to a specified period, perhaps 3 or 6 months before election day. Campaigning outside this run up to the election would be illegal.

Prosocial Effects

Virtually every argument that can be made for the harmful or negative effects of media can also be applied to the ability of media to do good. A sizable body of research clearly demonstrates that people, especially children, can and will model the good or prosocial behaviors they see in the media, often to a greater extent than they will the negative behaviors. Research on the impact of media portrayals of cooperation and constructive problem solving (Baran, Chase, & Courtright, 1979) and other "good" behaviors indicates that much more than negative behavior can be learned from the media.

Television's ability to serve prosocial ends is obvious in the public service messages we see sprinkled throughout the shows we watch. For example, dozens of national broadcast and cable channels simultaneously ran a 30-second PSA in August 1999 that featured President Clinton urging parents to talk with their kids about violence. And "this is your brain; this is your brain on drugs" is familiar to everyone who can operate a remote control.

But there is a movement among television writers and producers to more aggressively use their medium to produce prosocial effects by imbedding important cultural messages in the entertainment fare they create. The goal is to produce both micro-level individual effects and larger macro-level cultural effects. The ABC series *The Hughleys* is indicative. This 30-minute comedy opened its 1999 season with an episode centering on the show's lead, D. L. Hughley, discovering his children playing with a gun he keeps in the house. Neither pro- nor antigun, this "program-length PSA" was intentionally designed to keep alive the gun debate that began with the string of murderous school shootings than blighted 1998 and 1999.

Making good use of television's prosocial potential, UPN's *The Hughleys* opened its 1999 season with a "program-length PSA" intentionally designed to keep alive the gun debate that began with the string of murderous school shootings that blighted 1998 and 1999.

This "prime time activism" can be traced to Harvard professor Jay Winsten and his 1988 campaign to get Hollywood to push his novel designated driver idea, possibly the most successful of such efforts. We all know what a designated driver is. He or she is the person among a group of friends who is selected to remain alcohol-free during a get-together and drive everyone else home. The concept, much less the term, did not even exist until Professor Winsten, through the intervention of CBS executive Frank Stanton, contacted Stanton's friend Grant Tinker, then chairman of NBC, to ask for help. Intrigued by Winsten's plan to develop a new social norm, Tinker put his considerable clout behind the effort, writing letters to the heads of the 13 production companies that did the most business with the networks. If they did not reply

The Drew Carey Show and *Sports Night* were two of the shows that accepted the federal government's offer of reduced public service expectations in exchange for anti-drug messages written into story lines. These arrangements proved to be very controversial when they were later made public.

quickly enough to suit Tinker, he called them on the phone. Once everyone was in line, Tinker personally escorted Professor Winsten, director of Harvard's Center for Health Communication, to meetings with all 13 producers.

In the four network television seasons that followed these meetings, designated drivers were part of the story lines of 160 different prime time episodes seen by hundreds of millions of viewers. Did these story lines make a difference? Within one year of the introduction of the idea of the designated driver in these television shows, 67% of American adults said they were aware of the concept, and by 1991, 52% of adults under 30 years old said they had served as a designated driver. From 1988, the campaign's first year, to 1997, the number of drunk driving fatalities in the United States dropped by 32% (Cox, 1999).

The federal government, too, is convinced of television's ability to have prosocial effects. In a controversial program, it offered the national television networks a deal—embed "an aggressive antidrug message" in their programming and they are no longer bound by the legal requirement that they air specific numbers of public service announcements. This rendered them free to sell the then vacant spots to paying advertisers (Kurtz & Waxman, 2000). Among the shows involved were *Beverly Hills 90210, ER, Chicago Hope, The Drew Carey Show, Sports Night, Sabrina the Teenage Witch, General Hospital,* and scores of others. The controversy revolved around two issues. First, critics argued that licensed broadcasters should not have to be paid to do what is right and responsible. Second was the argument that the government has no business suggesting story lines, approving scripts, and inserting messages of any kind into entertainment fare, especially without public knowledge.

Not at all controversial is what some people are calling the "60 Minutes Effect." In May 1999 the venerable television news magazine aired a segment on Dress for Success, a New York City–based group. This charity provides job interview wardrobes for low-income women seeking work. Within a few weeks of the broadcast, Dress for Success received nearly 900 calls specifically mentioning *60 Minutes,* 150 of which were from people wishing to set up chapters in their home cities. In addition, scores of corporations offered their products, services, and—as in the case of Bank of America, which donated $75,000—cash (Werner, 1999).

Online Democracy

The Internet generates significant concern as well as promise. This concern and promise often find their expression in efforts to control or regulate the new medium, as we saw in some detail in Chapter 12. Here, however, we will focus on an effects issue frequently raised about the Internet—its impact on democracy and self-governance.

As we discussed earlier, the Internet has been characterized by freedom and self-governance. These, too, are the hallmarks of true democracy. It is no surprise, then, that computer technology is often trumpeted as the newest

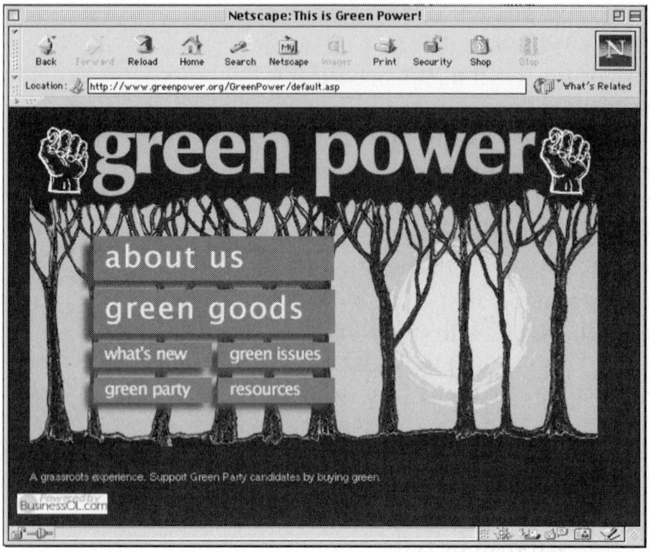

Virtually all political parties of any size have created a presence on the Web.

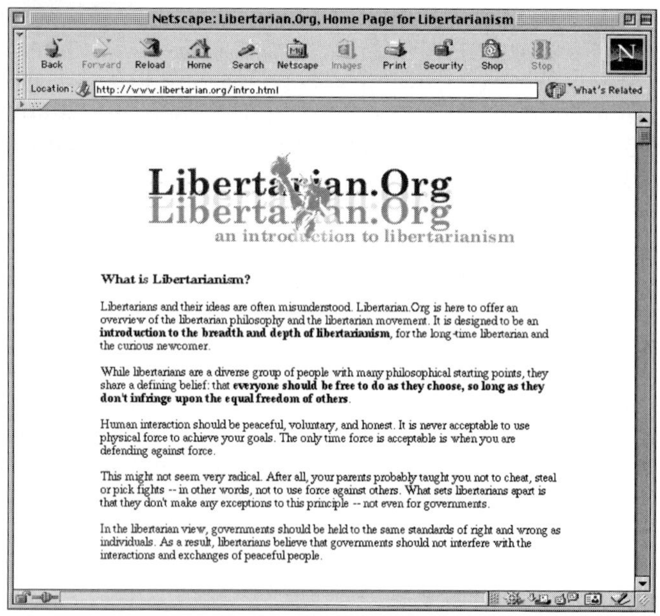

and best tool for increased democratic involvement and participation. Vice President Al Gore conducted history's first interactive, computer network *news* conference on January 13, 1994. It was not a *press* conference, in the traditional sense, because the people themselves rather than their representatives in the press could query Mr. Gore. Several states are studying the possibility of online voting. Among Newt Gingrich's first official acts after becoming Speaker of the U.S. House of Representatives in 1995 was to order that all

Congressional documents be available online on a Web site he dubbed Thomas (<http://thomas.loc.gov/>), after Thomas Jefferson. Republican presidential hopeful Steve Forbes launched his 1999 bid for his party's nomination on the Internet rather than at a more traditional press conference. The 1999 elections saw Shelton, Washington become the first city in the United States to allow voting by Internet. Virtually every politician of any standing, including the President, Vice President, Senate Majority Leader, and Speaker of the House, maintains at least an e-mail address if not a full Web page.

This enthusiasm for a technological solution to what many see as increased disenchantment with politics and the political process mirrors that which followed the introduction of radio and television. A September 3, 1924, *New Republic* article, for example, argued that overwhelming public interest in the broadcast of the 1924 political party conventions brought "dismay" to "the most hardened political cynic" (as cited in Davis, 1976, p. 351). The November 1950 *Good Housekeeping* claimed that television would bring greater honesty to politics because "television is a revealing medium, and it is impossible for a man or a woman appearing before those cameras to conceal his or her true self" (p. 359). In 1940 NBC founder and Chairman David Sarnoff predicted that television would enrich democracy because it was "destined to provide greater knowledge to larger numbers of people, truer perception of the meaning of current events, more accurate appraisals of men in public life, and a broader understanding of the needs and aspirations of our fellow human beings" (as quoted in Shenk, 1997, p. 60). Few would argue that these predictions have come to pass.

Some critics claim that the Internet will be no more of an asset to democracy than have been radio and television because the same economic and commercial forces that have shaped the content and operation of those more traditional media will constrain the new media just as rigidly. Communication scholar Everette Dennis (1992) condensed the critics' concerns into two overarching questions: (1) Will computer networks be readily accessible to all people—even if it means depending on institutions such as schools, churches, and community organizations—or only to some? (2) Once the technology is in place and people have access to it, are they going to know how to use it? What Professor Dennis is discussing are the technology and information gaps (Chapter 1). An important principle of democracy is "one person, one vote." But if democracy is increasingly practiced online, those lacking the necessary technology and skill will be denied their vote.

Ironically, the technology gap may be exacerbated by the sheer popularity of the Internet. Monthly **flat rate billing** has become the standard form of paying for Internet access. For example, in October 1996 America Online (AOL) converted its billing to a flat monthly rate of $20, changing from its previous practice of charging for time connected. It was forced to do so because many of its competitors were successfully employing this practice.

But many in the telecommunications industry argue that flat rate billing will eventually damage the Internet. With no financial incentive to limit the time they spend online, more people will spend more time on the Internet.

The Net therefore becomes more crowded and operates more slowly, limiting its efficiency and effectiveness and angering users. AOL, for example, experienced a number of system failures due to heavy use after its change in pricing. Calling flat rate billing "bad business," many industry insiders predict that **pay-as-you-go pricing** will eventually become necessary and, as a result, will become the industry standard. Those who make the most use of the Internet, the logic argues, should pay the highest toll.

Among the pay-as-you-go pricing plans being considered are fees based on the length of time connected (like billing for long-distance telephone calls), pay-as-you-go billing for use during peak times, and tiered or multilevel pricing. Users wanting to maintain flat rate pricing might be limited to certain hours of the day or night. Those willing or able to pay for time connected might have greater freedom. Any of these changes, argue technology gap critics, will shut out or at the minimum limit access for those who cannot afford the higher prices. The democratic ideal of one person, one vote will be reduced to one dollar, one vote, and the technology gap will grow even wider.

Another important principle of democracy is that a self-governing people govern best with full access to information. This is one reason our culture is so suspicious of censorship. The technology gap feeds a second impediment to virtual democracy, the information gap. Those without the requisite technology will have diminished access to the information it makes available.

Information, Knowledge, and Understanding

Some critics of the idea of **cyberdemocracy,** the online practice of self-governance, point to the amount of information available to contemporary citizens and the speed with which it comes as potentially troublesome. Add to these the difficulty in assessing the veracity of much online information (Chapter 12), and they argue that the cyberworld may not be the best place to practice democracy.

Advocates of cyberdemocracy see the Internet as a way to let citizens have more direct, quicker access to politicians. Elected officials should hear what the people have to say. But does democracy necessarily benefit when its leaders respond directly, maybe even impulsively, to public sentiment? Until there is no more technology gap, certain voices—the poor, the uneducated, the elderly—will have less access to their leaders than those who are connected. Moreover, claim cyberdemocracy critics, ours is a representative, deliberative democracy. It was intentionally designed to allow the public's representatives to talk to one another, to debate ideas and issues, to forge solutions that benefit not just their own but others' constituents as well. They claim that the political alienation that is felt by many citizens today is the product of politicians *listening too much* to the loudest voices (that is, special interests) and *being too responsive* to the polls. People often criticize politicians for "flip flopping" or "having no personal conviction." How can having elected officials responding daily to the voices in the electronic town hall improve this situation? Journalist Robert Wright (1995) wrote in *Time:*

> The Founders explicitly took lawmaking power out of the people's hands, opting for a representative democracy and not a direct democracy. What concerned them, especially James Madison, was the specter of popular "passions" unleashed. Their ideal was cool deliberation by elected representatives, buffered from the often shifting winds of opinion. . . . Madison insisted in the Federalist Papers on the need to "refine and enlarge the public views by passing them through the medium of a chosen body of citizens, whose wisdom may best discern the true interest of their country and whose patriotism and love of justice will be least likely to sacrifice it to temporary or partial considerations." (p. 16)

Critics also argue that cyberdemocracy, in and of itself, is antidemocratic. Not only is there now an additional medium to further fragment the audience, but by its nature the Internet solidifies people into smaller, more homogeneous, more narrowly interested groups. This cannot be good for democracy. The Internet and the Web encourage people to splinter into virtual communities based on a shared interest in some given information. This renders real world communities irrelevant. No longer "required" to coexist with other people in the day-to-day world, cybercitizens have little need to examine their own biases. They need not question their own assumptions about the world and how it works. There is little benefit to seeking out and attempting to understand the biases and assumptions of others outside the self-chosen virtual community.

Media critic and scholar Robert McChesney (1997) writes that among the criteria that must be met if democracy is to serve the needs of its people are "a sense of community and a notion that an individual's well-being is determined to no small extent by the community's well-being" and "an effective system of political communication, broadly construed, that informs and engages the citizenry, drawing people meaningfully into the polity" (p. 5). Where advocates of the new online communication technologies see them doing just this, many others, often likening the Net and the Web to "talk radio writ large," fear the opposite, that the Internet has already become, in the words of *Time* environmental writer John Skow (1999), "a stunning advance in the shoring up of biases, both benign (one's own views) and noxious (other views)" (p. 61).

Chapter Review

The debate over media effects focuses on a number of issues. Are media effects limited by the fact that audiences know content is only make believe? Is impact limited because media are used only as entertainment? Does media content simply reflect society as it is? Do media effects typically exist only in the form of reinforcement? Can media affect only unimportant aspects of people's lives?

One way to sort through the different answers to these questions is to distinguish between micro-level (or personal) and macro-level (or societal or cultural) effects. Another is to base questions about effects on either the transmissional perspective of media, which sees media as mere senders of information, or on the ritual perspective, which sees media as central to the maintenance of society.

The impact of mediated violence is the best known effects issue. The causal relationship between violent media content and subsequent aggressive behavior has been demonstrated for decades, giving rise to the stimulation and aggressive cues models of media's influence on aggressive behavior. Social cognitive theory, which argues that people model from media through the processes of imitation and identification, has been central to demonstrating the operation of both vicarious reinforcement and environmental incentives. Catharsis, the idea that viewing mediated aggression reduces people's innate aggressive drives, has been discredited.

Effects researchers have also made important discoveries about the relationship between media representations of drugs and alcohol and subsequent behavior that parallel those of the violence researchers. Disagreement exists, however, regarding the interplay of content and behavior. Those who adopt the wider, macro-level ritual view of mass communication more readily accept media's influence on people's behavior than do those who hold to more transmissional views.

Other effects issues that have attracted scientific attention are the impact of media portrayals of different groups of people (stereotyping) and the impact of media, especially television, on political campaigns. Again, links have been scientifically demonstrated. None of this, however, is meant to imply that media content cannot have good or prosocial effects. Research demonstrates this as well.

The Internet brings with it effects issues as well. Primary among them is its impact on democracy and self-governance. Proponents see the Internet as the great democratizer, giving voice to all. Others view it as spreading even wider the technology and information gaps and destroying people's larger sense of community.

Questions for Review

1. What is the early window? What is willing suspension of disbelief?
2. What is implied about media when it is said they hold a fun-house mirror to society?
3. What are micro-level media effects? Macro-level effects?
4. What is the transmissional perspective on media? The ritual perspective?
5. What is the stimulation model of media's influence on aggressive behavior? The aggressive cues model?
6. What is social cognitive theory? How does it differentiate between imitation and identification?
7. What are observational learning, inhibitory effects, and disinhibitory effects?
8. What is catharsis, and what is contemporary thinking about its operation?
9. What are vicarious reinforcement and environmental incentives? How are they related to the media violence debate?
10. What is desensitization?
11. What is stereotyping?

12. What does agenda setting theory say about media's influence?
13. What are pseudo-events?
14. What is cyberdemocracy? What are some of its benefits? Potential drawbacks?
15. What are flat rate billing and pay-as-you-go pricing? What are the benefits and drawbacks of each for cyberdemocracy?

Questions for Critical Thinking

1. Do media set the agenda for you? If not, why not? If they do, can you cite examples from your own experience?

2. Conduct your own mental content analysis of any of the effects issues we've discussed. How is violence portrayed on your favorite prime time drama? In your favorite video game? What beer and wine ads have you seen on television? How are alcohol and drinking presented?

3. Do you pay attention to alcohol advertising? Do you think it influences your level of alcohol consumption? The "value" you place on alcohol in social settings? Explain your answers.

4. How did you do on the Heartland/big city experiment? Why do you think you responded as you did?

5. Do you have less difficulty accepting the idea that media can have positive or prosocial effects than you do their ability to have negative effects? Why or why not? How would you resolve the broadcasters' dilemma in asserting powerful effects for the commercials they air but limited or no effects for other content?

Engaging Technology

1. Several major scholarly organizations house most of today's influential media researchers and theorists. Visit their Web sites to determine where they stand on the various effects issues we've studied. The American Communication Association is at <http://www.americancomm.org/>. The Association for Education in Journalism and Mass Communication is at <http://www.aejmc.sc.edu>. The International Communication Association is at <http://www.icahdq.org/>. And the National Communication Association is at <http://www.natcom.org>.

2. We've seen in this chapter that the Surgeon General's office can and does involve itself in important media effects debates. Visit its Web site at <http://www.surgeongeneral.gov> to see which effects issues are at the top of its agenda today.

3. Alcohol advertising and its effects are the subject of numerous Web sites. Here are two. The Center on Alcohol Advertising is at <http://www.traumafdn.org/alcohol/ads/>, and the New Zealand Drug Foundation is at <http://www.nzdf.org.nz/notices/messages/23.htm>. Visit these and as many others as you can find. Then determine for yourself the contemporary status of the debate.

Glossary

absolutist position regarding the First Amendment, the idea that no law means no law

account executive in an ad agency the person who serves as liaison between agency and client, keeping communication flowing between the two and heading the team of specialists assigned by the agency to the client; in a broadcast operation, someone who sells airtime or cable time

actual malice the standard for libel in coverage of public figures consisting of knowledge of its falsity or reckless disregard for whether it is true or not

addressable technology technology allowing a program provider to switch access to pay services on and off at the **hub**

ad hoc balancing of interests in individual First Amendment cases weighing several factors when determining how much freedom the press is granted

ad hoc network stations that come together or network only for specific broadcasts

adjacency in television when a commercial runs between two shows

advertising allowances computed as a proportion of a retailer's purchase order, the manufacturer's contribution to the merchant's buying of commercial time (often mistakenly called **co-op advertising**)

affiliates broadcast stations that belong to a **network**

aftermarket additional sources of revenue for television content after its cable or network airing (e.g., in videotape sales and rental)

afternoon daypart in television programming typically noon to 4 p.m.

agenda setting theory argues that media may not tell us what to think but that media tell us what to think about

aggressive cues model media portrayals of violence can indicate that certain classes of people are acceptable targets for real world aggression

AIDA approach the idea that to persuade consumers commercials must attract attention, create interest, stimulate desire, or promote action

all-channel legislation law requiring all sets imported into or manufactured in the United States to be equipped with both **VHF** and **UHF** receivers

alternative programming noncommercial radio content rarely heard on commercial stations

AM/FM combo two stations, one AM and one FM, having common ownership

amplitude the height and depth of a radio wave

amplitude modulation (AM) shaping radio waves' height and depth; occupies radio spectrum frequencies between 535 and 1705 kHz

applied ethics the application of values and **normative ethics** to very specific situations

ascertainment required broadcasters to ascertain or actively and affirmatively determine the nature of their audiences' interest, convenience, and necessity (no longer exists)

audience flow the carry-over or movement of audience members from one program to the next

audimeter device for recording when the television set is turned on, the channel to which it is tuned, and the time of day; used in compiling **ratings**

audion tube vacuum tube developed by DeForest that became the basic invention for all radio and television

authoritarian media system a national media system characterized by strict government control

automated station in radio computer-programmed stations that have no local personnel other than engineers and a sales staff

avails as yet unsold broadcast of cable commercial spots

average quarter-hour persons (AQH Persons) average number of people listening to a particular radio station for at least 5 minutes during a 15-minute period over a **daypart**

average quarter-hour rating (AQH Rating) the **AQH Persons** expressed as a percentage of the population being measured

awareness test advertising research technique that measures the cumulative effect of a campaign in terms of a product's "consumer consciousness"

bandwidth portion of the spectrum utilized to transmit video and audio signals; an indication of a channel's information carrying capacity

banners online advertising messages akin to billboards

barter in broadcast advertising sales the exchange of airtime for sponsors' products or services; in television syndication "selling" content to local stations for their promise to run the commercials it contains

barter syndication form of broadcast content distribution that enables stations to "pay" for content by promising to air commercial spots included with that content

basic cable "free" channels provided automatically by a cable operator to subscribing homes

bicycling distribution of broadcast content to stations by mail or other land-based delivery systems

billings total sale of broadcast airtime

Bill of Rights the first 10 amendments to the U.S. Constitution

Biltmore Agreement settled the press war between newspapers, which had refused their services to their electronic competitors, and radio

binary code information transformed into a series of digits 1 and 0 for storage and manipulation in computers

bitcasters "radio stations" that can be accessed only over the **World Wide Web**

blocking in programming network prime time, when a network programs a strong lineup of similarly typed programs in one evening (sometimes called **stacking**)

block programming large segments of related radio content scheduled at the same time each day or week, providing listeners with a consistent schedule but allowing stations to air several different formats

BOC regional Bell operating companies (often called Baby Bells) created when AT&T was forced to separate from local telephone service providers in 1984

bots online software that automatically searches the Web for a requested item or piece of information

brand awareness an advertising goal when a number of essentially similar brands populate a given product category

branding in telecommunications the creation of a specific identity (or brand) for a specific content or service provider

bridging television programming strategy that involves nontraditional use of a show's start time to attract and keep viewers away from opponents' programs that start at traditional times

broadcasters radio and television stations that transmit their signals over the airwaves

broadcast model of telecommunications basing a telecommunications operation's financial support on advertising

broadcast network group of broadcast stations linked together for the purpose of reaching large national or regional audiences

broadcast spectrum the range of electromagnetic frequencies on which the FCC grants permission to transmit

browser software program loaded on personal computers and used to **download** and view Web files

bundling delivering television, audio, high-speed Internet access, multiple telephone lines, and fax via **cable**

buttons graphically presented links on Web pages

cable delivery of audio and video content to consumers by wire

cable access modem connects user's computer to the **Internet** over cable television lines via the cable company's designated ISP

cable channels audio and video services carrying content acquired specifically for delivery by **cable**

carrier wave a transmitted wave sent at a constant **amplitude** and **frequency**

cash-plus-barter in television syndication the acquisition of a program license fee through a combination of cash and barter

catharsis the theory that watching mediated violence reduces people's inclination to behave aggressively

CD-ROM read-only discs that permit storage of large amounts of easily accessible, indexed digital information

cease-and-desist order demand made by a regulatory agency that a given illegal practice be stopped

chains the first radio networks

channel surfing moving through the television dial while commercials are on and returning to the original show when commercials are over (also called **cruising**)

checkbook journalism paying sources for information or interviewees for their interviews

checkerboarding in local television programming rotating different programs through the same time slot on different days of the week

churn in cable television turnover in subscribership whereby new subscriptions in a given period of time are offset by cancellations

clandestine stations illegal or unlicensed broadcast operations frequently operated by revolutionary groups or intelligence agencies for political purposes

clear time what local affiliates do when they agree to carry a network's program

click stream series of choices made by a user on the Web

click through rate in cyberadvertising the frequency with which people move from an ad to the advertiser's ordering site

clutter in broadcasting when many individual commercials share one commercial break

coaxial cable copper-clad aluminum wire encased in plastic foam insulation, covered by an aluminum outer conductor, and then sheathed in plastic

coincidental telephone measurement radio audience measurement technique in which listeners are asked what they are listening to at the time of the call

common carriers telecommunications companies that carry the messages of others with no power to shape or restrict them

communication the process of creating shared meaning

communist media systems a national media system in which media are seen as a "tool of the people"; because the people are the Communist Party, the government controls this tool

community antenna television (CATV) original name for what has developed into the current cable television industry

community radio noncommercial stations typically underwritten by community groups, local businesses, and listener donations

comparative analysis the study of different countries' mass media systems

comparative studies see **comparative analysis**

compensation network payments to **affiliates** for clearing content

compulsory license fee in cable television payments to a central collection agency to be distributed to copyright holders

conduction system wireless communication system using ground and water to carry signals

confidentiality the ability of media professionals to keep secret the names of people who provide them with information

conglomeration the ownership of media companies by even larger companies having media and nonmedia holdings

consumer culture where personal worth and identity reside not in the people themselves but in the products with which they surround themselves

consumer juries advertising research technique where people considered representative of a target market review a number of approaches or variations of a campaign or commercial

content analysis the objective, systematic, and quantitative description of media content

convergence erasure or blurring of distinctions among different forms of communication media

co-op advertising see **advertising allowances** and **cooperative advertising allowances**

cooperative advertising allowances matching funds arrangements between local retailers and manufacturers in which, up to a specified limit, the manufacturer will match dollar-for-dollar the local merchant's commercial time buys (universally called **co-op advertising**)

copyright identifying and granting ownership of a given piece of expression to protect the creator's financial interest in it

copy testing measuring the effectiveness of commercial messages by showing them to consumers

corporate telecommunications use of telecommunications technologies to serve very specific instructional, economic, public relations, or individual needs of organizations and businesses

corrective advertising a new set of commercials required by a regulatory body and produced by the offender that correct the original misleading effort

cost per rating point cost to an advertiser for reaching an audience equivalent to one percent of the population in a specified area

cost per thousand (CPM) cost to an advertiser of garnering 1,000 gross impressions

counterprogramming in programming network prime time scheduling content that is unlike that on the other networks at a given time

CPM See **cost per thousand**

crossownership rules rules limiting ownership of different media by a single entity

cruising see **channel surfing**

cultural imperialism the invasion of an indigenous people's culture, through mass media, by outside, powerful countries

cume persons total number of different people who listen to a station for at least 5 minutes in any one **daypart** over a specified period of time, usually one week

cume rating cume persons audience expressed as a percentage of all people estimated to be in the specified area

cyberadvertising placement of commercials on various online sites

cyberdemocracy the online practice of self-governance

cycle the movement of an individual wave past a specific point

dataveillance collection and distribution or sale of personal consumer data by online and other businesses

daypart time of the radio or television broadcast day

democracy government by the people

demographics the composition of the audience in terms of varying personal and social characteristics such as race, gender, and economic level

demographic segmentation advertisers' appeal to audiences composed of varying personal and social characteristics such as race, gender, and economic level

density in cable television the number of households per mile of cable

deontological moral reasoning duty-based moral reasoning in which decisions emphasize adherence or a commitment to principles that should be universally applied

deregulation relaxation of ownership and other rules for radio and television

desensitization the idea that viewers become more accepting of real world violence because of its constant presence in television fare

dial-up modem connects user's computer to an ISP via traditional telephone lines

diary sample ratings technique in which people are asked to maintain a written diary of their listening or viewing for a specified period of time

digital conversion of signals into 1s and 0s logged in millisecond intervals in a computerized translation process that permits near instantaneous and noise-free transmission and reception

digital audio radio service (DARS) direct home or automobile delivery of digital audio by satellite

digital audio tape (DAT) introduced in the early 1970s, offers digital quality sound purity in a cassette tape format

digital cable television delivery of digital images to subscribers' homes

digital compression reduction of signal size to permit multiple signals to be carried over one channel

digital computer computer that processes data reduced to a binary code

digital must-carry potential FCC rules that would require cable systems to carry both digital and analog channels offered by over-the-air broadcasters

digital recording recording based on conversion of sound into 1s and 0s logged in millisecond intervals in a computerized translation process

digital subscriber line modem (DSL) connects user's computer to the **Internet** via dedicated telephone lines passing through the ISP offering the DSL service

digital video recorder (DVR) television service that uses sophisticated digital software to put significant amounts of control over content in viewers' hands

digital watermarking information added to a computer file that identifies its owner or origin; a protection against piracy

direct broadcast satellite (DBS) direct delivery of video to home receivers via satellite

direct sales see **retail sales**

disinhibitory effects in social cognitive theory seeing a model rewarded for making prohibited or threatening behavior increases the likelihood that the observer will make that behavior

distance learning broadcast and online instructional programming, usually offered by colleges and universities for people who live too far away to regularly attend classes

distributors those who provide content to delivery systems such as cable channels and television stations

DMA an audience measurement area (designated market area); sampling area as defined by television ratings company A. C. Nielsen

DMX (Digital Music Express) home delivery of audio by cable

D-notice in Great Britain an officially issued notice of prior restraint

domain name an identifying name on the **World Wide Web** rather than a site's formal **URL,** which gives some indication of the nature of a site's content or owner

download when Internet users transfer files from an outside computer to their own computer

drive time a radio station's most expensive time, typically 6 a.m. to 10 a.m., when listenership is assumed to be the highest

drop cable the cable television line that runs from the feeder to people's homes

duopoly ownership of two or more radio stations of the same service (AM or FM) in a single market

DVD CD-sized digital video discs that contain video images

early evening daypart in television programming typically 6 p.m. to 7 p.m. or 7:30 p.m.

early fringe in local television programming typically 4 p.m. to 6 p.m.

early morning daypart in television programming typically 7 a.m. to 9 a.m.

early window the idea that media give children a window on the world before they have the critical and intellectual ability to judge what they see

e-book digital book having the appearance of a traditional book but whose content is digitally stored and accessed

e-commerce commercial use of the **Internet** for the sale of goods or services

economies of scale concept that relative cost declines as the size of the endeavor grows

electronic scanning in television dividing a scene through electronic means into an orderly pattern of transmittable picture elements that can be recomposed as a visual image

e-mail electronic mail communication via computer with anyone else online, any place in the world, with no long-distance fees

embedded system computer system that adds intelligence to other technologies

encryption coding online content at the source so it can only be read by an end user with appropriate decoding software

environmental incentives in social cognitive theory the notion that real world incentives can lead observers to ignore negative vicarious reinforcement

error term see **margin of error**

estimated sampling error see **margin of error**

e-tailers companies offering goods and services for sale over the **Internet**

ethics questions of behavior or moral principles that guide actions in given situations

Euro English the official language of the European Union

evergreen a classic off-network program and a guaranteed syndication winner

exclusive cume number of different people who listen to only one radio station during a given **daypart**

exogenous stations clandestine broadcast operations functioning from outside the regions to which they transmit

expanded basic cable in cable television a second, somewhat more expensive level of subscription

external service an international broadcasting service designed by one country to counter enemy propaganda and disseminate information about itself

extramercials columns of ad content in cyberadvertising that slide down over a Web page's content

Fairness Doctrine required broadcasters to cover issues of public importance and to be fair in that coverage (abolished in 1987)

fair use in copyright law instances where material can be used without permission or payment

features radio content such as traffic and business reports, newscasts, sports, and weather

feature syndicators distributors of short, standalone or individual radio programs or series

feeder cables the television cable that runs from the trunk cable into individual neighborhoods or areas

fiber optics signals carried by light beams over glass fibers

Financial Interest and Syndication Rules (Fin-Syn) FCC rules no longer in force that limited television networks' ownership in the entertainment fare they air

first-run syndication programming, primarily television, produced specifically for sale to local stations

flat rate billing flat monthly charge levied by a provider for **Internet** access

focus groups small groups of people typically interviewed to provide advertising professionals with detailed information

forced exposure advertising research technique used primarily for television commercials, requiring advertisers to bring consumers to a theater or other facility where they see a television program, complete with the new ads

forfeiting audience when a show in network prime time programming loses audience because it is not preceded by a strong **lead-in**

format a radio station's particular programming content

format doctor consultants who help radio stations improve their **format** or sound

format syndicators companies that provide both long- and short-form radio programming in a given **format** via satellite to local, usually automated stations

fraction of selection graphic description of how individuals make media and media content choices based on expectation of reward and effort required

franchise official authority to operate a cable television system

frequency (in advertising) average number of times a radio listener hears an advertiser's schedule of commercials

frequency (electromagnetic waves) the number of cycles of a wave that occur in one second

frequency modulation (FM) shaping radio waves' frequency; occupies radio spectrum frequencies between 88 and 108 mHz

FTP (file transfer protocol) program that allows copying files on the **Inernet** between computers

functional displacement when existing media survive the advent of emerging media by finding new functions to serve

geosynchronous Earth orbit (GEO) satellites in orbit around the Earth at a height of 22,300 miles

grazing watching several television programs simultaneously

gross impressions (GIs) total of the **AQH Persons** for all commercials in a specified period of time

gross ratings points (GRPs) total of all ratings points achieved for all the commercials in a specified period of time

group systems audio and video systems designed for videoconference sites engaged in meetings held in large rooms

guaranteed CPMs television networks' guarantee to their advertisers of a certain return for their money, computed in **CPM**

hackers people interested in technology, information, and communication through computers

hammocking placing a new or struggling prime time television show between two proven programs

HDTV digital compression of television signals enabling a single high-resolution signal to be sent in a standard television channel

headend the place or equipment necessary for cable television to receive and process a signal

Hertz the standard measurement of electromagnetic waves; one Hertz is one cycle per second

home page entry way into a Web site containing information and **hyperlinks** to other material

horizontal integration when companies own other similar companies in an attempt to control a larger portion of a specific market

hosts computers linking individual personal computers to the **Internet;** more commonly called **servers**

hub the cable television system operation itself, where signals are processed and boosted for distribution

HUT households using television; an audience measure

hypercommercialism with the concentration of media ownership outlets can increasingly commercialize their content without fear of consumer refusal

hyperlink connection imbedded in the **Internet** or Web site that enables instant access to other material in that site as well as to material in other sites

iconoscope tube first practical television camera tube, developed in 1923

identification a special form of imitation in social cognitive theory where observers do not exactly copy what they have seen but make a more generalized related response

imitation in social cognitive theory the direct replication of an observed behavior

importation of distant signals the ability of cable and DBS television to deliver channels from outside the receivers' area for the purpose of improved reception and to offer a wider variety of programming

in-band-on-channel (IBOC) digital radio technology that uses **digital compression** to "shrink" digital and analog signals, allowing both to occupy the same channel

indecency broadcasting language or material that depicts sexual or excretory activities in a way offensive to contemporary community standards

independents broadcast stations that remain apart from a **network**

indigenous stations clandestine broadcast operations functioning from inside the regions to which they transmit

industrial telecommunications see **corporate telecommunication**

infomercials see **program length commercial**

information gap diminished access to information available in telecommunications technologies because individuals do not have those technologies

information superhighway descriptive, collective expression used to describe worldwide digital data networks

inhibitory effects in social cognitive theory seeing a model punished for a behavior reduces the likelihood that the observer will express that behavior

in-tab rate proportion of people who agree to participate in the Nielsen ratings who do and do not complete their diaries accurately or who do and do not employ their **peoplemeters** properly on a daily basis

integrated circuit miniature computer component employing tiny transistors and other components etched onto a single silicon wafer

interactive cable the ability of subscribers to talk back to the cable system operator

interdiction technology technology that descrambles pay channel cable television signals outside the home

interlaced scanning television format in which electron beams sweep the picture tube twice, creating half the image's lines on the first pass and filling in the gaps on the second pass

intermercials attractive, lively commercials that run while people are waiting for Web pages to **download**

Internet global network of interconnected computers that communicate freely and share and exchange information

Internet service provider (ISP) company that offers **Internet** connections at monthly rates depending on the kind and amount of access needed

interstitial ads cyberadvertising images that seem to mysteriously appear and disappear on the screen as users click from one page to the next

island a television commercial that airs inside a show and stands alone, without other commercials

kinescope improved picture tube developed by Zworykin for RCA

Kranzberg's law technology is neither good, nor bad, nor neutral

LANs (local area networks) networks connecting two or more computers, usually within the same building

late fringe the 30 or 60 minutes in local television programming that come immediately after network prime time

late night daypart those hours immediately after television networks' late night programs

lead-in scheduling a strong show in programming network prime time so that a second, presumably weaker show inherits some of its audience

lead-off position the first television show in an evening's prime time lineup

libel false and malicious publication of material that damages a person's reputation

libertarian media system a national media system based on the philosophy of **libertarianism**

libertarian philosophy philosophy of the press that asserts that good and rational people can tell right from wrong if presented with full and free access to information; as such, censorship is unnecessary

licensing fee the fee paid to a syndicator by a local station for a given piece of content

line of sight signals that can only be received by sites that are in a straight line

liquid barretter first audio device permitting reception of wireless voices; developed by Fessenden

LISTSERV software program used to run **e-mail** mailing lists, bulletin boards, and discussion groups

local avails in broadcast and cable network/affiliate relationships, advertising time that can be sold by the local affiliate who keeps any income from that sale

local carriage rules see **must-carry rules**

local franchising authorities local bodies charged with granting and oversight of cable television franchises

local marketing agreement (LMA) arrangement allowing one broadcast station in a market to operate another without formally owning it

local sales selling broadcast time to advertisers through their local advertising agencies

log an FCC required record of the technical aspects of a radio station's broadcast day

long-form programming syndicator-supplied, complete radio shows, typically 4 hours or longer

low Earth orbit (LEO) satellites in orbit around the Earth at altitudes from 100 to 1,000 miles above the ground

macro-level effects media's wide-scale social and cultural impact

mainframe computer large, central computer to which users are connected by terminals

make-goods if a television network fails to deliver on a guaranteed CPM, additional commercials are aired to compensate the sponsor

margin of error the predictable built-in level of error that accompanies all **samples** of a given size

market exclusivity a guarantee that only one station in a given broadcast area can carry a syndicated program

mass communication creating shared meaning between those producing mass media messages and their audiences

mass media technologies (and the industries that utilize and support them) that carry messages to a large number of people

master antenna television (MATV) connecting multiple sets in a single building or location to a single, master antenna

mechanical scanning mechanically dividing a television scene into an orderly pattern of transmittable picture elements that can be recomposed as a visual image

media councils panels of people from both the media and the public who investigate complaints against the media and publish their findings

media department the department in an ad agency where decisions about placement and purchase of commercials are made

mediated communication communication conducted through the use of electronic and digital technology

member station in public broadcasting the equivalent of an **affiliate**

metaethics examination of a culture's understanding of its fundamental values

metro a broadcast audience measurement that includes a city (or cities) whose population is composed of that central city and the county (or counties) in which it is located

microbroadcasting low-power community radio stations that typically broadcast at a power from 10 to 100 watts

microcomputer small computer that uses a **microprocessor** to handle information (also called a **personal computer** or **PC**)

micro-level effects effects of media on individuals

microprocessor a computer chip carrying several thousand to millions of circuits

microwave relay system audio and video transmitting system in which super high frequency signals are sent from land-based point to land-based point

middle Earth orbit (MEO) satellites in orbit around the Earth at altitudes ranging from 1,000 to 22,300 miles

minicomputers relatively large central computer to which users are connected by terminals; not as large as a **mainframe computer**

MNA (multinetwork area) reports television ratings computations based on the 70 largest markets

modeling in social cognitive theory learning through **imitation** and **identification**

modem device that translates digital computer information into analog form so it can be transmitted through telephone lines

moral agent in an ethical dilemma the person making the decision

moral reasoning use of systematic thought in making ethical decisions

morning daypart in television programming typically 9 a.m. to noon

MP3 file compression software that permits **streaming** of digital audio and video data

MSA metropolitan statistical area; an audience measurement area defined by the federal government Office of Management and Budget

multichannel multipoint distribution services (MMDS) a form of "wireless cable" that distributes programming via microwave

multimedia advanced sound and image content available to **microcomputers** and on **videodiscs**

multiple systems operator (MSO) a company that owns several different cable television operations

multiplexing the practice of using one channel to transmit multiple forms of content; in FM radio by using extra channel space and in television and cable through signal compression

music licensing company organizations that collect fees based on recorded music users' gross receipts and distribute the money to song writers and artists

must-carry rules rules requiring cable television systems to carry the signal of every television station within a specified radius

national sales representatives broadcasting companies that sell stations' local time to national advertisers (commonly referred to as **reps**)

national spot sales selling local broadcast time to national advertisers

netlets smaller, newer national broadcast television networks such as WB and UPN

net reach number of different people reached by an advertiser's schedule of broadcast commercials

network centralized production, distribution, and decision-making organization that links **affiliates** for the purpose of delivering their viewers to advertisers

news cycle a rotation of radio news that might include national news, regional news, local news, weather, sports, and a leisure or lifestyle feature

news staging re-creation on television news of some event that is believed to have happened or which could have happened

Nipkow disc first workable device for generating electrical signals suitable for transmission of a picture

noncommercial broadcasting over-the-air radio and television stations that do not rely on the sale of commercial time for their financial support

nonduplication rule mid-1960s FCC ruling that holders of an AM and FM license in the same market must broadcast different content at least 50% of the time (since repealed)

normative ethics generalized theories, rules, and principles of ethical or moral behavior

normative theory an expression of how media should ideally operate in a given system of social values

NSI market-specific television sweeps data collected and reported by the Nielsen Company

NTI Nielsen Television Index; national ratings computed from metered markets

NTSC technical standards for U.S. television, mandating a picture composed of 525 lines, scanned completely 30 times per second

O&O a broadcasting station that is owned and operated by a **network**

obscenity unprotected expression determined by (a) whether the average person, applying contemporary community standards, would find that the work, taken as a whole, appeals to the prurient interest, (b) whether the work depicts or describes, in a patently offensive way, sexual conduct specifically defined by the applicable state law, and (c) whether the work, taken as a whole, lacks serious literary, artistic, political, or scientific value

observational learning the social cognitive theory that observers can learn new behaviors simply by seeing those behaviors performed

off-network broadcast industry term for syndicated content that originally aired on a **network**

off-network syndication syndicated programming that initially appeared on a network schedule

off-season in network television the weeks and months other than the 30 weeks running from late September to May

oligopoly concentration of control of media industries into an ever-smaller number of companies

ombudspeople internal arbiters of performance for media organizations

open access the requirement that telephone companies make their lines available for a fair price to competing telecommunications service providers; does not yet apply to cable operators

open directory a free, volunteer-created classified directory to **World Wide Web** sites

open source software free software that can be **downloaded**

operating system the software that tells the computer how to work

organizational telecommunications see **corporate telecommunication**

outright cash purchase when a station pays a fee, based on its local advertising rates, for a given piece of syndicated content

overbuilds establishment of a second cable service in an area

overnight dayparts television programming hours after late night until morning

overnights instant television ratings gathered from homes in several major urban centers

panel a group of Internet users deemed typical of the larger population whose Internet use is tracked as an audience measurement

parity products products generally perceived as alike by consumers no matter who makes them

participation when a television commercial airs inside a program

pass-by rate the number of homes passed by that have the potential to subscribe to a cable system's **cable**

pay-as-you-go pricing billing Internet users by their providers based on the amount of time online

pay cable cable television channels delivered to those viewers who pay a fee over and above their basic service charge

payola the practice of taking payments in exchange for playing specific pieces of music on the radio

pay-per-view cable television viewers pay a set fee for delivery of a specific piece of content (rather than an entire cable channel)

PC see **personal computer**

penetration the number of homes passed by the television cable that actually subscribe

peoplemeter remote control key-pad device and set-top box that records television viewing for establishing ratings

perishable inventories products and services that are salable for a relatively short or specific period of time

personal audience venue an advertising venue that people use individually, on their own rather than on the programmer's schedule

personal computer (PC) see **microcomputer**

piggybacking when a single sponsor presents two products in the same television commercial

pilot a sample episode of a proposed television program

piracy illegally recording and selling copyrighted material

pirate broadcasters unlicensed or otherwise illegally operated broadcast stations

pixels the smallest picture element in an electronic imaging system

platform in media synergy the individual delivery systems

playlist the predetermined selection of music to be played by a disc jockey

pocketpieces television ratings based on a national sample that are computed and reported every 2 weeks

pods groupings or clusters of commercials on television

point-of-purchase video promotion using monitors in retail settings to present satellite-delivered, product-specific audio and video

policy book delineates standards of operation for local broadcasters

pop-outs cyberadvertising that appears in a smaller window at the border of the Web page being read

population in audience research the group under scrutiny from which a **sample** is drawn

pornography expression calculated solely to supply sexual excitement

portal one-stop brand-name entryway to the Web offering users news, chat, search functions, original content, links to specific sites, advertising, and **e-commerce**

postbuy analysis an analysis to determine if the network has delivered the agreed upon number and type of viewers for a television sponsor's schedule of ads

premium tier the highest level of cable television subscription, which includes all programming from the lower tiers plus a selection of pay channels

prime access in local television programming typically 7 p.m. or 7:30 p.m. to 8 p.m.

prior restraint the power of the government to prevent publication or broadcast of expression

producers those who create content

program clock a visual representation of where the music, commercials, station identification, promotions, and news appear in a radio sound hour (sometimes called a **program wheel**)

program length commercial (PLC) 30- and 60-minute commercials touting the benefits of a single product or service (often called **infomercials**)

programmers those who schedule and organize content

program wheel see **program clock**

progressive scanning digital television format compatible with **personal computers** in which the entire picture is built line by line in one scan of the television's electronic beam

promos see **promotional spots**

promotional spots commercial-like announcements promoting broadcasters' and cablecasters' upcoming program schedules

protocols common communication rules and languages for computers linked to the **Internet**

pseudo-events events that have no real informational or issue meaning that exist solely to attract media attention

psychographic segmentation advertisers' appeal to consumer groups of varying lifestyles, attitudes, values, and behavior patterns

public access channels cable channels reserved on a first-come, first-served, nondiscriminatory basis for use by groups or individuals who maintain editorial control of their programming

public domain in copyright law the lawful use of material without permission once the **copyright** expires

public service announcement (PSA) spot promoting nonprofit causes and organizations aired for free

public service remit limits on advertising and other public service requirements imposed on Britain's commercial broadcasters in exchange for the right to broadcast

puffery the little lie or exaggeration that makes advertising more entertaining than it might otherwise be

push technology software permitting automatic delivery or downloading of Internet data even to sleeping computers

put agreement between a **producer** and a **network** guaranteeing that the network will order at least a **pilot** or the network must pay a hefty penalty

radiation system wireless communication system that enables reliable sending and receiving of signals sent through the air

random sample an audience research sample in which each member of the population has equal likelihood of being selected

rate card a broadcast station's statement of advertising fees and services

rate protection policy guarantee to a sponsor that the price of time agreed on for a given period will not be altered even if the station changes its fee schedule

rating an audience measure of the percentage of all available sets tuned in to a given program

ratings, the all inclusive, generic name given to various audience measurements

recall method radio ratings technique in which listeners are asked to remember what they had been listening to

recall testing advertising research technique where consumers are asked to identify commercials they find easiest to remember

recognition tests advertising research technique where people who have seen a given program are asked whether they remember seeing a particular commercial

reps see **national sales representatives**

repurposing reusing or reshaping a piece of content for use as original programming on network television

reruns network rebroadcasts of television shows originally broadcast by those same networks

research the objective search for knowledge

reserved channels broadcast frequencies set aside or reserved specifically for noncommercial broadcasting

retail sales in broadcasting the direct, one-to-one contact between station salespeople and advertisers in their signal area (sometimes called **direct sales**)

retransmission resending material on cable that was originally meant for broadcast

retransmission consent local television broadcasters' permission for the cable system to carry its signal

reverse compensation affiliates pay networks for programming rather than the more traditional situation where networks pay affiliates

reverse filtering web filtering technology through restrictive **portals** that allow access only to preapproved sites

ride-alongs the practice of allowing television reporters to accompany police in the conduct of their duty

ritual perspective the view of media as central to the representation of shared beliefs and culture

roadblock when a sponsor runs its spot at the same time on multiple radio or cable stations

roster-recall method radio ratings technique in which listeners, interviewed in person and often given a roster of stations, are asked to name stations listened to in the past day and past week

rotation the frequency with which different songs (or types of songs) are played on radio stations

safe harbor times of the broadcast day (typically 10:00 p.m. to 6:00 a.m.) when children are not likely to be in the listening audience

sample in audience research respondents drawn from a population who are considered representative of that population

satellite-delivered media tour satellite-facilitated "tour" in which spokespeople can be simultaneously interviewed by a worldwide audience connected to the on-screen interviewee via telephone

satellite master antenna television (SMATV) a type of minicable system serving a single large apartment building or complex

search engines Web or Internet search software providing on-screen menus (sometimes called **spiders** or **web crawlers**)

secondary service a radio station's second, or nonprimary, **format**

self-righting principle John Milton's articulation of **libertarianism**

semiconductor material that sometimes conducts electricity and sometimes doesn't

servers see **hosts**

sets in use an audience measure of the number of radio or television sets turned on

share an audience measure of the percentage of sets in use tuned to a given program

shield laws legislation that expressly protects reporters' rights to maintain sources' **confidentiality** in courts of law

short-form programming syndicator-supplied, individual radio programs or series

short-ordering television network practice of ordering only one or two episodes of a new television series

shortwave radio radio signals transmitted at high frequencies that can travel great distances by skipping off the ionosphere

Sixth Report and Order the fundamental blueprint for the technical operation of television (issued in 1952)

skip ability of radio waves to reflect off the ionosphere

sky waves radio waves that are skipped off the ionosphere

slander oral or spoken defamation of a person's character

snail mail unflattering name for the kind of mail delivered by the postal service

social cognitive theory the idea that people learn through observation

social responsibility theory normative theory asserting that media must remain free of government control but in exchange must serve the public

sound hour a specific hour of radio broadcast time

spam mass mailing of Internet messages to users whether they want them or not

spectrum scarcity broadcast spectrum space is limited, and those who are granted licenses must accept regulation

spiders see **search engines**

spot commercial sales selling individual advertising spots on a broadcast program to a wide variety of advertisers

stacking see **blocking**

Standards and Practices departments the internal content review operation of a television network

stereotyping application of a standardized image or conception to members of certain groups, usually based on limited information

stickiness in cyberadvertising an attribute of an ad that compels people to stay with a given Web page

stimulation model posits that viewing mediated violence can increase the likelihood of subsequent aggressive behavior

streaming simultaneously downloading and playing digital audio or video data

stripping television programming strategy that shows an episode of the same program five, six, or seven nights a week in the same time slot

stunting the practice of programming special events in network prime time to draw viewers away from other networks

subscription TV television's early experiments with over-the-air pay television

subsidiary communications authorization (SCA) use of "extra" signal space in radio to transmit a second, distinct form of content

super trunk cable the television cable that leads from the **headend** to the **hub**

surrogate service an international broadcasting operation established by one country to substitute for another's own domestic service

sweeps periods special television ratings times in February, May, July, and November in which diaries are distributed to sample households in all television markets

syndicated exclusivity rules rules prohibiting cable operators from importing programming from distant sources that might duplicate that aired by a local station

syndication sale of radio or television content to stations on a market by market basis

synergy using a piece of content or message across a number of different media **platforms**

T1 connection that downloads Internet data at a speed of 1.5 megabits per second

targeted keyword buys when a sponsor buys the right from an **Internet service provider** to have its ads pop up on the computer screen every time a user types in the sponsor's name

taste publics groups of people or audiences bound by little more than their interest in a given form of media content

technological determinist a person who believes machines and their development drive historical, economic, and cultural change

technology gap the widening disparity between communication technology haves and have-nots

telcos telephone companies

telecom model of telecommunications model that posits that audiences do not consume content prepared for them but use technology to create their own content

telecommunications the technologies and industries that facilitate **mediated communication**

telecourse college and university instruction delivered to students by television

telenovelas Spanish-language soap operas

teleological moral reasoning consequence-based moral reasoning in which the most moral action is that which results in the best outcome

Television Freeze 1948 freeze in authorization of new television stations while the FCC resolved a number of technical problems

Telnet Internet tool that allows users to log into other computers as if they were connected to them directly

tent-poling scheduling a proven winner in the midway spot of the network prime time schedule (typically 9:00 p.m.) in hopes of having it benefit the programs surrounding it

terminal user workstation connected to a larger, centralized computer

tier groups of cable television channels made available to subscribers at varying prices

time-shifting taping a show on a VCR for later viewing

time spent listening (TSL) estimate of the number of quarter hours an average person spends listening to the radio during a given period of time

time splits in television **barter syndication** the number of national commercial minutes that must be run plus the number of local minutes available for sale

toll broadcasting an early radio system in which anyone could buy airtime to deliver a message

toy-based children's television programming television programs aimed at children that feature characters that are also toys

traffic-cop analogy the idea that the FCC, as a traffic cop, has the right to control not only the flow of broadcast traffic but its composition

transaction journalism in cyberadvertising directly linking editorial content to sales

transistor miniature electronic switch employing **semiconductors**

transmissional perspective the view of media as senders of information for the purpose of control

transnational corporation (TNC) a large international conglomerate

transponders part of the satellite that receives and retransmits a specific telecommunications service

trunk cable the television cable that leads from the **headend** into the community

trustee model the idea that broadcasters serve as the public's trustees or fiduciaries

TSA the total service area, which is a geographical area encompassing an **MSA** and including additional areas outside it meeting certain listening criteria

UHF ultra high frequency, which are television channels 14 to 69 on the broadcast spectrum

unbundle when a cable operator make its lines available to competing phone and Internet providers

underwriting public broadcasting financial support for programming from outside, corporate sources

unified analysis the Nielsen practice of measuring the same samples of viewers a year apart to verify ratings accuracy

uniform resource locator (URL) the designation for each file or directory on the **host** computer connected to the **Internet**

unique selling proposition (USP) the aspect of an advertised product that sets it apart from other brands in the same product category

unique visitors a standard ratings measure for the Web that shows the number of people who visit a specific site in a specific period of time

Universal Licensing System (ULS) the FCC's online application renewal system

unwired networks collections of unrelated radio stations that are sold to advertisers as if they were in fact actual broadcast networks

upfront sales the sale of commercial spots on the next season's prime time network television schedule

uploading when Internet users transfer files from their own computers to others

USENET an internationally distributed Internet bulletin board system; also known as network news

user individual who uses the services of a telecommunications operation rather than only consuming its content

VALS advertisers' psychographic segmentation strategy that classifies consumers according to values and lifestyles

vast wasteland expression coined by FCC Chair Minow in 1961 to describe television content

vertical integration when a telecommunications company holds financial interest in two or more aspects (production, distribution, and exhibition) of its industry

VHF very high frequency, which includes television channels 2 to 13 on the broadcast spectrum

vicarious reinforcement in social cognitive theory the idea that observation of reinforcement operates in the same manner as actual reinforcement

videoconferencing meeting with people in different locations via satellite-distributed audio and video that permits full and instantaneous interactivity

videodiscs digitally recorded discs that contain audio and video images; some are read-only, some recordable

video news release preproduced report about a group or product distributed free of charge to television stations on satellite or by videocassette

video-on-demand the ability of cable viewers to call up any piece of content at any time and have it delivered immediately

viewing year in network television the 30 weeks running from late September to mid-May

virtue theories moral reasoning that requires a moral agent to evaluate his or her own character

WANs (Wide Area Networks) networks that connect several **LANs** in different locations

wavelength the distance between successive positive or negative peaks of an electromagnetic wave

Web crawlers see **search engines**

weblets newer national television networks striving to develop complete schedules along the lines of the more established competitors

Web radio delivering "radio" over the **Internet** directly to individual listeners

WebTV online delivery of high-definition television and Internet content to special home receivers

willing suspension of disbelief the audience practice of willingly accepting the content before it as real

wired radio centrally located loudspeakers in remote areas of many developing countries that deliver radio broadcasts

wireless Internet delivering Internet data to **users** through the air

wireless telegraphy early efforts at long-distance communication that led to the development of radio

wireless telephony early efforts at long-distance voice communication that led to the development of radio

World Wide Web tool that uses sophisticated transfer **protocols** as a means of accessing audio, video, graphic, and text files on computers connected to the **Internet**

year-round scheduling the practice of debuting network television programs outside the traditional fall premiere season

zapping using the remote control to switch to other content when a commercial appears

zipping fast-forwarding through taped commercials on a VCR

zone Internet address information typically either geographic or descriptive of the type of organization

References

Accas, G. (2000, January 10). The cost of spots. *Broadcasting & Cable,* 26–30.

A. C. Nielsen Company. (1996). Company history. [On-line]. Available: http://www.nielsenmedia.com.news.history.html.

Adalain, J. (1998, December 14–20). Ingenue survives hairy ride. *Variety,* pp. 1, 148.

Adams, M. H. (1996). The race for radiotelephone: 1900–1920. *The AWA Review, 10,* 78–149.

Adams, M. H. (1999, August 25). Public service essential to radio. *San Jose Mercury News,* p. 4A.

Adams, M. H., & Massey, K. K. (1995). *Introduction to radio: Production and programming.* Dubuque, IA: Brown & Benchmark.

Adams, W. J., Eastman, S. T., & Lewine, R. F. (1993). Prime-time network television programming. In S. T. Eastman (Ed.), *Broadcast/cable programming* (pp. 115–159). Belmont, CA: Wadsworth.

Advertising outlook: how the pie will be sliced in '99. (1999, January 4). *Broadcasting & Cable,* 26–50.

AFRTS. (2000). Mission statement. [On-line]. Available: http://www.eucom.mil/programs/afrts/index.htm.

Aiken, E. G., & Affe, R. B. (1993). Independent station programming. In S. T. Eastman (Ed.), *Broadcast/cable programming* (pp. 216–241). Belmont, CA: Wadsworth.

Albiniak, P. (1998, December 21). Gore panel takes middle ground. *Broadcasting & Cable,* 7.

Albiniak, P. (1999, May 3). Media: Littleton's latest suspect. *Broadcasting & Cable,* 6–15.

All together now. (1987, December 15). *Electronic Media,* 14.

Alterman, E. (2000, June 5). Disinterestedness dies. *The Nation,* p. 12.

Apar, B. (1997, Summer). DaViD meets Goliath. *Video Business DVD Supplement.* p. 10.

Arbitron. (1999). Survey process. [On-line]. Available: http://www.arbitron.com.

Assignment E! (1999, August 29). Too young Hollywood. E! cable television network. Los Angeles, CA.

Aumente, J. (1999, January/February). Cracking down: Yugoslavia's campaign against an independent media harkens back to the Cold War era. *American Journalism Review,* 40–43.

Baker, R. (1999, October 20). Profits vs. truth. *San Jose Mercury News,* p. 6B.

Bandura, A. (1965). Influence of model's reinforcement contingencies on the acquisition of imitative responses. *Journal of Personality and Social Psychology, 1,* 589–595.

Baran, S. J. (1999). *Introduction to mass communication: Media literacy and culture.* Mountain View, CA: Mayfield.

Baran, S. J., Chase, L. J., & Courtright, J. A. (1979). *The Waltons:* Television as a facilitator of prosocial behavior. *Journal of Broadcasting, 23* (3), 277–284.

Baran, S. J., & Davis, D. K. (2000). *Mass communication theory: Foundations, ferment, and future* (2nd ed.). Belmont, CA: Wadsworth.

Barber, B. (1998, June 15). When VOA talks, millions around the world listen. *Insight on the News,* 40–41.

Barlow, J. P. (1996). Selling wine without bottles: The economy of mind on the global Net. In L. H. Leeson (Ed.), *Clicking in: Hot links to a digital culture* (pp. 148–172). Seattle, WA: Bay Press.

Barnouw, E. (1968). *The golden web: A history of broadcasting in the United States, 1933–1953.* New York: Oxford University Press.

Barnouw, E. (1990). *Tube of plenty: The evolution of American television.* New York: Oxford University Press.

Barr, B. W., Brewer, D. F., & Eastman, S. T. (1993). Information radio programming. In S. T. Eastman (Ed.), *Broadcast/cable programming* (pp. 447–469). Belmont, CA: Wadsworth.

Barry, D. (1990, March 18). It's Howdy Duty time. *West Magazine,* 23.

Bayers, C. (1999, March). The inner Bezos. *Wired,* 115–121, 172–187.

Bergman, L. (2000, May/June). Network television news: With fear and favor. *Columbia Journalism Review,* 50–51.

Bierbaum, T. (1998, December 7–13). Diminished Peacock poised for demo dash. *Variety,* 33.

Big Blue vs. Big Brother. (1999, April 12). *San Jose Mercury News,* p. 6B.

Bing, S. (2000, February 21). Will we all be couch potatoes? *Time,* 58.

Bittner, J. R. (1994). *Law and regulation of electronic media.* Englewood Cliffs, NJ: Prentice-Hall.

Black, J., & Barney, R. D. (1985–86). The case against mass media codes of ethics. *Journal of Mass Media Ethics,* 27–36.

Black, J., & Whitney, F. C. (1983). *Introduction to mass communications.* Dubuque, IA: Wm. C. Brown.

Boddy, W. (1992). *Fifties television: The industry and its critics.* Champaign: University of Illinois Press.

Boddy, W. (1994). Alternative television in the United States. In H. Newcomb (Ed.), *Television: The critical view* (pp. 352–362). New York: Oxford University Press.

Boothroyd, J. (1999, Winter). Media Carta, the human rights battle of the Information Age. *Adbusters,* 16–27.

Brent, W. (1999, July 12–18). China says rats to sats. *Variety,* p. 35.

Brill, S. (2000, April). The mega threats. *Brill's Content,* 23–27.

Broadcasting & Cable Yearbook, 1997. (1997). New Providence, NJ: R. R. Bowker.

Broadcasting & Cable Yearbook, 1998. (1998). New Providence, NJ: R. R. Bowker.

Broadcasting & Cable Yearbook, 1999. (1999). New Providence, NJ: R. R. Bowker.

Broadcasting & Cable Yearbook, 2000. (2000). New Providence, NJ: R. R. Bowker.

Brown, C. (1999, June). Fear.com. *American Journalism Review,* 50–71.

Brown, L. (1979). *Keeping your eye on television.* New York: Pilgrim Press.

BRS Radio Consultants. (1999). Web radio survey. [On-line]. Available: http://www.radio-directory.com.

Buel, S. (1998, August 3). Measuring sticks fall short. *San Jose Mercury News,* pp. 1E, 5–6E.

Buel, S. (1999, March 30). Amazon.com to do battle with eBay. *San Jose Mercury News,* pp. 1C, 2C.

Burress, C. (1999, March 25). Kosovo teen e-mails her diary to the world. *San Francisco Chronicle,* p. A14.

Burstein, D., & Kline, D. (1995). *Road warriors: Dreams and nightmares along the information highway.* New York: Dutton.

By the numbers. (1999, December 13). *Broadcasting & Cable,* 115.

Carey, J. W. (1975). A cultural approach to communication. *Communication, 2,* 1–22.

Carman, J. (1998, November 17). The fall of network television. *San Francisco Chronicle,* pp. E1, E4.

Childs, K. (1998, October 24). Debate over world news ethics code. *Editor & Publisher,* 11.

Chin, F. (1978). *Cable television: A comprehensive bibliography.* New York: IFI/Plenum.

China adopts new Net curbs. (1997, December 31). *San Jose Mercury News,* pp. 1C, 4C.

Christians, C. G., Rotzoll, K. B., & Fackler, M. (1991). *Media ethics: Cases and moral reasoning* (3rd. ed.). White Plains, NY: Longman.

Colman, P. (1998, December 7). Keeping the customer, period. *Broadcasting & Cable,* 58.

Conniff, K. (1999, June). Bringing the conflict home. *Brill's Content,* 108.

Cooper, G. (1999, September/October). Darts & laurels. *Columbia Journalism Review,* 17–18.

Corporation for Public Broadcasting. (1999). Frequently asked questions. [On-line]. Available: http://www.cpb.org/content/faq.

The cost of regulation. (1999, June 28). *Broadcasting & Cable,* 13.

Cox, A. M. (1999, May/June). John Hockenberry. *Mother Jones,* 40–43.

Cox, C. (1999, September/October). Prime-time activism. *Utne Reader,* 20–22.

Crandall, R. W., & Furchtgott-Roth, H. (1996). *Cable TV: Regulation or competition?* Washington, DC: The Brookings Institute.

Crystal, D. (1997). *English as a global language.* Cambridge: Cambridge University Press.

Culture wars. (1998, September 12). *The Economist,* 97.

Davidow, B. (1998, December 6). Tools are making the rules. *San Jose Mercury News,* pp. 4P–5P.

Davis, D. (1993). *The five myths of television power: Or, why the medium is not the message.* New York: Simon & Schuster.

Davis, D. K. (1990). News and politics. In D. L. Swanson & D. Nimmo (Eds.), *New directions in political communication* (pp. 147–184). Newbury Park, CA: Sage.

Davis, L. J. (1998). *The billionaire shell game: How cable baron John Malone and assorted corporate titans invented a future nobody wanted.* New York: Doubleday.

Davis, R. E. (1976). *Responses to innovation: A study of popular arguments about new mass media.* New York: Arno.

Day, L. A. (1997). *Ethics in media communications: Cases and controversies.* Belmont, CA: Wadsworth.

Dennis, E. E. (1992). *Of media and people.* Newbury Park, CA: Sage.

Diamond, D. (1999, August 29). In brief: What is Linux, anyway? *SV Life,* 13.

Ditingo, V. M. (1999, January 4). Radio. *Broadcasting & Cable,* 39.

Dobson, W. J. (1998, July 6). Protest.org. *The New Republic,* 18–21.

Draper, R. (1986, June 26). The faithless shepherd. *The New York Review,* 14–18.

Duncan, K. (1998). Antitrust law and the media monopoly. *The Cultural Environment Monitor, 2,* 1, 8.

Eastman, S. T. (1993). Cable system programming. In S. T. Eastman (Ed.), *Broadcast/cable programming* (pp. 245–282). Belmont, CA: Wadsworth.

Effron, E. (1999a, February). The big blur. *Brill's Content,* 44–45.

Effron, E. (1999b, July/August). Journalism. commerce. *Brill's Content,* 56–57.

Effron, E. (1999c, November). Taking the sin out of synergy. *Brill's Content,* 47–48.

Eisenstein, E. L. (1979). *The printing press as an agent of change: Communications and cultural transformations in early-modern Europe.* Cambridge: Cambridge University Press.

Ellis, J. (1999, January 4). In tomorrow's economy, only the online will thrive. *San Jose Mercury News,* p. 7B.

Emert, C. (1999, May 4). Web ad spending passes outdoor medium. *San Francisco Chronicle,* pp. C1, C4.

Epstein, J. (1999, June 26). Sex and sleaze put Brazil's television viewers in a tizzy. *San Francisco Chronicle,* pp. A10, A12.

Eron, L. D. (1992, June 18). Testimony before the Senate Committee on Government Affairs. *Congressional Record.*

Fawcett, A. W. (1996, October 16). Interactive awareness growing. *Advertising Age,* 20.

Financial Times. (1999, April 6). Report to re-ignite Net privacy debate. *San Jose Mercury News,* p. 5C.

Fisher, D. E., & Fisher, M. J. (1996). *Tube: The invention of television.* Washington, DC: Counterpoint.

Fleming, H. (1996, December 16). Valenti delivers the V-chip code. *Broadcasting & Cable,* 6–7.

Fogarty, J. R., & Spielholz, M. (1985). FCC cable jurisdiction: From zero to plenary in twenty-five years. *Federal Communications Law Journal, 37,* 113.

Fowler, M. S., & Brenner, D. L. (1982). A marketplace approach to broadcast regulation. *Texas Law Review, 60,* 205–254.

Frankel, M. (1999, February 21). Save democracy first! *New York Times Magazine,* 27–28.

Franklin, N. (1999, July 12). Networking. *The New Yorker,* 82–85.

Friedman, T. L. (1998, April 14). The network is the power. *San Jose Mercury News,* p. 6B.

FTC faults Net on privacy. (1998, June 4). *San Jose Mercury News,* pp. 1C, 4C.

Future of English? Project. (1998). [On-line]. Available: http://www.english.co.uk/cgis/archive.cgi?data/gen/gen1.

Gaitskell, R. (1998, Spring). Managing technological change: The challenge of the twenty-first century. *IEEE Technology and Society Magazine,* 36–39.

Garay, R. (1988). *Cable television: A reference guide to information.* New York: Greenwood Press.

Garcia, E. (1999, February 9). Spanish TV adopts an American look. *San Jose Mercury News,* pp. 1A, 18A.

Garfinkel, S. (2000, February 28). Privacy and the new technology. *The Nation, 28,* 11–15.

Garreau, J., & Weeks, L. (2000, January 17). Visions of a world that's nothing but Net. *Washington Post National Weekly Edition,* p. 8.

Gates, W. (1995). *The road ahead.* New York: Viking.

Gaw, J. (1998, December 29). To online sellers, Santa is still in a giving mood. *Los Angeles Times,* pp. C1, C18.

Gay, V. (1999, September 3). Other minority groups may support TV boycott. *San Jose Mercury News,* p. 23A.

Gillmor, D. (1998, November 8). Net protocol: Possibilities are endless. *San Jose Mercury News,* pp. 1D, 7D.

Gillmor, D. (1999, February 23). Push technology has new pull as an idea that's truly practical. *San Jose Mercury News,* pp. 21C, 10C.

Gillmor, D. M., & Barron, J. A. (1974). *Mass communication law: Cases and comments.* St. Paul. MN: West.

Gitlin, T. (1983). *Inside prime time.* New York: Pantheon.

Glass, A. J. (1999, September 12). Blacks unite to erase "digital divide. " *San Jose Mercury News,* p. 16A.

Global Reach. (1999, October 1). [On-line]. Available: http://www.glreach.com.

Global Reach. (2000, July 1). Internet users, by language. [On-line]. Available: http:///www.glreach.com.

Goldstein, T. (1998, September/October). Does big mean bad? *Columbia Journalism Review,* 52–53.

Gomery, D. (1998, December). U.S. waits to tune in, turn on digital TV. *American Journalism Review,* 70.

Graser, M. (2000, January 24–30). Triumph of the tech toys. *Variety,* pp. 31–32.

Grossman, L. K. (1998, September/October). The death of radio reporting: Will TV be next? *Columbia Journalism Review,* 61–62.

Grossman, L. K. (1999, July/August). From Marconi to Murrow to—*Drudge? Columbia Journalism Review,* 17–18.

Hachten, W. A. (1992). *The world news prism* (3rd ed.). Ames: Iowa State University Press.

Haldi, J. A., & Eastman, S. T. (1993). Affiliated station programming. In S. T. Eastman (Ed.), *Broadcast/cable programming* (pp. 189–215). Belmont, CA: Wadsworth.

Haley, K. (1999, September 6). New direction. *Broadcasting & Cable,* 18–19.

Harper's Index. (1999, January 17). Annual number of Americans who turn 50 each day. *San Jose Mercury News,* p. 2C.

Harper's Index. (2000, March 12). Number of companies that controlled over half of all U.S. media outlets. *San Jose Mercury News,* p. 2C.

Head, S. W., Eastman, S. T., & Klein, L. (1993). A framework for programming strategies. In S. T. Eastman (Ed.), *Broadcast/cable programming* (pp. 4–40). Belmont, CA: Wadsworth.

Healey, J. (1998, October 27). Phone-via-cable service tested. *San Jose Mercury News,* pp. 1C, 2C.

Healey, J. (1999a, February 18). AT&T-TCI merger gets official blessing. *San Jose Mercury News,* pp. 1C, 2C.

Healey, J. (1999b, May 4). Demand for data: More, more, faster, faster. *San Jose Mercury News,* pp. 1F, 4F.

Healey, J. (2000, January 28). Lights, camera, online! Hollywood plans Web shows. *San Jose Mercury News,* pp. 1A, 23A.

Heilbrunn, L. (2000, February). Waking up with Bob Edwards. *Brill's Content,* 116–118.

Heyboer, K. (1999, January/February). Enhancing the integrity of electronic news. *American Journalism Review,* 14–15.

Hickey, N. (1996, July/August). What's at stake in the spectrum war? *Columbia Journalism Review,* 39–43.

Hickey, N. (1999, September/October). Beachfront property on the cyber sea. *Columbia Journalism Review,* 47–48.

Higgins, J. M. (1999, May 10). All for just $5,000. *Broadcasting & Cable,* 16–18.

Higgins, J. M. (2000a, May 1). The big three. *Broadcasting & Cable,* 24–52.

Higgins, J. M. (2000b, February 7). Dish dealer drama. *Broadcasting & Cable,* 10.

Hilliard, R. L., & Keith, M. C. (1996). *Global broadcasting systems.* Boston: Focal Press.

Hodges, L. W. (1999, October). Questions to ask. *American Journalism Review,* 4 (special section).

Hollins, T. (1984). *Beyond broadcasting: Into the cable age.* London: BFI Publishing.

Hoover v. Intercity Radio Co., Inc. 286 F. 1003 (1923).

Hopkins, M. (1999, July/August). A babel of broadcasts. *Columbia Journalism Review,* 44–47.

Hundt, R., & Levin, B. (1998, July/August). The digital TV shuffle. *Brill's Content,* 76–77.

Industry News. (2000, March). [On-line]. Available: http://www.itva.org.

Inglis, F. (1990). *Media theory: An introduction.* Oxford: Basil Blackwell.

Innis, H. A. (1972). *Empire and communications.* Toronto: University of Toronto Press.

Jamieson, K. H., & Campbell, K. K. (1997). *The interplay of influence: News, advertising, politics, and the mass media.* Belmont, CA: Wadsworth.

Janah, M. (1999a, April 4). Web gateways try mass marketing to build identity with consumers. *San Jose Mercury News,* pp. 1E, 6E.

Janah, M. (1999b, April 1). Yahoo maps its future, will buy Broadcast.com. *San Jose Mercury News,* pp. 1A, 10A.

Jarvik, L. (1997). *PBS: Behind the screen.* Rocklin, CA: Prima Publishing.

Jensen, E. (1999, March). Meet the Nielsens. *Brill's Content,* 86–91.

Jessell, H. A. (2000, March 6). A mess of clutter. *Broadcasting & Cable,* 14.

Jones, A. S. (1999, November/December). Facing ethical challenges: The Integrity/Judgment Grid. *Columbia Journalism Review,* 63.

Katz, R. (1999, January 25–31). PBS plugs in 24-hour digital kids channel. *Variety,* p. 56.

Katz, R., & Peers, M. (1998, November 30–December 6). Digital cable plexes muscle. *Variety,* pp. 19, 22.

Kava, B. (1999, January 29). FCC proposes licensing "pirate" radio stations. *San Jose Mercury News,* pp. 1A, 22A.

Kiesewetter, J. (2000, April 2). Consolidation, cookie-cutter approach drive radio industry. *Louisville Courier-Journal,* pp. E1–E2.

Knight Ridder. (1999, March 30). News service. *San Jose Mercury News,* p. F1.

Kong, D. (1998, December 22). GO puts brand onto the Net. *San Jose Mercury News,* pp. 1C, 8C.

Kong, D. (1999, June 7). Catering to cyber customers. *San Jose Mercury News,* pp. 1C, 8C.

Konner, J. (1999, March/April). Of Clinton, the Constitution & the press. *Columbia Journalism Review,* 6.

Kristula, D. (1997, March). *The history of the Internet.* [On-line]. Available: http://www.davesite.com/webstation/net-history.shtml.

Kunkel, D. (1997, January 31). Why content, not the age of viewers, should control what children watch on TV. *Chronicle of Higher Education,* pp. B4–B5.

Kuralt, C. (1977). *When television was young* [videotape]. New York: CBS News.

Kurtz, H. (2000, January 17). When the news is all in the family. *Washington Post National Weekly Edition,* p. 7.

Kurtz, H., & Waxman, S. (2000, January 14). Anti-drug shows pay off for TV. *San Jose Mercury News,* pp. 1A, 24A.

Lambert, R. (1998, November/December). Rebuilding trust. *Columbia Journalism Review,* 39–42.

Lawlor, J. (1998, August 30). More meetings being held via video conference. *San Jose Mercury News,* p. 4F.

Lazarus, D. (1999, December 28). Christmas cheer for e-tailers. *San Francisco Chronicle,* pp. A1, A11.

Leading countries in Internet users per capita. (1999). *Internet industry almanac.* [On-line]. Available: http://www.c-i-a.com.

Ledbetter, J. (1998, September/October). Steven Brill: The Mother Jones interview. *Mother Jones,* 44–46.

Ledbetter, J. (2000, January/February). Wanted: A way of counting that you can count on. *Columbia Journalism Review,* 66–67.

Let them eat vowels. (1999, January–February). *Utne Reader,* 33.

Levine, M. (1996). *Viewing violence: How media violence affects your child's and adolescent's development.* New York: Doubleday.

Levy, S. (1998, February 16). New media's dark star. *Newsweek,* 78.

Littleton, C. (1999a, January 25–31). Sinclair's way is pay-to-play. *Variety,* pp. 53, 57.

Littleton, C. (1999b, March 29–April 14). U.S. birds made to get local for TV businesses. *Variety,* p. 58.

Littleton, C. (1999c, March 15–21). Webs play cross-lingo bingo. *Variety,* pp. 25, 30.

Lo, C. (1998, October). Space jam. *Wired,* 142–143.

Loco for local fare. (1999, September 27–October 3). *Variety,* p. M2.

Lorando, M. (1999, March 4). Why not just cancel Nielsens? *San Jose Mercury News,* p. 7E.

The man with the golden gut. (1977, September 5). *Time,* 46–52.

Mann, C. C. (1998, September). Who will own your next good idea? *The Atlantic Monthly,* 57–82.

Markels, A. (1999, July/August). Low power to the people. *Mother Jones,* 68–71.

Marlow, E. (1989). *Managing corporate media.* White Plains, NY: Knowledge Industry Publications.

Marshall, J. (1997, September 16). Wireless Web surfing. *San Francisco Chronicle,* pp. C3, C10.

Matthei, H. (1997, May/June). Inventing the commercial. *American Heritage,* 62–68, 70, 72–74.

McAdams, D. D. (1999, May 3). Cable expects big spring. *Broadcasting & Cable,* 27.

McChesney, R. W. (1997). *Corporate media and the threat to democracy.* New York: Seven Stories Press.

McChesney, R. W. (1999). *Rich media poor democracy.* Urbana: University of Illinois Press.

McChesney, R. W. (2000, March/April). The Titanic sails on. *Extra!* 10–15.

McClellan, S. (1999a, July 19). NAACP decries TV "whitewash." *Broadcasting & Cable,* 16–18.

McClellan, S. (1999b, August 23). Nielsen sells for $2.7B. *Broadcasting & Cable,* 35.

McClellan, S. (1999c, May 3). Upfront. *Broadcasting & Cable,* 24–26.

McClellan, S. (2000, May 29). It's "mother of all upfronts" at $8.2B. *Broadcasting & Cable,* 4–8.

McCollum, C. (1999, March 15). How rude. *San Jose Mercury News,* pp. 1C, 2C.

McCollum, C. (2000, January 16). The mob's big hit. *San Jose Mercury News,* pp. 1G, 3G.

McConnaughey, J. W. (1998). *Falling through the Net II: New Data on the digital divide.* Washington, DC: National Telecommunications and Information Administration. [On-line]. Available: http://www.ntia.doc.gov/ntiahome/net2/falling.html.

McConnell, B. (2000, January 17). FCC powers up. *Broadcasting & Cable,* 10.

McConnell, B. (1999a, April 19). Big flap over small stations. *Broadcasting & Cable,* 27.

McConnell, B. (1999b, January 25). FCC tunes microradio. *Broadcasting & Cable,* 100.

McConnell, B. (1999c, September 6). Right to respond challenged. *Broadcasting & Cable,* 13.

McConnell, B. (1998, November 16). FCC moving, cautiously, on ownership. *Broadcasting & Cable,* 27–42.

McConnell, B., & Albiniak, P. (1999, January 4). House to get dereg Speaker. *Broadcasting & Cable,* 15.

McDowell, J. (1999, May 3). Firing up the imagination. *Time,* 70–72.

McNichol, T. (2000, March). Naughty word alert. *Wired,* 99.

McQuail, D. (1987). *Mass communication theory: An introduction.* Beverly Hills, CA: Sage.

Miller, R. (1995). To use or not to use: The role of English in cross-border marketing. *Direct Marketing, 58,* 20–21.

Miller, R. (1998, November 30). Glorious history of radio. *San Jose Mercury News,* p. 5C.

Molony, D., & Cukier, K. (1998, October 5). Industry pushes ITU into truce with Net task force. *Communications Week International,* 1, 33.

Moore, F. (1999, June 7). Free speech isn't free, reports PBS special. *Santa Cruz County Sentinal,* p. A11.

Morales, E. (1999, September 27). We ought to be in pictures. *The Nation,* 9.

Nachman, G. (1998). *Raised on radio.* New York: Pantheon.

National Association of Broadcasters. (2000). Radio statistics [On-line]. Available: http://www.nab.org.

National Cable Television Association. (1998, Spring). Cable TV developments [On-line]. Available: www.ncta.com.

National Cable Television Association. (1999, February 23). Cable TV developments [On-line]. Available: http://www.ncta.com/yearend98_6.html.

National Cable Television Association. (2000). 1998 yearend statistics [On-line]. Available: http://www.ncta.com/yearend98_6.html.

New York Times. (1998). *1999 Almanac.* New York: Penguin.

Ninety-Second Congress. (1972). *Hearings before the Subcommittee on Communications on the Surgeon General's Report by the Scientific Advisory Committee on Television and Social Behavior.* Washington, DC: U.S. Government Printing Office.

Noack, D. (1999, June 19). Gannett pushes ethics. *Editor & Publisher,* 9.

North, J. (1999, February 8). Media starvation diet. *The Nation,* 28–30.

Nova. (1992). *Can you believe TV ratings?* Boston: WGBH Educational Foundation.

Novello, A. C. (1991, November 4). Youth and alcohol—advertising that appeals to youth [Surgeon General's Press Conference].

Numbers. (1999, January 25). *Time,* 23.

O'Brien, J. M. (2000, March). Sort stories. *Wired,* 104.

Ogilvy, J. (1998, November). Dark side of the boom. *Wired,* 188–189.

Online ad revenues surge. (2000, January 19). *San Jose Mercury News,* p. 1C.

Out of the boardroom. (1991, June 7). *Variety,* p. S10.

Overholser, G. (1999, December 27). Here's hoping the media will take the broad view. *San Jose Mercury News,* p. 7B.

Pareles, J. (1998, November 15). With a click, a new era of music dawns. *New York Times,* pp. 1, 22, (section 2).

Patchen, R., & Kolessar, R. (1999, Summer). Preparing for audience measurement in the digital age: Pilot testing Arbitron's Personal, Portable Meter (PPM). *Feedback,* 42–49.

Pember, D. (1999). *Mass media law.* Boston: McGraw-Hill.

Peraino, V. (1999, August). The law of increasing returns. *Wired,* 144–147.

Peterson, L. C. (1998, November). The *Wired* world atlas. *Wired,* 162–167.

Petrozzello, D. (1999, February 22). It's a buyer's market. *Broadcasting & Cable,* 17–18.

Platt, C. (1999, June). Ethernet at 60,000 feet. *Wired,* 150–155, 208–209.

Plotnikoff, D. (2000, March 2). Heady dreams for diversity go unfulfilled on the Internet. *San Jose Mercury News,* pp. 1E, 11E.

Potter, R. B. (1972). The logic of moral argument. In P. Deats (Ed.), *Toward a discipline of social ethics* (pp. 93–114). Boston: Boston University Press.

Potter, R. B. (1999, October 3). *The origins and applications of "Potter Boxes."* Paper presented to the State of the World Forum, San Francisco.

Prime time champs. (2000, January 10–16). *Variety,* p. 97.

Pringle, P. K., Starr, M. F., & McCavitt, W. E. (1991). *Electronic media management.* Boston: Focal Press.

Public Broadcasting System. (1997). Revised national program funding standards and practices adopted by PBS Board of Directors on February 2, 1997. [Memo to affiliated stations].

Put away the checkbook. (1999, June 12). *Editor & Publisher,* 50.

Quain, J. R. (1999, September). Fast-forward, rewind—and take control. *Brill's Content,* 64–65.

Quittner, J. (1999, May 17). Coinless juke box. *Time,* 94.

Rathbun, E. A. (1998, December 14). "A good day for women." *Broadcasting & Cable,* 14.

Reel, A. F. (1979). *The networks: How they stole the show.* New York: Charles Scribner's.

Rense, R. (1999). Fred Silverman [On-line]. Available: http://www.siterating.com/emmy/hof/silverman.html.

Reporters Sans Frontiers. (1999, November 1). Enemies of the Internet [On-line]. Available: http://www.freemedia.at/rsf.htm.

Roman, J. W. (1983). *Cablemania: The cable television sourcebook.* Englewood Cliffs, NJ: Prentice-Hall.

Rosenberg, J. (1998, October 10). Old idea in a digital age. *Editor & Publisher,* 34–35.

Rosenberg, R. S. (1992). *The social impact of computers.* Boston, MA: Harcourt Brace Jovanovich.

Rosenwein, R. (1998, August). The difference is radio. *Brill's Content,* 35.

Rosenwein, R. (1999a, April). Super Bowl hype: It works! *Brill's Content,* 30.

Rosenwein, R. (1999b, July/August). Trafficking in news. *Brill's Content,* 94–97.

Rothenberg, R. (1998, December). Go ahead, kill your television. NBC is ready. *Wired,* 204–207, 256–264.

Rushkoff, D. (2000, February 14). When hackers attack. *San Francisco Chronicle,* p. A23.

Safire, W. (1996, August 29). Quo lingua? *New York Times Magazine,* 61–62, 64.

Sandage, C. H., Fryburger, V., & Rotzoll, K. (1989). *Advertising theory and practice.* New York: Longman.

Sandler, A. (1999, June 14–20). A song in retailers' hearts. *Variety,* p. 1.

Sarnoff, D. (1953, September 21). Address to NBC Radio Affiliates Committee in Chicago. *Broadcasting/Telecasting,* 108–112.

Savan, L. (2000, March). Truth in advertising? *Brill's Content,* 62–63, 114.

Schlosser, J. (1998, December 7). Why CBS loves "Raymond." *Broadcasting & Cable,* 12.

Schlosser, J. (1999a, December 6). CBS solos for NAACP. *Broadcasting & Cable,* 53.

Schlosser, J. (1999b, July 26). Nets plan meeting with NAACP. *Broadcasting & Cable,* 42–47.

Schneider, M. (1999, December 13–19). Dramatic shift wizens webs. *Variety,* pp. 63–64.

Schneider, M., &Adalian, J. (2000, May 22–28). Nets get it together. *Variety,* pp. 15, 70.

Schramm, W. (1954). *The process and effects of mass communication.* Urbana: University of Illinois Press.

Seltel. (1999, January 18–24). Up for grabs. *Broadcasting,* pp. N24–N28.

Shapiro, A. L. (1998, December 21). AOL. mergergame.com. *The Nation,* 7.

Shenk, D. (1997). *Data smog: Surviving the information glut.* New York: Harper Edge.

Shepard, A. C. (1999, March). Leaving it "out there." *American Journalism Review,* 28–29.

Shiver, J. (2000a, January 20). Broadcast "pirates" due for a break. *San Jose Mercury News,* p. 5A.

Shiver, J. (2000b, March 16). Study shows a digital divide in Web content, too. *San Francisco Chronicle,* p. A12.

Siebert, F. S., Peterson, T., & Schramm, W. (1956). *Four theories of the press.* Urbana: University of Illinois Press.

Sinton, P., & Carman, J. (2000, February 10). Battle for control at Channel 4. *San Francisco Chronicle,* pp. A1, A17.

Skow, J. (1999, April 26). Lost in cyberspace. *Time,* 61.

Soley, L. C., & Nichols, J. S. (1987). *Clandestine radio broadcasting: A study of revolutionary and counterrevolutionary electronic communication.* New York: Praeger.

Soundbites. (2000, June). Jean Dominique, 1930–2000. *Extra! Update,* p. 2.

Specter, M. (1996, April 14). World, wide, web: 3 English words. *New York Times,* pp. A1, A5.

Specter, M. (1999, December 6). Your mail has vanished. *The New Yorker,* 95–104.

Statmarket. (1999, October 1). The language of the Internet [On-line]. Available: http://www.statmarket.com.

Stead, D. (1997, January 5). Corporations, classrooms and commercialism. *New York Times Magazine,* 30–47.

Stein, J. (1999, November 8). Babe tube. *Time,* 133–135.

Sterling, C. H., & Kittross, J. M. (1990). *Stay tuned: A concise history of American broadcasting.* Belmont, CA: Wadsworth.

Stern, C. (1999a, September 13–19). Digital transition: An expensive joke. *Variety,* p. 8.

Stern, C. (1999b, June 28–July 11). Radio receives rivals by satellite. *Variety,* p. 5.

Sterngold, J. (1998, September 20). How cable captured the mini-series—and the high ground. *New York Times Magazine,* 86–87.

Stevenson, R. L. (1994). *Global communication in the twenty-first century.* New York: Longman.

Stone, M. L. (1999, April). Classified conundrum: Building brand names. *Editor & Publisher,* 18–19.

The Strategis Group. (2000, May 8). High-speed forecast. *Broadcasting & Cable,* 32.

Streeter, T. (1997). Blue skies and strange bedfellows: The discourse of cable television. In L. Spigel and M. Curtin (Eds.), *The revolution wasn't televised: Sixties television and social conflict* (pp. 221–244). New York: Routledge.

Surowiecki, J. (2000, June 5). Can the record labels survive the Internet? *The New Yorker,* 35.

Taylor, C. R., & Stern, B. B. (1997). Asian-Americans: Television advertising and the "model minority" stereotype. *Journal of Public Policy and Marketing, 26,* 47–61.

Tedesco, R. (1999, March 8). Who'll control the video streams? *Broadcasting & Cable,* 20–24.

Tessler, J. (1999a, December 27). Net sales soar past '98 mark. *San Jose Mercury News,* pp. 1A, 14A.

Tessler, J. (1999b, December 20). Season was bright for online retailers. *San Jose Mercury News,* pp. 1C, 2C.

Thomas, O. (1999, October 4). Rewind. *Time Digital,* 18–19.

Ticker. (1999, May). *Brill's Content,* 128.

Tighter online privacy urged. (2000, May 23). *Palo Alto Daily News,* p. 13.

Top 25 television networks. (1999, December 13). *Broadcasting & Cable,* 31–40.

Turner, C. (1998, July 1). 19 nations join to counter spread of U.S. culture. Madison (WI) *Capital Times,* p. 1A.

UNESCO. (1980). *Many voices, one world: Towards a new more just and more efficient world information and communication order.* New York: Unipub.

United Kingdom at a glance. (1998, December 14–20). *Variety,* p. 70.

U.S. v. Zenith Radio Corp. et al., 12 F. 2d 61b (1926).

von Soosten, J. (1993). Domestic syndication. In S. T. Eastman (Ed.), *Broadcast/cable programming* (pp. 80–111). Belmont, CA: Wadsworth.

Voukydis, A. (1998, October). Privacy: CNN v. MSNBC. *Brill's Content,* 36.

Warner, C., & Buchman, J. (1993). *Broadcast and cable selling.* Belmont, CA: Wadsworth.

Wartella, E. A. (1997). *The context of television violence.* Boston, MA: Allyn and Bacon.

Werner, E. (1999, September/October). The 60 Minutes effect. *Columbia Journalism Review,* 13.

Wilke, J., Vamos, M. N., & Maremont, M. (1985, August 5). Has the FCC gone too far? *Business Week,* 48–54.

The wired diaries. (1999, January). *Wired,* 97–101, 106–135.

Wolf, M. J., & Sands, G. (1999, July/August). Fearless predictions: The content world, 2005. *Brill's Content,* 109–113.

Wolper, A. (1999, July 31). CNN analyst forgoes $10K moderating fee. *Editor & Publisher,* 9.

World-class tax dodger! (1998, September/October). *Mother Jones Extra,* 3.

Wright, R. (1995, January 23). Hyperdemocracy. *Time,* 14–21.

Zachary, G. P. (1999, January/February). The world gets in touch with its inner American. *Mother Jones,* 50–56.

Zoellner, D. B. (1995, September/October). The truth about the Heartland. *Columbia Journalism Review,* p. 8.

Zoll, D. (1998, November 25). Bad reception. *San Francisco Bay Guardian,* 16–23.

Credits

Chapter 1 **p. 3,** Courtesy NuvoMedia, Inc.; **p. 6,** Kobal Collection; **p. 16,** © Bernard Gutfryd/Woodfin Camp and Associates; **p. 17,** © Corbis/Bettmann; **p. 20,** Reprinted with permission from NBCi.com, a service of NBC, copyright 5/23/00

Chapter 2 **p. 26,** © AP/Wide World Photos; **p. 27,** © The Granger Collection, New York; **p. 31,** © The Granger Collection, New York; **p. 34,** Everett Collection; **p. 42L,** © AP/Wide World Photos; **p. 42R,** Photo courtesy Richard W. Fatherly and the Robert H. Storz Foundation; **p. 50,** Reprint permission granted by XM Satellite Radio Inc.; **p. 52T,** Reproduced with permission of Yahoo! Inc. © 2000 by Yahoo! Inc. YAHOO! and the YAHOO! logo are trademarks of Yahoo! Inc.; **p. 52B,** Courtesy Spinner.com, "Where Music Lives Online"; **p. 55,** © Lynn Goldsmith/Corbis

Chapter 3 **p. 61L,** © Culver Pictures, Inc.; **p. 61R,** © The Granger Collection, New York; **p. 66,** Photofest; **p. 69,** © NASA/Liaison Agency, Inc.; **p. 71L,** Photo by Andrew Eccles/© 2000 ABC, Inc.; **p. 71R,** Photo by Steve Fenn/© 2000 ABC, Inc.; **p. 76,** Kobal Collection; **p. 78,** Courtesy Sony; **p. 79,** Courtesy Stanley Baran; **p. 81,** Courtesy TiVo, Inc.; **p. 82,** Courtesy ImaginOn, Inc., San Carlos, CA and Friffin P.R., New York, NY

Chapter 4 **p. 93,** Courtesy Broadcasting and Cable Magazine; **p. 99,** Everett Collection; **p. 102,** Courtesy NPR. Photo: Jim Graham; **p. 104,** © 2000 Public Broadcasting System; **p. 105,** Courtesy Annenberg/CPB; **p. 107,** Promotional materials provided by Adventist Radio Network Stations; **p. 108,** Courtesy WMUA, University of Massachusetts, Amherst, MA

Chapter 5 **p. 118,** Courtesy Maui Economic Development Board; **p. 120,** © R.W. Jones/Corbis; **p. 121,** Atmospheric, Ltd.; **p. 122,** © Robert W. Ginn/PhotoEdit; **p. 123,** Courtesy Stanley Baran; **p. 124,** © Hulton Getty Picture Library/Liaison Agency, Inc.

Chapter 6 **p. 138,** Courtesy Community Television of Santa Cruz County, Channels 25, 26, 27; **p. 141,** Courtesy New England Cable News; **p. 143,** Courtesy Stanley Baran; **p. 147,** Courtesy Paper Tiger TV, 339 Lafayette St., NY, NY 10012, http://www.papertiger.org

Chapter 7 **p. 163,** © AP/Wide World Photos; **p. 166L,** © Hunter Freeman. Photo courtesy of Apple Computer, Inc.; **p. 166R,** Photo courtesy of Apple Computer, Inc.; **p. 167T,** The AOL logo is a registered trademark of America Online, Inc., used with permission; **p. 167M,** Courtesy Prodigy Internet (TM); **p. 167B,** © 1999 EarthLink Network, Inc. and Sprint Communications Company L.P. Trademarks are property of their respective owners. Sprint and the Sprint logo are trademarks of Sprint Communications Company L. P. EarthLink, EarthLink TotalAccess, and the EarthLink logo are trademarks of EarthLink Network, Inc. All rights reserved; **p. 171,** © AP/Wide World Photos; **p. 181,** Courtesy Monster.com; **p. 182,** Courtesy MySimon.com; **p. 184,** iWon.com; **p. 188,** Courtesy BackWeb Technologies Inc.; **p. 189,** Courtesy S3 Incorporated; **p. 190,** Napster

Chapter 8 **p. 199L,** © AP/Wide World Photos; **p. 199R,** © AP/Wide World Photos; **p. 205,** Arbitron; **p. 208,** Courtesy AGB Media Services; **p. 211,** Everett Collection; **p. 220,** Courtesy Arbitron; **p. 221,** Courtesy Stanley Baran

Chapter 9 **p. 227,** © 1978 Sid Avery/MP & TV Photo Archive; **p. 228,** © Timex/Timexpo Museum; **p. 231,** Interep is the largest independent sales and marketing company specializing in radio, the Internet and new media, with offices in 15 cities and annual billings in excess of $1.25 billion dollars; **p. 233,** Courtesy New Wave Broadcasting, L.P.; **p. 235,** Courtesy Stanley Baran; **p. 238,** Everett Collection; **p. 243,** Courtesy of The Humane Society of the United States; **p. 247T,** Courtesy QDECOR; **p. 247B,** Courtesy Privada.com; **p. 254,** © 1999 Warner Brothers/MP & TV Photo Archive; **p. 256,** © The Media Foundation/Adbusters Magazine

Chapter 10 **p. 261,** © AP/Wide World Photos; **p. 264,** Courtesy KSJS, San Jose State University, San Jose, CA; **p. 265,** Courtesy KSJS, San Jose State University, San Jose, CA; **p. 270,** © 1996 Universal/WB/MP & TV Photo Archive; **p. 274L,** Everett Collection; **p. 274R,** Kobal Collection; **p. 276,** Everett Collection; **p. 278,** Everett Collection; **p. 279,** © 2000 ABC, Inc.; **p. 280,** © AP/Wide World Photos; **p. 283,** Everett Collection; **p. 289,** Everett Collection

Chapter 11 **p. 302,** © Hulton Getty Picture Library/Liaison Agency, Inc.; **p. 303,** © Reuters Newsmedia, Inc./Corbis; **p. 309,** © Peter Turnley/Corbis; **p. 312,** Everett Collection; **p. 318,** Globo TV, Brazil; **p. 319** © AP/Wide World Photos

Chapter 12 **p. 327,** © AP/Wide World Photos; **p. 330,** Courtesy John Frost Newspapers; **p. 332,** © Dan Dion/Liaison Agency, Inc.; **p. 345,** Courtesy Net Nanny Software

Intl. Inc.; **p. 351,** Courtesy Truste.org; **p. 352,** Courtesy Anonymizer, Inc.

Chapter 13 **p. 362,** Reprinted with permission from the New York Post. © 2000, NYP Holdings, Inc.; **p. 365,** © AP/Wide World Photos; **p. 367,** © AP/Wide World Photos; **p. 368,** Everett Collection; **p. 373,** Everett Collection; **p. 374,** © AP/Wide World Photos; **p. 376,** © Neal Preston/Corbis

Chapter 14 **p. 384,** © AP/Wide World Photos; **p. 387,** Courtesy Stanley Baran; **p. 391,** © AP/Wide World Photos; **p. 395,** The V-Chip Education Project—The Center for Media Education and The Henry J. Kaiser Family Foundation; **p. 398,** © AP/Wide World Photos; **p. 399,** © Corbis; **p. 403,** Everett Collection; **p. 404T,** Everett Collection; **p. 404B,** Everett Collection; **p. 406T,** Courtesy The Green Party; **p. 406B,** Courtesy The Libertarian Party

Index

Page numbers in **bold type** indicate a glossary term.